Research Methods

Guidance for postgraduates

Research Methods

Guidance for postgraduates

edited by
Tony Greenfield
Industrial research consultant

A member of the Hodder Headline Group
LONDON • SYDNEY • AUCKLAND
Copublished in North, Central and South America by
John Wiley & Sons Inc., New York • Toronto

First published in Great Britain in 1996 by
Arnold, a member of the Hodder Headline Group,
338 Euston Road, London NW1 3BH

Copublished in North, Central and South America by
John Wiley & Sons Inc., 605 Third Avenue,
New York, NY 10158-0012

British Library Cataloguing in Publication Data
A catalogue record for this book is available from the British Library

Library of Congress Cataloging-in-Publication Data
A catalog record for this book is available from the Library of Congress

ISBN 0 340 64629 2
ISBN 0 470 23618 3 (Wiley)

Typeset in Times and Helvetica and produced by Gray Publishing, Tunbridge Wells, Kent
Printed and bound in Great Britain by J W Arrowsmith Ltd, Bristol

Contents

List of Contributors ix

Preface to the first edition xi

PART ONE INTRODUCTION

Chapter 1 A view of research 1
Tony Greenfield

Chapter 2 The research process: four steps to success 7
Tom Bourner

PART TWO GENERAL

Chapter 3 Research proposals 15
David Williams

Chapter 4 Finding funds 20
David Williams

Chapter 5 Ethics of research 29
Tony Greenfield

Chapter 6 Documenting your work 38
Vivien Martin

Chapter 7 Reviewing the literature: use of library and information systems 47
Susan Frank

Chapter 8 Data handling on computers 52
R Allan Reese

Chapter 9 Buying your own computer 60
R Allan Reese

Chapter 10 Who can help? 66
Shirley Coleman

Chapter 11 Managing your PhD 71
Stan Taylor

PART THREE RESEARCH TYPES

Chapter 12 Randomised trials 87
Douglas Altman

Chapter 13 Laboratory and industrial experiments 95
Tony Greenfield

Chapter 14 Agricultural experiments 107
Roger Payne

Chapter 15 Surveys 115
Roger Thomas

PART FOUR DATA: MEASUREMENT

Chapter 16 Principles of sampling 127
Peter Lynn

Chapter 17 Sampling in human studies 137
Peter Lynn

Chapter 18 Problems of measurement 144
Jim Rowlands

Chapter 19 Sources of population statistics 152
Tony Rowntree

Chapter 20 Instrumentation for experiments 160
Andrew Penney

Chapter 21 Interviewing 169
Mark Hughes

PART FIVE DATA: ANALYSIS

Chapter 22 Elementary statistics 181
David Hand

Chapter 23 Further statistical methods 189
David Hand

Chapter 24 Data analysis by computer 195
C E Lunneborg and David Hand

PART SIX SPECIAL TOOLS

Chapter 25 Mathematical models and simulation 207
Andrew Metcalfe

Chapter 26 Deterministic models 211
Andrew Metcalfe

Chapter 27 Stochastic models *Andrew Metcalfe* 228

PART SEVEN PRESENTATION

Chapter 28 Writing the thesis *Tony Greenfield* 243

Chapter 29 Presenting your research: reports and talks *Paul Levy* 253

Chapter 30 Graphical presentation *Paul Levy* 269

PART EIGHT THE FUTURE

Chapter 31 Protecting and exploiting technology *Edward Nodder and Fiona Dickson* 281

Chapter 32 Career opportunities *Ralph Coates* 289

Index 295

List of contributors

Mr Douglas G Altman
ICRF Medical Statistics Group
Centre for Statistics in Medicine
Institute of Health Sciences
PO Box 777
Headington
Oxford OX3 7LF

Mr Tom Bourner
Centre for Management Development
University of Brighton
Mithras House
Lewes Road
Brighton BN2 4AT

Mr Ralph Coates
Assistant Director (Employer Services)
Careers Service
University of Newcastle
Newcastle upon Tyne NE1 7RU

Dr Shirley Coleman
Ellengowen, Preston Park
North Shields
Tyne and Wear NE29 9JL

Ms Fiona J Dickson
Bristows Cooke and Carpmael
10 Lincoln's Inn Fields
London WC2A 3BP

Dr Susan Frank
The Main Library
The University of Sheffield
Western Bank
Sheffield S10 2TN

Dr Tony Greenfield
Middle Cottage
Little Hucklow
Derbyshire SK17 8RT

Professor David J Hand
Statistics Department
Faculty of Mathematics
The Open University
Walton Hall
Milton Keynes MK7 6AA

Mr Mark Hughes
Department of Finance and Accountancy
University of Brighton
Watts Building
Moulsecombe
Brighton BN2 4GJ

Mr Paul Levy
Centre for Research in Innovation Management
(CENTRIM)
University of Brighton
Falmer
Brighton BN1 9PH

Professor Clifford Earl Lunneborg
Department of Statistics
University of Washington
Seattle
WA 98195, USA

Mr Peter Lynn
Deputy Director
Survey Methods Centre at SCPR
35 Northampton Square
London EC1V 0AX

Ms Vivien Martin
Director of Management Education
Salomons Centre
Broomhill Road
Tunbridge Wells
Kent TN3 0TG

Dr A V Metcalfe
Department of Engineering Mathematics
Stephenson Building
University of Newcastle
Newcastle upon Tyne NE1 7RU

Mr Edward J Nodder
Bristows Cooke and Carpmael
10 Lincoln's Inn Fields
London WC2A 3BP

Professor Roger W Payne
Statistics Department
IACR-Rothamsted
Harpenden
Hertfordshire AL5 2JQ

Mr Andrew Penney
National Instruments
21 Kingfisher Court
Hambridge Road
Newbury
Berkshire RG14 5SJ

Mr R Allan Reese
Computer Centre
University of Hull
Hull HU6 7RX

Dr Jim Rowlands
Industrial Statistics Research Unit
Stephenson Building
University of Newcastle
Newcastle upon Tyne NE1 7RU

Mr J A Rowntree
12 Chester Road
Chigwell
Essex IG7 6AJ

Dr Stan Taylor
Staff Development Unit
Porter Building
University of Newcastle
Newcastle upon Tyne NE1 7RU

Mr Roger Thomas
Director
Survey Methods Centre at SCPR
35 Northampton Square
London EC1V 0AX

Dr David M Williams
Research Services Unit
University of Newcastle
Newcastle upon Tyne NE1 7RU

Preface to the first edition

The government proposed in 1994 in their White Paper *Realising our Potential* that all graduates who wish to study for doctorates should first take a one-year master's course in research methods. Several universities have since introduced such courses and more are planned. This book is a response to that development. It is not intended to be a deeply detailed textbook, rather a set of notes for guidance, to nudge the student's mind into useful avenues, to tell him or her what help is available and to show how he or she can help themselves. This guidance includes many references for further study. As a set of notes it should be useful to all researchers, those studying for doctorates as well as for masters' degrees, for their lecturers too and, indeed, for anybody in any field of research even if a higher qualification is not expected.

The breadth of the subject ruled out a single author: none but the most arrogant would pretend to such ability. The publishers and I therefore decided that we should seek contributions from many authors. This posed difficulties of recruitment, of meeting deadlines, of agreeing a common philosophy and adhering to it, and of imposing an editorial style without causing offence to the authors. These difficulties were resolved because there was one clear bond between all the authors: an enthusiasm to help young people to plan, manage, analyse and report their research better than they may otherwise. All of them are busy and successful as researchers and as teachers. I believe that all readers of this book will appreciate how much time and effort, as well as knowledge and experience, the contributors have devoted to its production.

Unusually, this preface is titled *Preface to the first edition*. This is because I have no doubt that there will be subsequent editions. The situation will change with the introduction of more courses on research methods and experience will accumulate. I invite all readers to tell me how it can be improved: what should be added, what should be omitted, and what should be rewritten. But if you like any of it, please write and tell me. I shall forward your comments to the authors. They deserve your praise.

Tony Greenfield

Thanks

I have long recognised a general need for guidance in research methods: my own need as well as that of others. While I was with BISRA (the British Iron and Steel Research Association) that need was expressed by many of the research staff who sought my help with the design and analysis of their experiments and with the writing of their reports. I thank them for encouraging me with their trust. But I also thank two of the directors of BISRA in particular. One was George Wistreich who persuaded me that creativity could be encouraged and indeed taught. Part of Chapter 1 is based on notes that I made at his lectures. The other was Eric Duckworth who insisted that project managers must include detailed experimental design in their research proposals before they could start work on the projects. They needed my help to do this so I was forced to be involved with their work and to learn about their areas of science.

I ran a research methods course for several years in the medical faculty of Queen's University, Belfast, and quickly discovered the need for several contributors. The subject is too wide for a single teacher. The medical statistics and computing staff all contributed as did staff from the library, the university computing service, various departments in the medical faculty and the dean himself, Gary Love. The course students were postgraduate doctors from first-year house officers to senior consultants and professors. I thank them all, lecturers and students, for their continuing encouragement and suggestions for improvements all of which led to the structure of this book.

Nicki Dennis is the commissioning editor at Arnold (as well as being the director of applied science and technology publishing). When the White Paper *Realising our Potential* was published I suggested to Nicki that she should produce a book for guidance of graduates who were starting their research studies. She agreed and asked me to edit the book. I thank her for that decision and for her continuing help and encouragement.

My search for authors revealed a common enthusiasm in a multitalented team. Their individual specialisms have combined into a successful whole and I thank them all: their contributions are great. Most specially, however, I thank Andrew Metcalfe who, as well as writing the whole of Part Six about models, arranged meetings with potential authors and put me in touch with others.

Part One

INTRODUCTION

1
A view of research

Tony Greenfield

Introduction

Research, depending on your viewpoint at the time, may be described as some combination of:

- a quest for knowledge and understanding
- an interesting, and perhaps, useful experience
- a course for qualification
- a career
- a style of life
- an essential process for commercial success
- a way to improve human quality of life
- an ego boost for you
- a justification for funds for your department and its continued existence.

To me, research is an art aided by skills of inquiry, experimental design, data collection, measurement and analysis, by interpretation, and by presentation. A further skill, which can be acquired and developed, is creativity or invention.

This book is mainly about the former set of skills, inquiry to presentation. Further useful topics are described, such as: how to find funds; how to protect your intellectual property; and how to find a job when your research is concluded. There is little mention of creativity so I shall briefly discuss it in this chapter.

But first, a few words about the origin of this book. It was inspired by the government's proposal in 1994 that all graduates who wish to study for doctorates should first take a one-year master's course in research methods. Whether or not you agree with this, you may agree that some notes for guidance for postgraduate research students would be useful. Some universities have already followed the government's proposal and have created research methods courses. There is already a place for this book. Whether you are studying a master's course in research methods, or doing some research for a master's degree or a doctorate, you can be guided by this book.

However, research is a big subject and it would not be possible to write a single volume about it in any depth. This book is intended to be a general reference on all aspects of research methods and should be used as notes for guidance. Its content is intended to be fairly simple and easily intelligible to most readers. There are references to more substantive texts.

The many viewpoints and components of research methods persuaded me that several contributors would be needed. Fortunately, there are enough qualified people in universities, consultancy and industry, who volunteered eagerly to write one or more chapters each. I asked them to write in a light style that could be read easily with a view to the reader picking up the general themes. I believe that between us we have achieved this but leave it now to you, the reader, to judge.

If there are parts that you don't understand, or that could be expressed more clearly, or if there are important omissions, please write to me or the publishers. Everything can be improved, especially the first edition of a book, and your opinions will help us.

Contents

These notes for your guidance have been divided into eight parts, with several chapters in each. Look through the contents list and see how the topics have been grouped. You may feel that some of the chapters are not for you. For example: do you know how to use the library? Of course you do! But do you? I suspect that many people who believe that they know how to find the right text at the right time will be pleasantly surprised to discover how much easier the task becomes when qualified guidance is given. Surely you will want to know how to find funds for your research but 'Ethics? Ethics has nothing to do with my research', you might say. It has. It has something to do with all research. Read the chapter and learn.

You will run into difficulties. You will find problems with management, resources, and people. There's a chapter telling you who can help. I suggest you read it before you meet these problems. There are chapters too on planning your work, about keeping documents, and examining your research process and keeping it on course.

There are several types of research and we have classified them as: clinical trials; laboratory and industrial experiments; agricultural experiments; and surveys. These may seem to be distinct, but there is a general philosophy running through all of them, expressed in different ways by the four writers. You may think that because your research fits into one class of research you can ignore the other three chapters in that part. Please make the effort and read those other three chapters. You will be stimulated to discover a new slant on your research.

Glance quickly at the part on data analysis and you may think 'I can leave that until much later, when I have some data to analyse'. Scientific method is about observing the world and collecting information so that you can understand the world better. The way in which you do this must surely depend on how you process the information once you have collected it. The data you collect will depend on how you analyse that data. Analysis is an essential feature of research and your progress will be easier the more you understand analysis. To some people it is hard and daunting. They would prefer to ignore it. To other people it is a challenge. Whichever is your viewpoint, make it a challenge and face it now; the more you understand about how you will analyse and interpret data, the better your planning and management of the way you collect it will be. The design of a good experiment depends on how the data from the experiment will be analysed.

Mathematical modelling and simulation may seem to be remote from the reality that you want to investigate in your research. They are powerful tools in many situations. Social, medical, economic and political systems, as well as physical, chemical and biological, can be described as mathematical models which can then be used in computers to predict the behaviour of those systems under various conditions. This is a useful approach to many types of research. While you read through the examples included in Part Six keep asking yourself how each example may relate to your research project.

Whatever research you do you must present your results in a thesis or dissertation, in reports and published papers, or in stand-up talks to live audiences. There are many books about presentation and some are recommended. Three chapters summarise the most useful points.

Other chapters offer good advice about how to: buy and use computers and instrumentation; sample from populations and interview people; protect your intellectual property; and progress in your career.

Creativity

Liam Hudson, in *Contrary Imaginations*, presents evidence that intelligence and creativity, as features of the human mind, are negatively correlated. That is, the average creativity of people with high intelligence, as measured by the researcher, was found to be lower than the average for people at the low end of the intelligence range. However, there is great variation about the average and there are some fortunate people who are both intelligent and creative. This combination must be desirable in research where we need both logic and imagination, where we need vision as well as the ability to plan and manage. It is perhaps encouraging to realise that people with exceptionally high IQ's do not have a monopoly on creative ideas.

Jacob Bronowski, writing about John von Neumann, said a genius is a person with two great ideas. One will do for a research project!

But what is creativity?

You are planning your research. You believe that every step on the way must be taken rationally. Indeed, that is the essence of most of this book: to guide you rationally through your work. But if you look at the most outstanding of creative leaps in the history of science you will see that many were founded on a quirk of thought. Well-known examples are: Watt's invention of the separate condenser for the steam engine as he strolled in the country; Poincaré's theory of Fuchsian functions as he boarded a bus; Kekulé's discovery of the benzene ring as he dozed by the fireside.

Nevertheless you can bring to bear some methods of intellectual discovery:

- **Analogy** – look for similarity between your problem and one for which the solution is known.
- **By parts** – break the problem into a series of sub-problems which you hope will be more amenable to solution.
- **By random guesses** – Edison used it extensively and brain-storming is a modern version of it.
- **Generalise** – if a specific problem is baffling, write a general version of it which includes its essential features, simplified as much as possible. An algebraic model leads to simplified solutions compared with tackling complicated arithmetic head on.
- **Add** – a difficult problem may be resolved by adding an auxiliary sub-problem.
- **Subtract** – drop some of the complicating features of the original problem. This is a trick used in simulation to make it more tractable.
- **Particularise** – look for a special case with a narrower set of conditions, such as tackling a two-dimensional example of a three-dimensional problem.
- **Stretch or contract** – some problems are more tractable if their scale or the range of variables is altered.
- **Invert** – look at the problem from the opposite viewpoint. Instead of 'when will this train arrive at Oxford?' ask 'when will Oxford arrive at this train?'.
- **Restructure** – in clinical studies we do not ask if a treatment will cure a disease, but will an inert treatment fail to cure the disease.
- **Analogy** – electrical circuits are imagined as water flowing through tanks, pipes, pumps and valves; some aspects of brain function are studied by comparison with computers. The more remote your analogue is from your problem, the more creative will be your solution.
- The method of **Pappus**: assume the problem is solved and calculate backwards.
- The method of **Tertullus**: assume a solution is impossible and try to prove why.

Check each of these approaches, asking yourself how you might bring it to bear on your problem. Then, if you need any more stimulation, read the following.

The *Art of Scientific Investigation,* a book by W I B Beveridge published in 1950 but still, half a century later, stimulating to read;
G Polya's *How to Solve It* offers practical recipes; and
Arthur Koestler's *The Act of Creation,* for a discussion of the working of the mind.

2
The research process: four steps to success

Tom Bourner

Most research projects take a long time to complete. A research degree, for example will usually take at least two years of full-time research. Completing a *part-time* research degree will take much longer. Sometimes the whole business of research can seem daunting to those who are new to it. At the start, research can seem like an ill-defined mishmash of activities littered with hidden pitfalls. When you are in the middle of your research it is sometimes difficult to see the wood for the trees. In this chapter I suggest a map for keeping in perspective your research project as a whole. The map is designed to give you an overview of the whole research process as though you were in a helicopter looking down on it so that you can keep it in view from start to finish.

Let's start by considering a problem with which you will probably be familiar. Suppose that you are a young person wanting to get a flat. How would you go about it? Well, if you're like most people, you would collect some literature from estate agents, local newspapers or letting agencies to get an overview of what is available. You would compare the features (such as price, size, location, amenities) of the different flats and make a shortlist of those most likely to meet your needs. Perhaps you would visit those on your shortlist. After making your decision you would sit in your new home and reflect on the process: the extent to which your flat met your original aspirations, what your first-hand experience had told you about the housing market, and what you had learned from the experience.

Now if you take off your flat-hunter's spectacles and put on a pair of researcher's spectacles you will observe some similarities between the processes of flat hunting and research. First, you did a literature review (local newspapers and estate agents' blurbs) to get an overview of the field. Second, you developed a theory of which of the available flats would be to your requirements (your shortlist). Third, you tested the theory by inspecting those on your shortlist. Finally, you reflected on the experience and your results. Stated formally, the process contains four parts:

- part 1: reviewing the field
- part 2: theory building
- part 3: theory testing
- part 4: reflecting and integrating.

Perhaps this sequence seems familiar. Perhaps you recognise it from other significant decisions you have made in your life, for example choosing a college, buying a personal computer, or choosing a job.

With some decisions it's not possible to go through all the stages. For example, when you choose a job the final test of your theory that you have chosen the right job is by doing the job. Unfortunately, this is possible only after you've committed yourself to the job. Perhaps that's why so many unsatisfactory job decisions are made.*

*The literature on labour turnover often refers to the period immediately following recruitment as the 'induction crisis' when job expectations are tested by the job realities. The propensity to quit is particularly high in this period. It is common for more job quits to occur within the first six months of recruitment than within any subsequent six-month period.

Once you recognise that you are already familiar with each of the major parts of the research process through your experience of making the larger decisions of your life you will have a valuable resource to draw on. Reflection on those experiences will also give you an indication of the possible pitfalls.

The four-part process can help you to put what you are doing into a broader picture when you start to get bogged down in the detail of research. It can also be useful in designing your research project.

Let us examine the parts of the process in more detail.

Part 1: reviewing the field

Many research projects arise from a study of current thinking in a field. The research project follows from identifying a gap in the literature. Most other research projects arise from awareness of a problem that is worth solving. In either case, a good start is an overview of current thinking in the field.

In case you are impatient with this part of the process and want to start immediately with fieldwork, here are some reasons for spending time and effort on a review of the field:

- to identify gaps in current knowledge
- to avoid reinventing the wheel (at the very least this will save time and it can stop you from making the same mistakes as others)
- to carry on from where others have already reached (reviewing the field allows you to build on the platform of existing knowledge and ideas)
- to identify other people working in the same and related fields (a researcher network is a valuable resource)
- to increase your breadth of knowledge of your subject area
- to identify the seminal works in your area
- to provide the intellectual context for your own work, enabling you to position your project relative to other work
- to identify opposing views
- to put your own work in perspective
- to demonstrate that you can access previous work in an area
- to identify information and ideas that may be relevant to your project
- to identify methods that could be relevant to your project.

Part 2: theory building

Theory building is the most personal and creative part of the research process. To some people it is the most exciting and challenging part.

In some cases data collection precedes theory building and in other cases it follows. Have you ever bought a used car? If so, you may have identified some possibles before narrowing down to a few probables. You collected data and then formed a theory about which of the cars would best meet your needs. Theory building followed data collection.

The process of developing a theory by inspecting individual cases has a special name – **induction**.

Our flat-hunting example is another illustration of induction. If each time you are sent the details of a flat in a certain area of town you notice that it is more expensive than you can afford, you may form the theory that all the flats in that area are too expensive for you. Acting on that theory you may ask the estate agents to stop sending details of flats in that area. That is the process of induction at work again – forming a theory from information about specific instances. Induction is a type of generalisation.

The alternative to induction is **deduction** which is the process of reaching conclusions about specific instances from general principles.

Here is an example of deduction: 'I can't afford to live in Mayfair so don't bother to send me the details of any flats in that part of town'. In this example 'I can't afford to live in Mayfair' is the generalisation and deduction leads me to the conclusion about any specific flat in Mayfair that I can't afford it.

Induction is a thought process that takes you from the specific to the general. Deduction is a thought process that takes you from the general to the specific.

We have seen how a theory can emerge from the data. However theory can also emerge from armchair theorising; introspection; deduction following a review of the literature; personal experience; chance remark; a brainstorm; a fortuitous metaphor; or pure inspiration.

I said earlier that data collection can precede theory building and that it can follow it. In the case of induction, data collection comes first. When data collection follows theory building then it is usually for the purpose of testing, or refining, the theory. That is part of the research process that we turn to next.

Part 3: theory testing

> *Experience has shown each one of us it is very easy to deceive ourselves, to believe something which later experience shows us is not so.*
>
> Rogers (1955)

When we were flat hunting we wanted to check if those apartments, described so attractively by the estate agent, would really meet our needs. Likewise when we are doing research we will want to check if the theory (or theories) that we have formulated fulfil our hopes and expectations.

The sort of theory testing we do will depend on our ambitions and claims for our theory. If we want to claim that our theory applies *generally** then we may want to use statistical methods (known as inferential statistics) which have been developed to enable us to make claims about whole populations based on information about a sample from a population.

If, however, your claims are only about the accuracy of your theory in the context of a particular situation† then theory testing may involve checking your conclusions (theory) from other perspectives. You may have looked at estate agents' brochures and now you want to look at the flats themselves, talk to the neighbours and inspect local schools, shops and transport. In research in the social sciences, the term **triangulation** is used to describe the process of checking if different data sources and different methods allow you to reach the same conclusions.

Testing theory can take many forms. At one extreme, you may simply invite the reader of a research report to test the conclusions against his or her own experiences. The test is whether the reader says: 'Aha! I can now make sense of my own experience in a new and convincing way'? But if the reader is unlikely to have first-hand experience for testing the researcher's theory, or if the claims being made involve a high level of generality, then the theory testing stage will be more formal and elaborate. In any case, at some level the testing of theories is likely to be part of any research process.

*For example: 'All two bedroom flats in Mayfair are more expensive than all two bedroom flats in Leytonstone'.

†For example: 'The flat that suits me best among those whose details have been sent is No. 10 Railway Cuttings'.

Part 4: reflection and integration

> *Knowledge doesn't exist in a vacuum, and your knowledge only has value in relation to other people's.*
>
> Jankowitz (1991)

Reflection and integration is the last stage of the research journey. There may be many things on which you want to reflect: what you have learned about the process of research; what you could have done differently; what you have learned about yourself. But there is one matter for reflection that is a crucial part of the research process itself. It will affect how your research is judged and the impact of your research. You must reflect on *how your research findings relate to current thinking* in the field of your research topic.

Your reflection on how your research results relate to current thinking will include your assessment of where your research *fits into* the field of knowledge. It will contain your assessment of *your contribution* to the field. In this part of the research process you are likely to return to your review of current thinking that you made at the outset and reassess it in the light of your results. It's as if the current thinking in your field of study is a partially complete jigsaw puzzle and you are detecting where your own new piece of the jigsaw fits in.

Relating the outcomes of your research to current thinking in the field may simply involve showing how it *adds* to what is already known in the field. This would be the case when you have filled a gap in the literature or found a solution to a particular problem in the field. It may involve seeking connections with current thinking. It may involve challenging some parts of the map of the current thinking in the field, so that you will be proposing some reconstruction of that map. It may involve asking 'what if?' questions of your research findings.

Any of these ways of relating your research findings to current thinking in the field may present further questions and new avenues to explore. The final chapter of a research report usually offers suggestions for further research.

A good practical question to ask yourself is: 'what are the implications of my research results for our understanding in this area?'. Here are some possible implications:

- You may have filled a gap in the literature.
- You may have produced a solution to an identified problem in the field.
- Your results may challenge accepted ideas in the field (some earlier statements in the literature may seem less plausible in the light of your findings).
- Some earlier statements in the literature may seem *more* plausible in the light of your findings.
- Your work may help to clarify and specify the precise areas in which existing ideas apply and where they do not apply (it may help you to identify domains of application).
- Your results may suggest a synthesis of existing ideas.
- You may have provided a new perspective on existing ideas in the field.
- Your results may suggest new ideas, perhaps some new lines of investigation.
- You may have generated new *questions* in the field.
- There may be implications for further research.
- Your work may suggest new methods for researching your topic.

Most of all, this last stage in the research process is about seeking to integrate the fruits of *your* research with current thinking in the field.

Summary and conclusions

It is sometimes difficult to keep in mind the whole research journey when all of your attention is focused on crossing some particularly difficult ground. My purpose in this chapter is to help you to keep the whole research process in perspective when you are engaged in a particular research activity. I have done this by giving you an overview map on which the whole journey is plotted in outline. I hope this will help you to *plan* your research journey.

I have related the process of research to the way that you find information needed for the larger decisions in your life. You already have much experience that is relevant to planning and doing your research.

I have suggested a four-part research process: (1) reviewing the field; (2) theory building; (3) theory testing; and (4) reflecting and integrating.

Your four parts may not follow this sequence strictly. For example, after you have reviewed the literature you may want to monitor developments in current thinking *while* you are doing the fieldwork. You may engage in some parts of the research process more than once. For example, you may find that data collected for theory building may enable you to test statements discovered in the literature. Or data collected to test a theory may suggest a new theory so that it becomes an element of theory building.

You may not want to spend the same amount of time and energy on each of the four parts of the process. For example, theory building may be only a token part of your research project if your main contribution lies in testing a theory that you found in the literature. On the other hand, you may direct most of your effort towards theory *building*, so that theory *testing* may be little more than establishing the plausibility of your theory in the light of the data you've collected.

The four parts will be present in almost all research projects, at least conceptually. If one of the four parts seems to be missing from your own research project, you should discuss it with other researchers and, if you are registered for a research degree, with your supervisor. If you intend to omit one of the parts from your own research project you must be able to state clearly why it has no role.

References

Jankowitz A D (1991) *Business Research Projects for Students*. Chapman & Hall, London.

Rogers C (1955) Persons or science: a philosophical question. *American Psychologist*, Vol. 10. Reprinted in *On becoming a Person*. Constable, London, pp. 267–279.

Part Two

GENERAL

3
Research proposals

David Williams

Introduction

A good research proposal is one that gets funded. With intense competition for funds, having a good research idea is not enough; it has to be well presented and clearly aimed at meeting the objectives of the funding source if it is to stand out from all the other applications. Each proposal will therefore be customised for a particular sponsor, but below there are some general principles to bear in mind when drafting any proposal.

Plan the proposal

Proposals that are put together at the last minute are seldom successful. Care needs to be taken to identify the most suitable sponsor, checking that what you propose falls within the remit of the sponsor, and preferably in a priority area. Most sponsors will produce handbooks or annual reports which lay out the objectives of the sponsor and the research areas or projects currently being funded. It is worth reading these carefully to get to know the sponsor's policy. If you are not sure whether you have an appropriate project, make a few preliminary enquiries by telephone to the sponsor.

Find out about the application procedures:

- What are the deadlines?
- Must the application be in a particular format?
- Are there application forms?

Most sponsors will issue guidelines for applications to the schemes they offer. The research councils send application forms and information booklets to each university. Your departmental or central administrative office should have bulk supplies. For other sponsors you may have to contact them directly for information.

Writing the proposal

Follow carefully any rules and formats specified by the sponsor. They may appear bureaucratic or silly, but the sponsor probably has a good reason, and it is their money. If you do not follow the rules your application may be rejected at the first stage before the research programme has even been assessed.

You will address two audiences in the proposal:

- the **specialist academic referees** who will assess the detailed research case you have presented; and
- the **non-specialists** who are on the board or panel to assess the wider context of the work proposed. It is important to address this second audience also.

Wherever possible, try to avoid using technical language which non-specialists would not understand, particularly in sections of the proposal which deal with the objectives and wider benefits of the work proposed. In all proposals you should try to impart the following information.

What are you trying to do?

Sponsors have finite budgets and cannot make open-ended commitments. They like projects that have a recognisable conclusion, even if the area of research may be continued by follow-up proposals. You need to define the scope of your application by giving the **aim** of the proposal and some specific **objectives**. These will be the criteria by which you, and the sponsor, will be able to judge the success of the research you do.

Why do this research?

Why is this research important? State why you believe it meets the objectives *of the sponsor* and why they should fund it. Give the **background** to the proposal, outlining the intellectual problem that has to be solved. Show you are really up to date in the field by stating what previous work has been done in the area by you or others, particularly work funded by that sponsor, and particularly by any researchers who may be referees or on the panel assessing the proposal.

What will be the benefits if the research is successful? Industrial sponsors have always been required to evaluate the financial benefits of research, but there is an increasing trend for publicly funded research to justify its contribution to wealth creation and the quality of life.

Why you?

You may have made a strong case for the research to be carried out, but you need to show the sponsor that you are a suitable or, since most major sponsors receive many applications in similar areas, the *most* suitable researcher to do it. Demonstrate your expertise by citing relevant publications or works, and projects that you and any collaborators have completed successfully.

What will you do?

Describe the **research programme**. This is usually the main section of the application and you should state clearly what are the various tasks within the work programme and who will do them (collaborators, assistants, technicians). Define the **methodology** that will be adopted; types of surveys, techniques or concepts to be used. Identify appropriate **milestones** at significant points during the research. These may be the production of some deliverables such as reports, prototypes, etc., or events such as seminars or meetings of the collaborators. The sponsors may wish to evaluate the progress of the research through the timely achievement of such milestones.

What resources will you need?

The resources you will require should be based on the programme of work you have defined. It is in no one's interest to submit a proposal with resources inadequate to carry out the project successfully, and top-up grants can be hard to come by. The particular resources that may be charged to the project vary considerably from sponsor to sponsor, so it is imperative to read carefully the financial terms and conditions of the particular sponsor scheme to ascertain what costs may be included in the proposal and what costs must be found from the applicant's institution or other external sponsors. Your finance office or equivalent should be able to give you advice. For all resources requested, there must be a justification in the application. They fall into the following general headings.

Staff

Not all sponsors will pay for the time of academic staff (such as lecturers) on the project, but most sponsors will pay all the costs of employing research support staff on the project (such as research associates, technicians). Explain how you will use such staff. For example, if you are asking for a senior researcher rather than a research assistant, state why a person of this grade is necessary for the project. Such staff are generally paid on nationally agreed terms and conditions which include annual pay rises and increments. You should check whether the sponsor will automatically supplement the grant for pay rises, or whether such rises need to be built into the costs of the proposal. This can be a significant sum for projects lasting several years.

Travel and subsistence

You should include realistic costs for travel and subsistence for the purposes of the project. Not all sponsors will pay for you to attend conferences overseas.

Consumables

Consumables include general running costs and materials used in the project. Such items generally need not be itemised but you should justify the funds requested in the proposal.

Equipment

You should specify large items of equipment that you require, giving the detail of the particular item, supplier and cost or quotation. Justify the purchase of the equipment rather than rental, and indicate the time the equipment will be used on the project. Note that some sponsors will not necessarily pay the full cost of equipment purchased.

Other items

Think carefully about any other resources that you may need and other costs that may be charged to the sponsor, such as recruitment and advertising costs.

Indirect costs (overheads)

You should include costs for services provided by the institution but which cannot be attributed specifically to the project. These include personnel costs, finance costs, staff facilities, training, library and other central facilities, and departmental services. These are real costs which will be incurred if you proceed with the project, and will have to be met by your department or institution if you do not receive funding from the sponsor. Much of the indirect costs result from the employment of additional staff. They are usually expressed as a percentage of total staff costs, and are generally close to 100 percent of such costs. That is, the indirect costs of a project are usually about the same as the total staff cost on the project.

It is therefore important that you recover as much of the indirect costs as possible from the sponsor. Some sponsors, such as the research councils, have a fixed level of 40 percent for such costs. Most universities have their own policies on charging for indirect costs, your finance office will be able to advise you on the appropriate costs at your institution.

What is the price?

When you have determined the *cost* of the resources you will need to carry out the research, you must decide on a *price* to charge the sponsor. The research councils and some other sources of publicly funded research have detailed financial conditions for grants which largely determine the price you should charge. For other projects, particularly from industrial or commercial sponsors, there is more flexibility.

There are many factors which may influence the price, but perhaps the most important is the long-term benefit that you or your institution may expect to receive, either financially or in other ways. If the research is long-term or basic and may be expected to generate academic publications or future research projects, then there may be an agreement to share the costs with a sponsor by, for example, not charging the full indirect costs. If the research is commissioned by a sponsor which places restrictions on publication or requires transfer of intellectual property to the sponsor, then you should not be expected to subsidise such research and may wish to set a price covering the full costs of the project and also opportunity costs for loss of intellectual property rights.

When you quote a price to a sponsor you should always make it clear that it is exclusive of VAT. Your finance office will tell you if the project is liable for VAT.

Can you manage the project?

Sponsors will want to ensure that their money is well spent and will look in the proposal for a plan for **project management**. This should include details of:

- who is in charge of the overall project
- who is in charge of the various tasks or sections of the programme
- how all the activities will be coordinated
- time scales
- milestones.

A diagram of the project plan will help and can save space where the length of application permitted is limited. For example, you could use a simple bar chart (Gantt chart) which shows the list of tasks and the time required to complete them; or, for more complex projects, you could include a network (PERT) diagram.

Before you submit

- Check again that the proposal meets all the necessary criteria of the sponsor. You may overlook the criteria when you revise the proposal with collaborators.
- Get someone else in your department or someone familiar with the research area to read and comment on the proposal.
- Check how many copies need to be submitted to the sponsor and the deadline for receipt.
- Allow plenty of time for final typing and printing or photocopying of the proposal.
- Many application forms require a signature of an administrative authority. Find out in advance who this is at your institution and allow plenty of time. Administrators are busy people too.

If you are unsuccessful

Try to get feedback from the sponsor as to why your proposal was not funded. If it was because of some aspect of presentation, or because of a minor technical reservation that you feel able to address, or simply that the sponsor did not have sufficient funds at that time, then it may be worthwhile revising the proposal in the light of feedback and resubmitting it. If it was rejected because it did not address the sponsors objectives or priorities then you might be better rewriting the proposal and submitting it to a different sponsor.

If you are successful

It is not uncommon to be awarded less than you asked for. It is also not uncommon for the sponsor to still want the full research programme. You must decide whether or not you can do the research with the resources offered, or if you need to curtail the research programme and outputs. If you think there is a problem, discuss it with the sponsor. It is in no one's interest for a research project to reduce its chance of success because it has inadequate resources.

From many sponsors you will receive a contract to be signed. This will contain various financial and administrative conditions as well as terms for doing the work. Conditions will include:

- the start and end dates of the contract, when the money will be paid
- if it is a fixed amount or if you will be reimbursed for only what you spend
- who owns the intellectual property rights
- what happens if you or the sponsor terminate the project early.

These are important conditions which can affect the way you do the work and you should read them carefully. Do not sign any contract before seeking advice from your finance office. If necessary, negotiate with the sponsor to achieve mutually acceptable terms.

A useful document summarising contract terms and their significance is *Sponsored University Research: Recommendations and Guidance on Contract Issues*, produced by the CVCP, June 1992. Copies, priced £5, may be obtained from:

Committee of Vice Chancellors and Principals
29 Tavistock Square
London WC1H 9EZ
Tel: 0171-387 9231
Fax: 0171-388 8649

The future

Do not wait until the project is near to finishing before you submit further applications. Plan in advance, particularly for those sponsors who are already funding you. Keep in touch with them. Send them reports from your current research. If your current research is producing results consult them about how you both may get good publicity from the work. Keep them interested by offering to put in position papers for possible future research areas or topics. And start planning your next application.

4
Finding funds

David Williams

Trends in external sources of research funding

Increasing importance

The level of research funding to universities provided by the funding councils through the block grant has remained static in recent years. This level is now set to decline by three percent each year over the next three years. There is, therefore, considerable financial pressure on universities to attract research income from external, non-block grant, sources in order to maintain research activity and provide a research infrastructure. Universities have been successful in identifying a wide range of funding sources and increasing the proportion of total research funds from these sources.

The research component of the block grant is allocated via the research-assessment exercise and is influenced by the level of external research funding and the income, publication and research staff it brings.

Increasing competition

These financial pressures, combined with increasing research activity from the new universities and many research organisations formerly funded by government departments now having to find a significant proportion of their income from external sources, have resulted in much greater competition for funds.

More directed funding

The recent White Papers on science and technology and competitiveness emphasised the requirement for research to underpin wealth creation and the quality of life. The *Technology Foresight* exercise further defined those areas of research of strategic importance. Funding from public sources has become more directed towards specific technologies and application areas, rather than academic disciplines. Increasingly, the relationship between government and the university is that of *customer–contractor* than of *sponsor–beneficiary*.

Increasing collaboration

Many of the most exciting advances in research are taking place at the interface between disciplines, such as engineering and biology. As the scope of expertise required and cost of state-of-the-art equipment increase, research programmes are becoming larger and involving more teams, often in different countries. Funding for personal scholarship is therefore reducing, particularly in the science and engineering fields.

Forms of research support

The diverse nature of research has led to a wide variety of forms of support. Each funding agency will have its own particular set of schemes with its own particular eligibility requirements. For example, some schemes may only be open to researchers currently employed as lecturers at a UK university. The handbooks and guidance notes for the funding agencies need to be read closely. In general, schemes fall into three types depending on the period of the research and the resources required.

Project funding

Project funding is perhaps the most common. Projects typically last one to three years and funding is provided for the employment of one or more research assistants together with the costs of equipment, consumables, travel, and so on, necessary to complete a defined programme of work. Some funding agencies may support longer-term research of major research groups by **rolling project support** where a particular project may be continued for several years subject to periodic review, or through **programme funding** which supports a portfolio of related projects, typically over a four-to-ten-year period. Support for focused research over a short period may be available, particularly from private funding agencies, as a **consultancy**.

Infrastructure

The larger funding agencies may support their major research groups by providing separate **infrastructure support** for buildings or equipment. This is generally given to research groups that already have extensive project funding, and may assist in the development of a specialist centre for research in a particular area.

Personal support

Some funds are available to support the work of a particular individual rather than a specific project and are related to the training and career development of the individual rather than focused on a particular piece of research. Such funding generally comes from the research councils or major charities. Research studentships provide funding for three years for a PhD. Fellowships are typically for one to five years to provide individual support. The longer-term fellowships are usually intended as a precursor to an academic position.

Funding bodies

Table 4.1 shows the source of funding for the UFC (old) university sector 1993/94.

Table 4.1

Source of funding	*£m*
Research councils	433
UK charities	293
UK central government	151
UK industry	124
European Union	101
Other overseas	53
UK health authorities/hospitals	30
Other	45
Total	**1230**

Government

Table 4.2 splits up sources of government spending.

Table 4.2 Sources of government spending

Department/agency	*Funding 1994/95 (£m)*
Research councils	1165.4
Higher education funding councils	941.6
Ministry of Defence (MoD)	2480.0
Civil departments	937.1
Office of Public Service and Science	23.9
Contribution to European Union programmes	305.0
Total	**5853.1**

Research councils

The research councils were reorganised on 1 April 1994 into six councils funded by central government through the Office of Science and Technology. Table 4.3 shows the split.

Table 4.3

Research council		*Funding 1994/95 (£m)*
Biotechnology and Biological Sciences Research Council	BBSRC	141.8
Economic and Social Research Council	ESRC	52.5
Engineering and Physical Sciences Research Council	EPSRC	345.0
Medical Research Council	MRC	268.6
Natural Environment Research Council	NERC	146.8
Particle Physics and Astronomy Research Council	PPARC	180.0
Total research council funding		**1165.4**

Each new council still has responsibility to support basic research within its remit, but there is increased emphasis towards the transfer of knowledge and skills, and working with industry and other public funding agencies to define and support strategic areas of research.

The research councils have a wide range of project and programme grant schemes, including special schemes to support young researchers, and award fellowships and studentships in most of the areas within their remit. Application forms and handbooks for these schemes are sent to central administrative offices within universities from where potential applicants can obtain copies.

Councils are particularly keen to support research in collaboration with industry. In addition to industry involvement in the project awards there are some special schemes, such as *Realising Our Potential Awards* (ROPAs). These are advertised in the press.

Each council produces handbooks, annual reports, corporate plan and 'forward-look' documents describing its research priorities. In addition, the main committees within each council usually produce annual reports. For further information on each council contact the appropriate press and information office:

Biotechnology and Biological Sciences Research Council (BBSRC)
Polaris House
North Star Avenue
Swindon SN2 1UH
Tel: 01793 413200
Fax: 01793 413201

Economic and Social Research Council (ESRC)
Polaris House
North Star Avenue
Swindon SN2 1UJ
Tel: 01793 413000
Fax: 01793 413001

Engineering and Physical Sciences Research Council (EPSRC)
Polaris House
North Star Avenue
Swindon SN2 1ET
Tel: 01793 444000
Fax: 01793 444010

Medical Research Council (MRC)
20 Park Crescent
London W1N 4AL
Tel: 0171-636 5422
Fax: 0171-436 6179

Natural Environment Research Council (NERC)
Polaris House
North Star Avenue
Swindon SN2 1EU
Tel: 01793 411500
Fax: 01793 411501

Particle Physics and Astronomy Research Council (PPARC)
Polaris House
North Star Avenue
Swindon SN2 1SZ
Tel: 01793 442000
Fax: 01793 442002

Government departments

An increasing proportion of government funding for research and development (R&D) is now open to competitive tendering rather than restricted to government research establishments. There are, therefore, increasing opportunities for universities to undertake research for UK government departments and agencies. The departments and their R&D expenditure for 1994/95 is shown in Table 4.4. The proportion of research funds from civil departments going to UK universities has increased to just under 50 percent, but of the MoD less than one percent goes to universities.

Table 4.4 Net government R&D expenditure 1994/95 (est.)

Department	*Funds 1994/95 (£m)*
Defence (MoD)	2480
Agriculture, Fisheries and Food (MAFF)	137.5
Trade and Industry (DTI)	248.1
Environment (DoE)	96.0
Overseas Development (ODA)	91.8
Health	61.5
Employment	54.2
Transport	38.7
Education	30.3
Health and Safety (HSC)	21.9
Home Office	19.5
National Heritage	11.8
Northern Ireland, Scottish and Welsh Offices	103.5
Others	22.3
Total	**3417.1**

Overall expenditure over the next three years is expected to remain static in cash terms, marking a fall in real terms. Much of this is taken up by cuts in the MoD, but there are also substantial cuts in the DTI and ODA.

Further details of the research priorities in each department is given in *Forward Look of Government-funded Science, Engineering and Technology* (HMSO), published annually; 1995 issue ISBN 0-11-430131-X (£29).

The forward-look document contains the contact addresses for each department. Most departments publish calls to tender for research programmes. Many departments are involved in LINK programmes to promote collaboration between universities and industry in areas of strategic research. Further details of the LINK schemes can be obtained from:

LINK Secretariat
Office of Science and Technology
Office of Public Service and Science
Cabinet Office
Room 171
Queen Anne's Chambers
28 Broadway
London SW1H 9JS
Tel: 0171-210 0556
Fax: 0171-210 0557

Other government-funded sources

Royal Society. The Royal Society is the UK academy of science. As such, it pursues an independent programme of activities and also receives funding through the OPSS to promote activities in the natural sciences, mathematics and engineering. It awards project grants, travel grants and fellowship schemes for both young and experienced researchers. For further details of the Society's schemes, contact:

The Royal Society
6 Carlton House Terrace

London SW1Y 5AG
Tel: 0171-839 5561
Fax: 0171-930 2170

British Academy (BA). The BA is a learned society which supports research in the humanities and some areas of social science. In the absence of a research council for the humanities, the BA established a humanities research board which fulfils a similar function. Government funding is provided through the Department for Education to make the board a major provider of funds and studentships for humanities research in the UK. A booklet describing the research funding and studentship schemes may be obtained from:

The British Academy
20–21 Cornwall Terrace
London NW1 4QP
Tel: 0171-487 5966
Fax: 0171-224 3807

British Council. The role of the council is to promote cultural, scientific, technological and educational cooperation between the UK and other countries. It does not fund major research projects, but does operate schemes to promote travel and short working visits for researchers to and from the UK which may be used to develop further research collaborations. Many schemes require an application to the council representative in the relevant country, but an address list and other information can be obtained from the UK headquarters:

The British Council
10 Spring Gardens
London SW1A 2BN
Tel: 0171-930 8466
Fax: 0171-839 6347

European Union (EU)

The EU is an increasingly important source of research funding. Income to universities from the EU in 1993/94 was £101 million; the comparable figure for 1988/89 was £23 million. The main source of research funds from the commission is a rolling four-year programme termed *Framework* within which are specific programmes aimed at areas of application. Details of the *Framework IV Programme* (1994–1998) are given in Table 4.5.

Funds within framework are specifically intended to promote the industrial competitiveness of Europe and to promote cohesion within the union. Each programme therefore has a detailed work programme specifying the areas to be studied. To be successful, applications must:

- specifically address a priority area within the work programme
- involve at least five partners from three member states of the EU
- show the relevance to improved industrial competitiveness or cohesion of the EU
- involve industry where relevant.

The requirement for a genuine collaboration with several partners in other member states means that preparing the application is time consuming, and actually carrying out the research will involve close liaison with overseas researchers and considerable travel.

Table 4.5 Framework IV, breakdown and amounts

Individual programmes	*Funding (ecu, millions)*
Telematics	843
Communication technologies	630
Information technologies	1932
Industrial and materials technologies	1707
Measurement and testing	288
Environment and climate	852
Marine sciences technologies	228
Biotechnology	552
Biomedicine and health	336
Agriculture and fisheries	684
Non-nuclear energy	1002
Nuclear fission safety	414
Controlled thermonuclear fusion	840
Transport	240
Targeted socio-economic research	138
Cooperation with third countries and international organisations	540
Dissemination and optimisation of results	330
Stimulation of the training and mobility of researchers	744
Total	**12,300**

A useful summary of the framework programme is contained in *Europe: Funding from the Fourth Framework Programme for Research and Technological Development (1994–98)* published by the Office of Science and Technology. Information on EU programmes and projects is available from the Internet site *http://www.cordis.lu/*.

In addition to the framework programme, sections of the European Commission may commission studies to assist in the development of policy issues. These are published in the *Journal of the European Communities* which is available in European documentation centres. Most universities subscribe to the *UK Research and Higher Education European Office* (UKRHEEO) based in Brussels and which produces a monthly bulletin and frequent e-mail communiqués on funding opportunities as they arise. You should contact the European officer or equivalent at your university to check the availability of these documents within your institution.

Charities and trusts

There are a wide range of charities and trusts that award grants for research. They range from the very large, such as the Wellcome Trust that has an annual turnover in excess of £200 million, to the very small local charity that may have only a few hundred pounds as a result of an individual donation or legacy. Collectively such organisations provided £293 million of funding to UK universities in 1993/94, much of it in medical research.

A comprehensive guide to charitable organisations providing funds for research is contained in the publication *Directory of Grant Making Trusts*. This publication provides details of trusts by principal fields of interest and by geographical area. It is updated annually and copies (~£50) can be obtained from:

Marketing department
Charities Aid Foundation
48 Pembury Road

Tonbridge
Kent TN9 2JD
Tel: 01732 771333

The Wellcome Trust is the largest such organisation with an annual turnover in excess of £200 million. It offers a range of project, programme, career development and travel grants for research into the biomedical sciences (clinical and non-clinical) and history of medicine, and is a major source of funding for research in these areas. A booklet giving details of the schemes is available from:

The Wellcome Trust
183 Euston Road
London NW1 2BE
Tel: 0171-611 8888
Fax: 0171-611 8545

A useful summary of details of medical research charities that make awards by peer-review of applications is contained in a handbook issued by:

Association of Medical Research Charities
29–35 Farringdon Road
London EC1M 3JB
Tel: 0171-404 6454
Fax: 0171-404 6448

Industry

Research support from industry and commercial organisations is focused on supporting its business priorities. Some major companies have their own research grant or studentship schemes for supporting strategic research. These schemes are generally an open competition for which applications are invited through advertisements in the press. For most companies, and particularly small companies, this will generally mean applied research in the form of short projects or consultancy on a specific problem.

Collaboration with industry can bring many advantages to universities. In addition to direct funds for research costs and fringe benefits, such as travel to conferences, there may be access to equipment and facilities not otherwise available. Advantages to industry include access to experienced researchers and facilities at relatively low financial risk and costs. The 1994 White Paper *Realising our Potential* emphasised the importance of wealth creation as an objective of future government funding for research. Schemes which combine government and industrial funding for academic research are likely to become increasingly important for both funding council and research council income.

LINK

LINK encourages companies and universities to work together on precompetitive research relevant to industry. There are about 30 LINK programmes covering a wide range of research fields, each with typically £5 million to £10 million funding from research councils, government departments and industry. Further details of the programmes and application procedures may be obtained from:

LINK Division
Office of Science and Technology
Office of Public Service and Science
Cabinet Office
Room 171
Queen Anne's Chambers
28 Broadway
London SW1H 9JS
Tel: 0171-210 0556
Fax: 0171-210 0557

Teaching Company Scheme (TCS)

TCS programmes involve a university participating in a company development programme, usually involving the implementation of new technology. Young graduates spend most of their time in the company supervised jointly by an academic and industrial supervisor. It is therefore more of a technology transfer scheme than a research scheme, but can be an effective way of developing a good relationship with a company. For more information contact:

The Teaching Company Directorate
Hillside House
79 London Street
Faringdon
Oxon SN7 8AA
Tel: 01367 242822
Fax: 01367 242831

Other research council schemes

Each research council is encouraging industry participation in its standard grants, and also developing some special schemes. Details may be obtained from the individual research councils above. All the research councils participate in the ROPAs scheme. This is an annual scheme for research groups having an annual support from industry of at least £25,000. Such groups are eligible to apply for ROPA awards, expected to increase to £70 million in 1996. The schemes are advertised in the press.

Finding an industrial partner can be difficult, particularly in a time of recession. For experienced researchers who have a high profile in academic or professional journals, industry may make the first contact. Usually, it is up to the academic to initiate the contact. Attendance at conferences, particularly those on the commercial applications of their research field may be valuable, as is a direct telephone call or visit to a relevant company. The industrial liaison officer (ILO) or equivalent at your institution may be able to help you identify such companies. You should take care to listen to the particular problems of the company, introduce your own research interests and expertise, and then identify a suitable problem that benefits you both before seeking possible government support through the schemes identified above.

5
Ethics of research

Tony Greenfield

Kant's wonder

Two things fill my mind with ever-increasing wonder and awe the more often and the more intensely the reflection dwells on them:
the starry heavens above me
and
the moral law within me

Critique of Pure Reason
Immanuel Kant (1724–1804)

Introduction

This chapter is about the ethics of well-designed and executed research; it is about honesty in analysis and reporting of results.

It is *not* about moral questions relating to projects such as experiments using aborted foetuses or the release of genetically altered viruses for the control of crop pests or the development of weapons of war.

The ethics of medical research have rightly demanded attention which has led to legislation, international agreements and declarations, regulatory authorities and local committees empowered to approve and monitor research projects. This emphasis, which arises from human concern, has extended to research about animals and the environment. There has been a lot of publicity and debate in all media about the ethics of research in these areas.

If you are researching in some area of medicine, either human or animal, or into some aspect of the environment you will almost certainly have thought about ethical aspects of your intentions. But if your research is in some other area such as sociology, education, physics, chemistry, materials, electronics, computing, mechanics, industrial manufacturing, you may think that there are no ethical questions for you to consider.

You would be wrong to think that.

Fraud is an obvious ethical matter but, surprisingly, so are experimental design, planning, management and execution; and so is publication.

If you know yourself to be thoroughly honest you must be confident that you will never be deliberately unethical. Unfortunately, no matter how good a person you are and how well intentioned, there is the possibility, indeed it is very likely, that you will be inadvertently unethical, insomuch as you infringe the accepted code of research behaviour. Anybody who embarks on research is at risk of such inadvertent unethical behaviour. Avoidance demands good advice at all stages. Where will you find that advice? Start here and follow the leads.

Background

We start with some definitions and, in the rest of the chapter, look at some good principles and bad behaviour.

Ethics, in its widest sense, as the principles of good human behaviour, is one of the issues for which philosophers have striven to provide guidance. Plato, in about 400 BC, proposed that there were *forms* of all things, including a *form of the good*. We could never experience true forms, but could at least approach them through knowledge.

For Plato, bad behaviour was the result of ignorance. Despite the enormous influence of his ideas, especially on religious belief, few people today would accept them in their original guise. For example, philosophers of the post-modern school hold the view that there are no absolute standards, and that morality can only be culturally determined.

There have been many philosophers, and theories, in the intervening years. Kant emphasised the *will* and the importance of intention. His *categorical imperative* is stated by Russell (1946) as: 'act as if the maxim of your action were to become through your will a general natural law'. The utilitarians, Jeremy Bentham and others, concentrated on consequences, and the 'greatest-happiness to the greatest-number' principle. Their ideas had a great, and generally highly beneficial, influence on the UK government during the middle of the nineteenth century, and probably still exert their influence today. Nietzsche's intense individualism was in stark contrast to this. He argued that such paradigms would stifle creativity.

The works of the major philosophers are not usually easy reading and, given the other demands on your time, you may think a more appealing understanding of ethics was given by Charles Kingsley in his children's adventure *The Water Babies*:

She is the loveliest fairy in the world and her name is Mrs Doasyouwouldbedoneby.

If philosophers cannot agree on the basic principles of ethics, and commentators cannot always agree about the correct interpretation of their work, it is hardly surprising that there is even more diversity of opinion about the practical application of those principles. Some philosophers, such as Nietzsche, had their ideas grotesquely misrepresented, and then reinterpreted in a more generous light. Hollingdale's translation of Nietzsche's *Thus Spake Zarathustra* (1969) sold well in the late 1960s. Despite all the controversy, there is enough common ground to establish codes of conduct which are generally accepted.

Codes of conduct

Most professional organisations have their own codes of conduct which are largely about the ethical standards that are expected of members. One of the best known of these codes of conduct is embodied in the *Declaration of Helsinki*.

Even if your research may be far removed from 'biomedical research involving human objects', which is what the Helsinki declaration is about, your should read it. Many of the points can be interpreted more widely. One that clearly applies to all research, without exception, and that includes yours, is:

It is unethical to conduct research which is badly planned or poorly executed.

The declaration was first adopted by the World Medical Assembly, Helsinki, in 1964. It has since been amended at assemblies in Tokyo, Venice and Hong Kong.

Here is a further selection of points:

- ... research ... must conform to generally accepted scientific principles ... based on adequately performed ... experimentation and on a thorough knowledge of the scientific literature.
- Every ... research project ... should be preceded by careful assessment of predictable risks in comparison with foreseeable benefits. ...
- In publication of the results of ... research ... preserve the accuracy of the results. Report of experimentation not in accordance with the principles ... should not be accepted for publication.
- The research protocol should always contain a statement of the ethical considerations involved.
- Special caution must be exercised in the conduct of research which may affect the environment.

Since all research involves the collection, analysis, interpretation and presentation of data, some points from the codes of conduct of statisticians are worthy of mention.

The Royal Statistical Society declares:

- Professional membership of the Society is an assurance of ability and integrity.
- ... within their chosen fields ... have an appropriate knowledge and understanding of relevant legislation, regulations and standards and ... comply with such requirements.
- ... have regard to basic human rights and ... avoid any actions that adversely affect such rights.
- ... identities of subjects should be kept confidential unless consent for disclosure is explicitly obtained.
- ... not disclose or authorise to be disclosed, or use for personal gain or to benefit a third party, confidential information ... except with prior written consent.
- ... seek to avoid being put in a position where they might become privy to or party to activities or information concerning activities which would conflict with their responsibilities.
- Whilst free to engage in controversy, no fellow shall cast doubt on the professional competence of another without good cause.
- ... shall not lay claim to any level of competence which they do not possess.
- ... any professional opinion ... shall be objective and reliable.

Some points from the code of conduct of the Institute of Statisticians (recently merged with the Royal Statistical Society) were:

- The primary concern ... the public interest and the preservation of professional standards.
- Fellows should not allow any misleading summary of data to be issued in their name.
- A statistical analysis may need to be amplified by a description of the way the data were selected and the way any apparently erroneous data were corrected or rejected. Explicit statements may also be needed about the assumptions made when selecting a method of analysis. Views or opinions based on general knowledge or belief should be clearly distinguished from views or opinions derived from the statistical analysis being reported.
- Standards of integrity required of a professional statistician should not normally conflict with the interests of a client or employer. If such a conflict does occur the public interest and professional standards shall be paramount.

None of these points needs elaboration. You can judge which apply to your research. However, thinking in terms of medical research, the ethical implications of statistically sub-standard research may be summarised as:

- *misuse of patients*
  - — put at risk or inconvenience for no benefit
  - — subsequently given inferior treatment
- *misuse of resources*
  - — diverted from more worthwhile use
- *misleading published results*
  - — future research misdirected

It is worth remembering that:

> ... *precise conclusions cannot be drawn from inadequate data.*
>
> *Biometrika Tables for Statisticians*
> Pearson and Hartley

Politics

Facts are sometimes distorted for political advantage. The ways in which this is done may be applied also in scientific research so some discussion of them is appropriate with a warning to be on your guard.

There is no official code of conduct about 'official statistics': those tables and graphs that are published by government departments and reach the public through newspapers, radio and television. But there is wide concern in the UK, and in most countries in the world, about the way that governments handle the figures. In the UK, for example, we are told that unemployment figures are expected to fall again, that the economy is recovering, that the poor are better off and that more is being spent on the National Health Service. Can we believe such statements?

Some tricks of official statistics:

- burying unfavourable statistics in a mass of detail
- changing definitions (what constitutes a major hospital project; items included and method of calculation of the retail prices index; who is unemployed)
- discrediting authors of unfavourable reports.

Cutting corners

Some of these guidelines are illustrated in the following anecdote from my own experience.

Pharmaceutical companies are naturally eager to conclude clinical trials quickly and favourably. This eagerness constitutes a commercial pressure on clinical research departments or agencies, and all others involved. This is fair enough provided no corners are cut and the highest ethical standards are maintained. Generally this is so, but sometimes statistical analysis reveals that it is far from the case.

What should I, the statistician, do then?

I believe that I must state my opinion firmly, without fear of loss of business or even of a libel action. I should do this just as if the trial had been conducted properly and the results had been entirely favourable with the expectation that the company would respect and honour my work and opinion.

The following example is of a trial that was designed and conducted by the pharmaceutical company. The data, already coded and entered into a computer file, came my way for analysis because the company was in difficulties, some of which will be revealed.

The trial was an open, randomised, phase two, multicentre study (see Chapter 12). The protocol specified that 150 eligible patients would be recruited by 12 investigators.

In fact 32 patients were recruited by five investigators, one of whom recruited only one patient.

Of those 32, only 21 patients were clinically evaluable and only seven were microbiologically evaluable.

The general conclusion was that nothing useful emerged from this study except for a strong message to the pharmaceutical company that they must pay closer attention to the design, planning, management and execution of trials than had been exhibited in this case.

Scientific integrity demanded comments on these aspects of this study. Recall that the *Declaration of Helsinki* (1975), states (Section 4.2):

It is unethical to conduct research which is badly planned or poorly executed.

It is generally accepted, by medical research ethics committees, that if the number of patients is too small to obtain a useful and significant result, then patients will have been submitted fruitlessly to inconvenience, discomfort, doubt of outcome, and to risk. Such a trial would be unethical.

The writers of the protocol assumed that if there was a 75 percent evaluability rate, approximately 112 of the 150 patients would be eligible for efficacy analysis. In fact 21 of the recruited 32 patients were clinically evaluable (65 percent).

The assumptions of a cure rate of 85 percent for the better of the two treatments and a 23 percent difference between the two with a significance level of 0.05 and a power of 0.8 indicated that 112 patients would be sufficient to detect that difference. In fact, the total cure rate was 43 percent and there was far from enough information to test for any differences of outcomes between the two treatment groups. No differences were indicated.

It was improper to embark on this trial without confidently expecting 150 eligible patients to be present. There was nothing in the protocol to show that the necessary number would present in a specified time. In fact a time was not specified.

Having embarked on the study it was not ethical to stop it, without clear evidence that one treatment was inferior, before the specified number of patients had been recruited.

The data-collection form provided for the collection of 1488 items of information on each patient. Much of this information, particularly relating to return visits, was returned as blanks.

Catch-all data forms may have a semblance of thoroughness to the uninitiated, but they demonstrate a lack of forethought and an absence of scientific planning.

Apart from demographic data collected to demonstrate the success of random allocation of patients to treatments and general homogeneity of the sample, all other data collected in a clinical trial should be related in some way to clearly stateable hypotheses.

The only hypotheses implicit in the protocol, although not explicitly stated, were:

1. There is no difference in the clinical improvement rates between the treatments.
2. There is no difference in the microbiological responses between the two treatments.
3. There is no difference in the incidences of adverse events between the two treatments.

These are straightforward hypotheses which may have been tested if 150 patients had been recruited. If any of the many variables of haematology, blood chemistry, medical history, age, sex, race, height and weight, surgical procedures and other medications may have influenced the outcomes, then the relevant hypotheses should have been stated. It could be left to the statistician to decide how to use these extra variables and what multivariate techniques to apply. However, the expected relationships should be stated in advance so that they could be taken into account in sample size determination.

A medical research ethics committee may have been misled by the protocol into believing that the study was well designed and would be well executed. The section on statistical methods contained an 'outline of statistical analysis plan' which appeared to be thorough. However, a responsible and careful committee would also look at the data-collection form and question its potential, not only for collecting the necessary data, but also for facilitating data processing and statistical analysis. They may question first the desirability of collecting so much information and how it was proposed to use it all in testing hypotheses. The plan suggested tabulations and complete listings but without any indication of how these would be interpreted. While clinical judgement may be needed to assess the effect of a treatment, it is not an appropriate tool to use when data recorded from clinical trials are interpreted. Clinical judgement is not necessarily reproducible between investigators, whereas formal statistical analysis is reproducible.

The most striking feature of the data-collection form was that there was no indication as to how the data were to be coded and entered into computer files. The consequence of this failing was that the data were entered into the computer files in formats which are very difficult to manage.

Pharmaceutical companies should understand that it is usual to have a pilot study for testing data-collection forms for their suitability for:

1. use by investigators
2. coding for data entry
3. statistical analysis.

The investigators had not generally completed the forms properly. This may be because the forms were badly designed.

Haematology and clinical-chemistry data called for individual assessment of 'significant abnormality' which was not defined. If it had been intended to be 'outside normal ranges', this could be left to calculation provided the laboratory normal ranges were given, although it is well known that these are contentious. The normal ranges were provided for the various centres but the data-collection forms had not been designed to include this information.

Because the data-collection forms were poorly designed, they could not be expected to encourage cooperation by the investigators who responded by leaving many questions unanswered, or improperly answered.

If the regulatory authorities were aware of the nature of this study, it is likely that they would admonish the company for conducting trials unethically.

The results of this trial were not satisfactory. This is because the trial was poorly designed and inefficiently and incompletely executed, and because the data forms were badly designed and unsuitable for data coding, computer entry, and statistical analysis.

The unavoidable conclusion was that the conduct of this trial was not ethical.

As the statistician responsible for analysing the data and reporting the results, that is what I told the pharmaceutical company.

Fraud

While much unethical science is inadvertent, caused mainly by poor management, there is a long history of scientific fraud reaching back several centuries. Charles Babbage, who was Lucasian professor of mathematics at Cambridge University (a chair held by many great scientists including Isaac Newton and Stephen Hawking), published a book in 1830 entitled *The Decline of Science in England.*

Read that again. The date was *eighteen* thirty.

One chapter in his book was about scientific fraud under which he described four methods of fraud: hoaxing; forging; trimming; and cooking. To these I would add obfuscation. For the first four, I cannot do better than quote him directly.

Hoaxing

In the year 1788, M Gioeni, a knight of Malta, published an account of a new family

> of Testacea of which he described, with great minuteness, one species. It consisted of two rounded triangular valves, united by the body of the animal to a smaller valve in front. He gave figures of the animal, and of its parts; described its structure, its mode of advancing along the sand, the figure of the tract it left, and estimated the velocity of its course at about two-thirds of an inch per minute ... no such animal exists.

There have been many more hoaxes since Babbage's day, including the saga of the Piltdown man.

Forging

> Forging differs from hoaxing, inasmuch as in the latter the deceit is intended to last for a time, and then be discovered, to the ridicule of those who have credited it; whereas the forger is one who, wishing to acquire a reputation for science, records observations which he has never made. ... The observations of the second comet of 1784, which was only seen by the Chevalier D'Angos, were long suspected to be a forgery and were at length proved to be so by the calculations and reasoning of Encke. The pretended observations did not accord amongst each other in giving any possible orbit.

Statistical methods now exist to discover forged data. Examples may be found in industrial research and in clinical trials. If you are tempted to forge your data, be warned. A good examiner will detect your forgery and you will be humiliated.

There can be great pressure on a student to complete a research project within the time specified by the university rules or before his or her grant expires. Under such pressure the student may be tempted to forge data which was never observed. Or, if he or she has made some measurements but they don't properly meet expectations, there may be a temptation to cook the results. Cooking is described later on by Babbage.

Trimming

> Trimming consists in clipping off little bits here and there from those observations which differ most in excess from the mean, and in sticking them on to those which are too small ... the average given by the observations of the trimmer is the same, whether they are trimmed or untrimmed. His object is to gain a reputation for extreme accuracy in making observations ... . He has more sense or less adventure than the cook.

Cooking

> This is an art of various forms, the object of which is to give to ordinary observations the appearance and character of those of the highest degree of accuracy.
>
> One of its numerous processes is to make multitudes of observations, and out of these to select those only which agree, or very nearly agree. If a hundred observations are made, the cook must be very unlucky if he cannot pick out fifteen or twenty which will do for serving up.
>
> Another approved receipt, when the observations to be used will not come within the limit of accuracy, is to calculate them by two different formulae. The difference in the constants, employed in those formulae has sometimes a most happy effect in promoting unanimity amongst discordant measures. If still greater accuracy is required, three or more formulae can be used.

> It sometimes happens that the constant quantities in formulae given by the highest authorities, although they differ amongst themselves, yet they will not suit the materials. This is precisely the point in which the skill of the artist is shown; and an accomplished cook will carry himself triumphantly through it, provided happily some mean value of such constants will fit his observations. He will discuss the relative merits of formulae ... and with admirable candour assigning their proper share of applause to Bessel, to Gauss, and to Laplace, he will take that mean value of the constant used by three such philosophers which will make his own observations accord to a miracle.

Obfuscation

Obfuscation means 'to make something obscure'. It is a deliberate act which is intended to convey the impression of erudition, of being learned, of great scholarship. Hence it is fraudulent. There is a style of academic writing, increasingly common in recent years, that is long-winded with long paragraphs, long sentences, long words, passive statements and tortuous structures (see Chapter 26). It is intended to deceive and it does so easily because the reader, even an examiner, is tempted to skim such verbosity and subsequently fears to confess he or she has not understood every word.

It is a trick that is apparent today in many academic papers and theses but it was not uncommon 100 years ago.

> *The researches of many commentators have already thrown much darkness on this subject, and it is probable that, if they continue, we shall soon know nothing at all about it.*
>
> Mark Twain

Perhaps some people can't help writing obscurely but if a postgraduate research student does so we should be suspicious.

> *People who write obscurely are either unskilled in writing or up to mischief.*
>
> Peter Medawar

Unnecessarily esoteric mathematics should be avoided. For example, it is not necessary to preface straightforward calculus, as applied to an engineering problem, with references to Hilbert spaces and sigma fields. Simple numerical examples can be a great help to your readers.

Advice

How can you, an inexperienced student, know how to avoid any of the problems, to be sure that your research is ethical? Only by seeking advice. The librarian is there to help you (Chapter 7); your supervisor is there to help you (Chapter 7); there are statisticians.

Why statisticians?

A statistician is objective. Although statisticians may know little about your special subject, they can advise you about how to do things fairly, achieve balance, measure and record information (which is what research is about), analyse data (Chapters 22 to 24), design your experiments (Chapters 12 to 14), avoid making too many measurements, be sure you are making enough, avoid bias, achieve high precision, present your results clearly and succinctly.

Anything qualitative is open to abuse, to subjective judgement and misinterpretation, either inadvertent or deliberate.

When you can measure what you are speaking about and express it in numbers, you know something about it: when you cannot measure it, cannot express it in numbers, your knowledge is of a meagre and unsatisfactory kind. It may be the beginning of knowledge, but you have scarcely in your thought advanced to the stage of science.

Lord Kelvin

The researcher's prayer

Grant, oh God, thy benedictions
On my theory's predictions
Lest the facts, when verified,
Show thy servant to have lied.

May they make me BSc,
A PhD and then
A DSc and FRS,
A Times *Obit. Amen.*

Oh, Lord, I pray, forgive me please,
My unsuccessful syntheses,
Thou know'st, of course – in thy position –
I'm up against such competition.

Let not the hardened editor,
With referee to quote,
Cut all my explanation out
And print it as a note.

Proceedings of the Chemical Society, January 1963, pp. 8–10
(Quoted in *A Random Walk in Science*,
an anthology published by the Institute of Physics in 1973)

References

Babbage C (1830) *The Decline of Science in England*. Fellowes, London.
Nietzsche F W (1969) *Thus Spake Zarathustra: A Book For Everyone and No One*, translated with an introduction by Hollingdale R J. Penguin Books, Harmondsworth.
Russell B (1946) *History of Western Philosophy*. George Allen & Unwin, London.

6
Documenting your work

Vivien Martin

Whatever type of research you do, you will need to keep records of what you do, how, when, where and why. You may not think that this is important, you may even think that you will easily remember everything and can write it up later. You won't and you can't!

Why document?

Your perception of what you do and why you are doing it will change in subtle ways as your research progresses. As you become clearer about some aspects you forget earlier doubts. As your findings accumulate you form firmer ideas and if some findings do not confirm hesitant proposals you are likely to reform or forget them and will concentrate on those that seem to offer interesting results. Much of the richness of your original thinking and planning is lost and hesitant directions which do not look immediately rewarding may be prematurely closed. Without good records you will forget earlier ideas.

Research is often talked about more for its results than for its processes. In planning your research you will have studied research methods and taken time to make a plan of your proposed work. You may think in terms of the broad question you plan to address and the ways in which you will explore that question. You will consider the research methods and try to choose appropriate ones for your intended study. You will probably plan the process carefully to ensure that your data collection is suitably rigorous. You will expect to write your 'methods' chapter explaining how you have planned. You may not have thought about how you will show that you did work in a methodical way. Many people declare their intentions in the 'methods' section and then jump straight to describing results. You would expect to have to substantiate your results, to offer supporting evidence for everything you claim to have found – so why not for your methods – for the process you have used to collect your data and analyse it. Experienced researchers know that the plan is only the starting point and that many annoying and illuminating hitches will occur between plan and report. Much of the learning for you and your reader will be in the detail of the process planned and that which really happened, why changes were made, what could have been anticipated and what could not, what caused time to slip and which expectations were unrealistic.

The value in having records of your process is in being able to supply and use the detail of the 'whole story' whenever you may need it. If all goes reasonably well, you can use the detail to substantiate your discussion of the planned and the actual methods used. If anything goes wrong you can use the detail to explore and explain. Some researchers despair if their original idea or hypothesis seems to be either unprovable or even disproven – but with good records of the process there is still much to say about why this might have happened and perhaps evidence to recommend a different approach another time.

One more good reason for methodical record-keeping is that at least you are writing something and you won't have to face the intimidating blank sheet when you start to write up. You will have a lot to start with and it will contribute to the introduction of why you are doing what you are doing, to the background for the study, the context, the choice of methods, the report of findings and possibly to your discussion of strengths and weaknesses in your study. How can you ignore such potential value from what is really a mundane discipline?

How to document?

Consider using different methods for different stages or different aspects of the study so that the method used fits as naturally as possible with the way in which you are working. Some methods you might consider are given below.

Diary

Keep a diary of the research from first idea to completion of write-up. The diary might be handwritten in a hardback book or in several small notebooks which you date and number as you fill them. You might prefer to use file paper and ring files so that you can rearrange the pages as themes emerge (but if you do this, number or date the pages so that you remember where they originated). You may prefer to keep the diary in a computer program, but this would restrict you to making entries only when you have access to the computer.

You may also consider using the diary as a major part of your research and analyse its contents occasionally. There are software programs with search facilities which can help you to do this. Other ways of pulling out themes and recurring interest once there is enough material in your diary include highlighter pens, applying different coloured Post-Its to significant pages, colour-coding entries with stick on dots or using coloured pens, or, more drastically, photocopying and cutting and pasting to make up reconstructed pages.

In some research you may ask your respondents to keep diaries. You might do this to elicit recurring issues, maybe to identify critical incidents in their work or lives, perhaps to identify their problem areas and responses to problems.

A useful source of further information about this approach is in Judith Bell's book, *Doing Your Research Project* (1993).

Recording observations

In a study involving observation you will need a way to record the issues relating to your research rather than everything that is going on. Are you an outsider observing a setting or are you a participant observer? What are the implications of each position?

Your note-making will need a mixture of writing, diagramming, mapping and drawing. The setting will need to be recorded in terms of anything which may impact on your data collection – this might include a map of the setting (showing physical features like doors, windows, furniture, if it is a room). If the research concerns movements of people or animals this will need to be shown, probably using diagrams to show paths taken and timing, perhaps coded to indicate who moved or how the movement was made. Devise grids to tick as things occur or checklists of things to look for and mark off. Observations which include listening to speech or sounds will need to be recorded in some way, perhaps with tape recording or perhaps with diagrams showing frequency of speech and types of interaction. Video recording might be sensible in some studies, but similar issues arise as for photography.

Judith Bell introduces methods suitable for recording interaction between people. In any study of this sort you will need pilot studies to derive a good method for recording observations and to be sure that you record what you intend and that you do not change the setting too much.

Could you benefit from structured observation, like activity sampling with a strict framework for when and how observations will be made and recorded? More information about how to do this can be found in *Management Research* by Mark Easterby-Smith *et al* (1991).

To some extent the method you choose for recording observations will relate to your approach as a researcher and it would be wise to be aware of the approaches used traditionally by your discipline or related disciplines. Anthropology field notes record observations as do the interaction diagrams used by social scientists and the activity samples of organisational behaviour practitioners. Your choice of method should relate to the traditions and expectations of your disciplinary area.

Laboratory notes

Laboratory notes apply more to experimental research. If you set up an experiment to prove your hunch, you will need to keep very accurate and methodical records to defend your findings. You should be careful to establish a method of note-keeping from the beginning of your studies so that your records are consistent. You must even record mistakes and omissions, like missed entries or lost information. The planning is very important, so consult literature from your discipline, for example Plutchik (1975) on experimental research in psychology. Also be sure that you really do measure what you need to measure. Without this even the most meticulous records would be worthless.

Recording voices

If your research involves interviews or focus groups how will you record what people say? People usually speak more quickly than a researcher can write, so although it is often possible to make notes, catching 'verbatim' quotes is more difficult. You may think your notes are sufficient, but they will be your *précis* of what you have heard, a brief analysis of what was said, not what was actually said. One way to record what was actually said is to explain to the interviewee that you need to write down every word and pauses will be needed. This is often acceptable if the interview raises questions which require thoughtful answers. It is sometimes more acceptable than tape recording.

Tape recording is the obvious way to record sound but it does have some hazards. The first consideration is the effect of having an obvious tape recorder between you and your interviewee or group. It can inhibit conversation and it can present difficulties of confidentiality if respondents might be recognised and would rather not be. If you are tempted to record but not to tell respondents, consider the ethical issues and the constraints your covert approach might bring in terms of the use of your findings and what would be likely to happen to you or your subjects if you were found out. If you think that it would not present particular problems for your interviewees it is best to explain why you want a recording, how you will use it and then to record openly. Plan to have an appropriate time available on the tape so that you don't have to disrupt the interview to change it. Choose the tape recorder with care and make sure that it can record effectively in the setting, particularly in group work. Learn how to use it confidently before you are in the interview. Be cautious about using mini-tapes and the very small recorders because you might well want to copy the tape or edit bits from it, so you will need more versatile equipment and standard tape size. Consider the advantages and disadvantages of using voice-activated equipment as the pauses between voices might be significant and important to record. You are likely to find a pause facility useful on the machine when you play back and try to make notes. Some researchers recommend using variable speed play-back to help at the analysis stage.

If you are not a good note-maker, consider using a personal Dictaphone to record your research process; the thoughts and additional ideas that occur and what actually happens as you go along. This is essentially the diary approach, but keeping an oral diary instead of a written one. You will need to write something eventually if you are making a research report, but you could then refer to your recordings and quote passages as appropriate or you could have a typed transcript made as you complete tapes.

Whenever you use tapes do be sure to label them as you use them with date and contents. If they are crucial to your research consider making backup copies.

Card index

For some record-keeping the card index is ideal. It may seem a bit old fashioned or associated too much with office files, but there is a very practical advantage in its portability and the portability of packs of file cards. Researchers can carry a few cards around with them and use them wherever they are to make an instant record which can be filed in a system once back at base with the box or boxes. An example of this sort of use is in using cards to record the sources in a literature search, perhaps to note key quotes. The cards can be arranged in a helpful sequence at the point when you write the literature review then rearranged when you compile your bibliography.

Computer database

The main advantage of a card index over a computer index used to be in its portability, but now laptop computers have filled this gap. If you have access to a laptop check out the software now available to help researchers at various stages of research. There are some very useful packages that will link stages of the literature review in the way described above as a manual system with card files. If your research involves collection of a considerable amount of data from which you will make a content analysis, consider putting the whole data collection into a package which will subsequently help with the analysis using keywords and phrases. You may be able to do this directly with a laptop if you can key in your data as you collect them. With a desktop model it may be less convenient and you should consider the advantages and disadvantages carefully if it will mean transcribing written notes which might be used effectively without computerising.

Again, a computer can be used very effectively for diary-keeping and for logging progress through your research process.

Mapping and making diagrams

Some sorts of research will involve complex ideas which can be described and recorded visually as maps. This can be useful if there is not yet an obvious sequence or priority, so making lists may be less appropriate. It may also allow ideas to be linked in ways that sequential writing does not.

Mind-mapping is one approach which works very well for some people. It can be used to develop detailed thoughts around a central theme, by writing the theme in a central bubble then drawing branches out from it as thoughts arise. Main thoughts are the lines

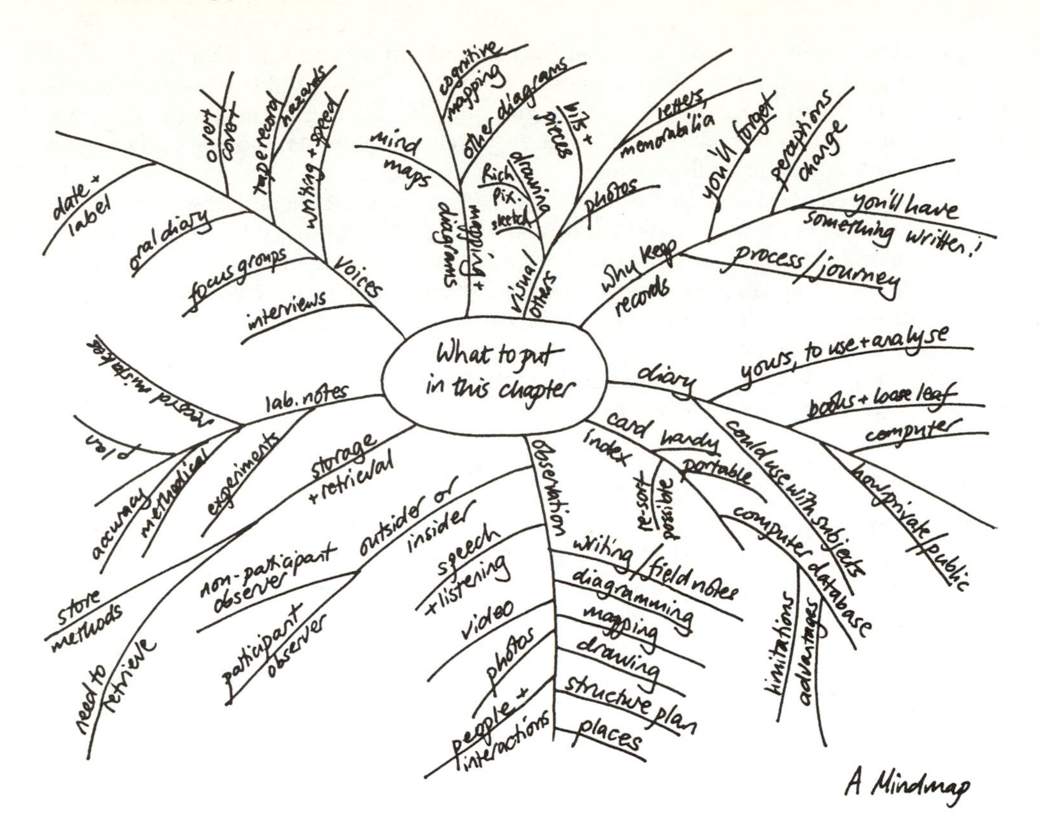

Figure 6.1 A mind-map.

out from the bubble and branches from these represent aspects of the main idea (see Fig. 6.1). More information about this method can be found in *Use Your Head* (1989) and other books by Tony Buzan. The method is useful at the early stages of generating ideas and connecting them, almost as a personal brainstorming. It is also very useful in planning the writing of chapters and mapping out how to report various aspects of research studies. It can also be used to make a quick record of a day or an event if it is more important to catch the elements and associated thoughts and feelings than to make a linear time-related record.

More conventional mapping may be useful in research records to record where something is located or to record differences over time when the research relates to physical changes in an environment or the use people or animals make of an environment. More comment on this method is made under recording observations.

Another recording system for group work is cognitive mapping, familiar in strategy development workshops in business and management research. The idea is based on mapping perceptions of the setting and has connections with the **repertory grid** technique. Groups can work with coloured cards and walls of flip-chart paper to map out issues and relationships and the group map can be the basis of subsequent planning. More information about how to use this can be found in *Management Research* (Easterby-Smith *et al*, 1991).

Remember that diagrams can save the use of a lot of words and explanations. Consider use of flow diagrams, Venn diagrams, multiple cause diagrams, fishbone diagrams and force-field diagrams.

Drawing

This is similar in some ways to mapping but is less formal. You might make a drawing of something as a record, as many archaeologists, palaeontologists and other researchers have done traditionally.

A loose way of using drawing is as a projective technique, for example, I might ask you to make a drawing of your research as you see it at the moment. You might draw yourself struggling to climb a mountain or disappearing down a black hole, or perhaps more cheerfully relaxing on a beach in the sun. The drawing could be the basis of a discussion about why you chose the images and what this means for you in terms of the progress of your research. The drawing would record the initial stimulus for the discussion and could be part of your record-keeping of the process of your research. This has been described in words and clearly words could be used for this sort of process instead of drawing, for example, in using metaphor to liken your feelings as a researcher to something else. ...

A use of drawing with individuals or groups is the compiling of a **rich picture** (see Fig. 6.2). The idea of this is to capture a situation in as much of its complexity as possible, showing all its component groups and individuals, sites and connections, communications, conflicts, inputs and outputs, messy areas, etc. It is drawn with pictures, symbols and connecting lines, in any way that makes sense to the individual or the group. It is drawn with discussion if it is a group work and the drawing captures the discussion and as many aspects of the situation as members of the group can come up with. For an individual it captures the thought process and the personal perception. These can be used only as a record, but are more usually the first stage in making an analysis of problem areas and muddled systems, in its most formal form a process called **soft system analysis** as described in the book by Peter Checkland, *Systems Thinking, Systems Practice* (1981).

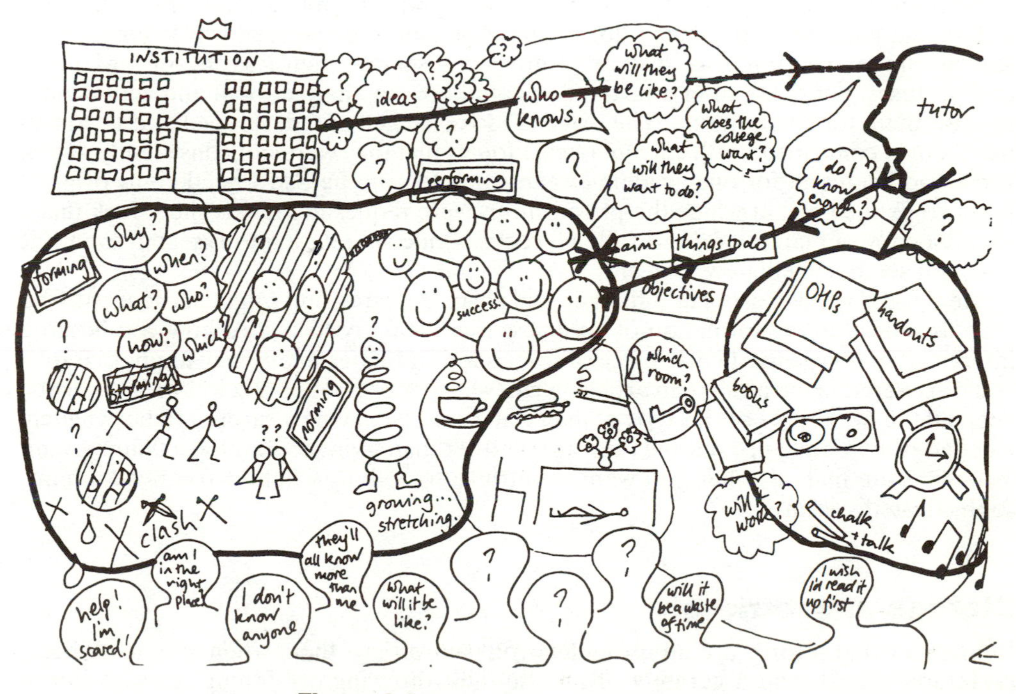

Figure 6.2 A 'rich picture' of a workshop.

Photographs in record-keeping

If your research concerns something which can be communicated effectively in visual records then photographs may be appropriate. Consider all the usual research issues in deciding how and when to use them as choice of viewpoint, span of view and selectivity are all choices made by the recording photographer. If you take photographs to support your research in any way, make sure you record when you took it (date and time), where from and looking at what, and the reason why you took it. The last point is important because of the delay in processing and the possibility that you will look at the prints and wonder what one or two were supposed to be about. When you look at the subject you photograph you know exactly what you are looking at, but the camera will record everything evenly unless you are very skilled in ensuring that you focus on your subject and reduce the importance of everything else. If you are not a skilled photographer you will need your notes to make good use of the prints. If you use the prints in writing up your research you will need to reference them and link them with your text. An example of the use of photographs in research can be found in a research into shop sites in high streets, looking at which areas had the most people shopping at what times of the day. Photographs taken regularly from the same spots were used with maps of streets and shops and people counts. The photographs added visual information which was richer than the other methods could produce alone.

Memoranda and correspondence

As soon as you start to research you will produce and receive all sorts of related notes, phone calls, letters, comments, and so on. Consider right from the beginning how you might keep these in an accessible form in case they are useful later. It is easy to lose the more trivial things, scraps of paper with notes and phone numbers, handwritten memos, letters that you write requesting information. A personal example of why keeping these somewhere is important was in some of my own research when I wrote off for college prospectuses, put them all in a box and debated how to analyse them and actually did not use them for several years. The original idea changed and developed and I went to use them in comparison with current ones. I found that they were not consistently dated, some had been sent for one academic year, some for another, so I needed my original letter to fix the point at which the prospectuses were requested. It was sheer luck that I found a copy of that letter. Now I file all correspondence relating to research even if it does not seem immediately relevant.

Memoranda of all sorts can influence and shape research and might subsequently be seen as crucial to the formation of your research idea. Try to record how you became interested in doing a particular piece of research. If it relates in any way to photographs you discovered in the attic or an article in a newspaper or a set of receipts or bus tickets, keep all the evidence. For this type of material, I suggest having a cardboard box vaguely labelled research and just putting things in it rather than trying to develop a filing system. You will soon find out when you want to retrieve something whether you need a more sophisticated system.

Storage and retrieval

Remember that records are no use unless you can retrieve them when you need them, preferably quickly and accurately. If all else fails, throwing everything relevant into a cardboard box will at least preserve the material, but the task of retrieval may be so daunting that you never actually get round to it. Some thought given to storage of records and retrieval of appropriate memoranda will pay off later! Consider storing in date order, in topics, in themes, in labelled envelopes, in transparent envelopes in ring files, in labelled boxes, in card files, in computer files ... much depends on your material and your own preferred methods. There is little in research more annoying than having to retrace your steps to find the exact reference for a quotation or the exact source of a piece of information.

Summary

Why document?

- *Your ideas will change about*
  - your aims
  - your methods
  - your knowledge
- Your plan will change
- Your interpretation of results will develop over time
- Abandoned side tracks may be worth later study
- Remember difficulties and why you lost time
- Other people's suggestions should be acknowledged
- You will create an assembly of written material to serve as first draft of your final report.

How to document?

- *Diary*
  - handwritten (use highlighters, coloured pens, sticky labels)
    - one large hardback notebook
    - several small pocketbooks (date and number)
    - loose pages in ring file (easy to reorder)
  - computer (there are programs that date, index, search and select items).
- *Making notes*
  - your position:
    - outsider looking in
    - participant observer
  - type of note:
    - writing
    - diagramming
    - mapping
      - relative positions
      - routes
      - movements
        - time
        - mode
      - sounds
    - drawing
    - photographs
    - video
  - bits and pieces
    - correspondence
    - memoranda
  - finding notes again
    - storage
    - some sort of order.

References

Bell J (1993) *Doing Your Research Project*. Open University Press, Milton Keynes.
Buzan T (1989) *Use Your Head*. BBC Books, London.
Checkland P (1981) *Systems Thinking, Systems Practice.* John Wiley, London.
Easterby-Smith M *et al* (1991) *Management Research, An Introduction*. Sage, London.
Plutchik R (1974) *Foundations of Experimental Research*. Harper & Row, London.

7
Reviewing the literature: use of library and information systems

Susan Frank

Introduction

Efficient and effective information retrieval underpins all successful research programmes. Searching techniques, especially in science, technology and medicine, now rely largely on electronic forms of information delivery such as compact disk and on-line databases, and on-line catalogues, although there are still important printed sources. The situation is changing very rapidly, both in response to developments in information technology, and to rapid growth in student numbers which demand less staff-intensive methods of information training and delivery.

It is therefore essential, before you start to search for information, to find out how facilities are organised within your institution. The traditional split between the library and the computing service may no longer exist, instead there might be a combined information service which supplies all the information/computing needs of the university. There is usually some degree of cooperation between the library and the computing service

You may need to improve your computing skills, as library services now assume a certain basic level of 'computer literacy'. Therefore, before you start to search for information, ensure that you have registered with your library and computing service, and have attended any basic computer courses that they offer. Owing to pressure of student numbers, these courses are usually supplemented by a wide range of publications on all aspects of computing skills, backed up by help on specific problems from an advisory service.

Universities are now starting to offer integrated training programmes in information management for their graduate schools. You should always take advantage of such programmes, as they will give you a thorough training which will be useful to you both during your time at university and beyond. Employers are looking for highly developed transferable skills, and an ability to retrieve and exploit information effectively is one of the most useful for a wide range of jobs.

I have not given you a long list of information sources, as these could never be comprehensive and would soon become outdated. This chapter is intended as a framework, within which you can fit the services available in your own university, in the most effective way. As a scientist, I am aware that postgraduate students have little time available for information retrieval, so I have limited myself to methods which will yield the most useful information as quickly as is reasonably possible. Your subject librarian is the person to see if you have problems at any stage with information retrieval: he or she will also be able to direct you to special sources relevant to your research topic.

Defining your search

All searches for information are useless unless you first give time to defining exactly what you want to look for. Do you want a general introduction to a topic, or more advanced, subject-specific references? Is currency critical? Do you really need to search back over a large number of years? The 'comprehensive search' can yield so many references that you will never be able to evaluate them all. Always search backwards from the latest information available: you can then stop when you have sufficient material and still ensure that what you have collected contains the most up-to-date references.

Many database searches lead to a reference to an article or book, not the item itself. What are the arrangements within your institution for obtaining material not in stock? University libraries nowadays are in severe financial difficulties, and may well not hold much of the material to which you find references. In such cases, what are the arrangements for borrowing from other libraries, how long will they take, and how much will you have to pay? All these factors need to be considered before you start to look for information, and they will influence the sources that you use.

You also need to take great care in the preparation of your search strategy. The way in which you word your questions is crucial. There is a general feeling that computers somehow compensate for inadequate queries on the part of their human users, but it is still a case of 'rubbish in, rubbish out'. It is a good idea to consult your subject librarian, who has a lot of experience in formulating search strategies and pointing out the pitfalls.

Many databases require you to do **free language** searching. This means that you have to work out all the likely ways in which the author may have described the topic and mirror this in your search strategy. You need to look for both singular and plural forms of keywords, US and UK spellings and synonyms. Thinking of all the alternatives takes practice: it requires both a knowledge of your subject and an ability to think laterally. A good subject encyclopaedia, introductory text, or talk with your supervisor can help here, by giving you a framework for your search and suggestions for subject keywords.

One mistake that many researchers make is to think that they will retrieve all references to members of a group by putting in the name of the group. For example, if you are interested in heavy metals, you will also need to search for lead, copper, zinc, and so on. You may find that your database does some of the work for you by offering fixed sets or hierarchies of keywords: if so, take advantage of these, as they can improve your search results.

When you have worked out your search strategy, try it out and then examine the results. Use these to modify your strategy and run the search again. You may find that the useful references contain keywords that you have not thought of. You will also find that you always get a certain number of references that are not useful, although they contain all your chosen keywords. Words have various meanings, and can occur in combinations that you did not anticipate. For example, in the search above, you might be interested in heavy metals in the soil acting as pollutants. A typical search statement might be:

SOIL/S AND (LEAD OR COPPER OR ZINC OR HEAVY METAL/S)

(The computer will look for any reference which mentions soil or soils in combination with any of the words/phrases in brackets.)

You will pick up references where lead, copper, and zinc act as soil pollutants, and also those on plant nutrition. In this case, you could narrow the search by putting in further search terms such as POLLUTION/POLLUTANT/S. Unfortunately, although this will find more of the references that are relevant to your subject, it increases the chance that you will miss other relevant references. This is because the concept of pollution can be described in so many different ways. With practice, you will become more skilled in balancing 'relevance' against 'recall' to carry out a satisfactory search.

Searching for books

The need to use books varies from subject to subject, but as a new postgraduate you will often wish to research a topic new to you, or to study the background to a particular area. This type of information is often found in books.

Most university libraries now have on-line catalogues covering their own book stock. Systems differ, but you will find that help is available, on the system itself, or from explanatory leaflets available in the library. If you find that queues to use the library terminals are excessive, check whether you can gain access in other ways. Most universities now have their own networks, and the library catalogue is often available from networked terminals around the system. If you need to know about books in other university libraries, many of their catalogues may be accessed from terminals on the network: check with your library.

You will find that on-line catalogues allow very flexible retrieval of material. You can search by author, title, subject keywords, and combinations of these. In addition, there may be special catalogues giving access to such things as university theses, audio-visual collections, and journal holdings.

Searching for journal articles

For researchers in science, technology and medicine, the most important source of information is the journal literature, because it is here that you will find the latest work on your subject. Journals are also a major source of information for most other subject areas. However, your library will only take a very small proportion of the journals of interest to your subject area, and in order to search the literature comprehensively you need to use specialised indexes to individual journal articles published world-wide.

These indexes are prepared by taking the individual papers from a wide range of journals in a particular subject area, and then providing access in various ways: searching by author or subject keywords is the most common. Some databases are very sophisticated: for example the major chemistry database allows structure searching by graphical input, and many databases have 'dictionaries' of subject headings which help you to formulate better search queries.

The output from these databases is in the form of bibliographical references to individual items in the journals which the database covers. In addition, these references may contain abstracts, summaries of the contents of the paper. Abstracts can be very useful in deciding if you want to go on and obtain a copy of the actual paper.

Nowadays, nearly all the important databases are available in electronic format. You may have access to these in a variety of ways, as compact disk (CD-ROM) databases on stand-alone machines or on your university network, as locally networked databases, or as a remote database made available via JANET (the UK's academic network). The means of delivery is not really important. You will be able to obtain advice on what is available, and means of access, from your library.

One of the most important general set of databases is provided by the BIDS service.

BIDS-ISI

The BIDS-ISI service is an on-line service provided via the JANET network linking UK universities. The database itself consists of three indexes in the sciences, humanities, and social sciences, and an index to scientific and technical proceedings.

BIDS-ISI indexes thousands of journals published world-wide, but even so, it can only cover a small proportion of the total number available. It is thus useful for a wide-ranging search of the literature, but it lacks the in-depth coverage of more specialist databases. BIDS-ISI is updated weekly, and, as well as doing searches over a period of time, it is also possible to search for the titles of articles in the latest issues of any specified journal, thus allowing you to keep up to date with what is being published in your field. The system is menu-driven, which makes it very easy to use.

You can view references (many containing abstracts), and have them e-mailed to you, downloaded or printed (depending on the facilities that you have available). There is also a document ordering service, if you do not wish/are not able to find the item in your library.

Specialised databases

You will usually find that your library holds a range of specialised databases for you to do your own searches. It is not possible to give a list here – new databases are continually appearing, and the ability of universities to acquire these varies. In general, science, technology and medicine are more comprehensively covered than the humanities. *Biological Abstracts* (biological sciences), *Medline* (medical and biomedical sciences), *Compendex* (engineering), *Inspec* (electrical and electronic engineering, computers and control, physics), and *Chemical Abstracts* are examples.

Other people's theses, in all subject areas, are a useful source of information, much of which may never appear in any other format. CD databases are available for British, other European, and North American theses.

Databases are not limited to those giving access to bibliographical references. You can find material as diverse as the *Oxford English Dictionary*, a multimedia encyclopaedia in the biological sciences, world maps, census data, and research in progress, all on CD-ROM. Your department/library may offer access to very specialist sources of information such as protein and nucleic acid sequence databases, and econometric data sets.

Your subject librarian may also be able to access databases held outside the library and not available on your network, and carry out searches for you. This is especially useful if you are working in a specialised area, such as patent searching, where the library will not usually provide user access to a suitable information source. Although there is a very large number of these databases, university libraries tend to make use of them much less than in the past, as more and more databases become available for direct searching by staff and students.

Obtaining books and journals

If your library does not hold the items that you want, you will need to use the interlibrary loan service. This may be free, or, increasingly, you may have to pay part or all of the cost. Do not put in a request for an item and expect it to appear by the next day: your library will advise you on the time which you may expect to wait. You may be able to take advantage of various rapid delivery services, usually at an extra cost to yourself.

Some electronic database services offer the option of document delivery once you have found the references that you want. You can input your credit-card number and have the costs debited to your account. This option is available for the BIDS-ISI databases, and also for the BIDS-Inside Information service. BIDS-II provides access to references from about 10,000 journals of the British Library's most requested titles. Currency is a major aim of this database, so it can act as a current awareness service, alerting you to the latest developments in your subject.

Reference management

You will have put a lot of time into finding your references. If you have recorded them in electronic form by downloading to floppy disk from a CD or on-line database you have a reasonable guarantee of their completeness and accuracy. If you record them from other sources (printed indexes, books, colleagues) remember to record all the details in full (authors, article title, full journal title, publication year, volume, first and last pages). Also note where you found a reference – it may be necessary to go back later to recheck a detail.

You may wish to use word processing or reference management software to edit your references and to incorporate them into a personal database. There are many different packages available from which you should be able to find one to suit your purposes. Your department may favour a particular package – ask your supervisor for advice. Universities may support a package/packages on their networks, but investigate them carefully – there is no one package suitable for all subject areas/purposes.

Future developments

Networking in universities is proceeding rapidly, radically affecting the way in which libraries offer their services. The network is the gateway to a great range of information: you may now be able to access the *Internet*, via such network tools as *World Wide Web*. University departments act both as consumers of information, and contributors, making their own material available to the wider electronic community. **Electronic journals** are becoming much more common, and in some areas, such as high-energy physics, conventional printed journals have become almost a thing of the past as news of the latest discoveries is quickly passed between interested researchers on networks around the world.

However, whilst researchers are under pressure to publish in recognised (usually printed) journals, and depend upon these publications for departmental recognition and career progress, the printed publication has still an important part to play in the information process. Electronic sources cannot be browsed in quite the same way as collections of printed volumes, where it is often possible to find useful papers on subjects quite different from the one that you were originally looking for.

Many universities, faced with escalating materials costs, especially for journals, are now moving from a *holdings* to an *access* (to information) policy. However, electronic sources of information are not cheap. There are also considerable financial implications in making these sources available: CDs have to be used in conjunction with specialist hardware, and if databases are made available on the university network, hardware, software, licensing and staffing costs have to be found. There are no free sources of information. Thus you will find yourself living in an exciting, but constrained information environment. It is your responsibility to make this work for you whilst you are carrying out your research.

8
Data handling on computers

R Allan Reese

Computers in research

The name **computer** still carries unfortunate overtones of calculation and arithmetic. You must disregard these, as the computer is now a ubiquitous piece of office equipment used for storing and processing information of all types. The vogue term **information technology** (IT) reflects this.

Until recently, research data would probably have been stored on paper, and may have been processed by shuffling and cross-referencing pieces of paper or card. The computer on your desk (or lap) can replace bulky piles of paper and gives you far faster access to any piece of data. Information handling with a computer is so much faster and more convenient, that you can perform analyses which would be unthinkably tedious by hand. However, the computer is only a mechanical slave; it works without understanding or thought. *You* must remain in control of the method of analysis and the interpretation of the output.

Data stored in a computer can relate to numbers, text or graphics. Inside the machine they may all look alike: text characters are stored as numbers (in one scheme 65 for the letter A) and a picture might be described as intensities of colour as each point. Indeed, at the fundamental level everything reduces to pulses of electricity or magnetism corresponding to binary (on/off) values, but you can usually remain indifferent to this. The processes of coding and decoding are automatic whenever you store or retrieve data: you press the key labelled A on the keyboard and the letter A (or a) is displayed on the screen. Just sometimes you need to be aware of the coding, as the way you choose to hold data may affect the type of analysis you can do.

A computer is distinguished from many other machines by being general purpose. One computer can run many different programs to perform different tasks. The term **hardware** relates to the physical part of the computer, while **software** is the particular set of instructions (or **program**) that is read and obeyed. Hardware has become more standard in recent years, conforming to choices made by the industry leaders. Software in contrast is amazingly diverse. Although many software writers attempt to match the appearance of perceived market leaders (as in 'look and feel'), the diversity of applications means that each program must have unique features. Any researcher must be prepared to devote time to learning the relevant software; this is as fundamental as learning any other research technique.

To have a sensible conversation about computers, whether with your supervisor, an adviser or a shop assistant trying to sell you one, it's essential to consider what you want to achieve, the type and amount of data you will use, and hence the choice of software and the type of hardware it will run on. It's best if you try to maintain strictly this order of consideration. Starting from the premise 'we have a model X computer, what can we do?' is almost certainly going to lead your research astray. When seeking advice you need to know what hardware and software you are using: questions such as, 'how can I print this file?' are meaningless in isolation.

Many people regard the jargon surrounding computers as a barrier and an irritation, but you will have to accept that this is a technical area where specific information can be conveyed concisely and accurately by the correct words. On the other hand, the use of technical words without inquiring as to their definition may lead to confusion. As a simple example, there are several types of floppy disks in use (double, high and quad densities) which differ in their physical properties and hence the types of signal used to write or read them. Some makes of computer do not detect that the wrong type of disk has been inserted. They will write data using the wrong format, so that disk may not be readable on other makes of computer and may not be reliable on the same machine.

Common English words may gain specific connotations. A **document** in a word processor is usually a special type of file which can be used only by that program. Putting text *into* a document is usually called 'entering' or 'typing'; *outputting* the file is called 'printing' though sometimes the output is not direct to a printing machine but to a different sort of file. Thus, 'how can I type this document?' and 'how can I print this document?' may evoke different answers.

Computer hardware divides roughly into the games and home entertainment market, office desktops and corporate systems. The divisions are arbitrary and movable: quite large organisations now base all their data processing on high-powered *desktop* machines. The best definition I have heard of a *personal* computer is 'one that can be hidden in a middle manager's budget'.

If we disregard the home machines, which may be equally powerful but do not have the range of software, it is likely that you will have access to either a **PC** or **Mac**. PCs are computers that run the same programs as the IBM PC though they may be made by any firm and the majority are not made by IBM. Macs, or Macintosh computers, are all made by the Apple Corporation.

PCs and Macs are different. They run different programs or different versions of the same program; disks and data files are not immediately interchangeable between PCs and Macs. Either type of machine can be linked to a **network** which is like a telephone system. Through the network you can access many other computers to run programs on them or to fetch data. The computers you access may be any make: typically they fall into the corporate description.

Software is far more diverse and many individuals offer specialist or novel programs. At the other extreme are well-established companies who sell polished and packaged software. The benefits of a major package include being able to assume that the program has been tested through use by many people, so it is unlikely to contain major **logical errors (bugs)**. Also you can get advice from experienced users, either at your own establishment or through user groups on the networks.

Programs may be aimed at **vertical markets**: groups of individuals with well-defined sets of needs and distinctive types of data values or structures. It is well known in computer science that any program with a basic set of operations can solve any (soluble!) problem, but you would be foolish to deliberately choose an inappropriate tool. For example:

- **Spreadsheets** were introduced to serve accountants, who deal mainly with tables of figures which they add across rows or columns or compare between sheets. Spreadsheets allow you to move rapidly round your data making adjustments and 'what if?' calculations to evaluate formulae. They have sometimes been recommended for teaching statistical methods as they allow students to see all stages of a calculation. As general programming tools, however, they soon become cumbersome and error-prone. They are not a replacement for a custom-written statistical package.
- **Word-processing packages** were introduced to serve typists, who were able to correct and improve text on screen. This saves retyping. Later versions made it possible to copy material directly from a spreadsheet, so that calculated tables could be incorporated into reports.

Some users, when shown this, assume that *any* table should be typed as a spreadsheet file and merged, rather than typed (as a typist would) using tabbed columns.

- **Graphics** packages are invariably sold with the slogan 'a picture is worth a thousand words'. Unfortunately most graphics (including many in professional publications) come nowhere near this ideal. As a first step in using any graphic software you must ask yourself if this graphic is being drawn to aid your analysis and understanding, or to present a particular message to an audience.

To return to a point above, a sensible question might be, 'how can I type a table into *WordPerfect* on a PC running *Windows*?', as opposed to Microsoft Word on a Macintosh.

Numeric data

Numeric data covers counts and measurements but also items that might be coded as numbers. Numbers may have implicit meaning (such as temperature) or at least imply an order (such as voting preferences) or may equate with named values (for example, female/male might be coded 0/1, 1/2, -1/1 or any other pair of numbers).

One common use is for strength of opinion, commonly recorded on seven-point **Likert** scales.

Nominal alternatives (technically called **factors**) code such items as the type of car owned by each respondent. Factors values (**levels**) are often conveniently recorded as names or mnemonic codes (such as f/m for sex). Most of the programs mentioned allow levels to be input as text codes; they can be converted into numbers if necessary by the program.

Whole numbers hold no problems of coding or processing. Fractional numbers are constrained both by the computer having a fixed space to hold each value (finite precision) and by the internal coding being binary (powers of two rather than powers of ten). This means that a number easily written in decimal notation (such as 0.1) will be held internally as an approximation:

$$\frac{1}{16}+\frac{1}{32}+\frac{1}{256}+\ldots$$

This can occasionally lead to problems when values are compared. Do two expressions give *exactly* the same result internally? Converting values between their external and internal coding is called formatting. You can choose how to write the values in your data but must then instruct the software of your choice of format so that it reads the values correctly. When you instruct the software to write values you can similarly control their appearance (not the value calculated) by choosing a format.

There are many computer packages well able to summarise and analyse data from questionnaires and experiments.

- **SPSS** and **Minitab** are widely available in research centres,
- **SAS**, another market leader, is more widely used in commercial organisations,
- **Genstat** and **Glim** are packages favoured by professional statisticians.

All such packages provide the standard analyses described in Chapters 22–24 of this book. Your role is to choose an appropriate analysis, describe your data and specify the analysis (usually with a short command), and interpret the output. The computer can be assumed to have done the calculation accurately. If it reports an error or produces unbelievable results, you almost certainly mis-specified the data or chose an inappropriate analysis.

For example, here is the set of commands to read data from a file into SPSS and tabulate the responses to two questions: the respondent's sex and their choice of newspaper. The `value labels` command is used to annotate the output; the computer does *not* understand the semantics of the data.

```
data list file = 'news.dat' free / gender paper.
value labels sex 1 'male' 2 'female' /
             paper 1 'Times' 2 'Telegraph'
                   3 'Guardian' 4 'Independent' 5 'Sun'.
frequencies variables = sex paper.
descriptives variables = paper.
```

and the output contains sensible tables. But what possible meaning could there be to the 'average paper' value?

Unlike a human assistant, the computer does what it's told, not what you would like or expect.

That little example also raises an important point of data collection and analysis. While each person has only one sex at a time, many people read more than one newspaper. It is easy to phrase a questionnaire so that respondents are constrained to give a single response, but they may resent this and it may reduce both your response rate and the quality of the data. If *multiple responses* are allowed, you must take account of this in your coding scheme and in the analysis. The software permits it, so this is not a computing but a research issue.

A good rule is to collect and record data in exactly those terms most familiar and comfortable to those providing the data; be they general public, trained professionals, paid technical assistants or yourself. Then copy those data verbatim into a computer file. If you must change or group values, do this on a copy within the computer. Computers are exactly suited to accurate and repetitive manipulations; people are not. If you try to recode data at the same time as recording or copying values, you will make mistakes and these will be more difficult to correct later on.

Numerical work without data

Computers can do huge numbers of calculations in seconds, and this had led to an emphasis on numerical (number-based) solutions. Don't be brainwashed into the view that computers can be used in only this way. They can be used for formal manipulation of mathematical models: **algebra**. You can include at a crude level the widely practised use of spreadsheets to study the value of a function over a range of input values, generally known as 'what if?' computing. A true algebraic package provides far more.

One use of computer algebra packages is as a check on conventional methods. Most computing uses a fixed-size pigeon hole for each value, so fractional values are truncated and held as approximations. When values are used in many calculations, the effect of truncation must be taken into account to ensure the accuracy of the answers. This is called **numerical analysis** and has been an important aspect of the growth of software.

Most **algorithms** shown in textbooks are derived from pure algebra and are *not* suitable for direct use in computers. That's another reason for not writing your own programs, especially as naive coding in a spreadsheet.

Computer algebra packages use special techniques to hold values as exact representations, whether they be whole numbers, fractions or irrational numbers. A coded algebraic formula will therefore give you an accurate answer, though perhaps take a long time to compute. The answer can be displayed exact or truncated. It's very simple for example to print out π to thousands of places of decimals: a calculation that may take a lifetime by hand.

At the other extreme, such packages can be used for sophisticated modelling and visualisation since they are often designed round the generation of graphical output. Among widely used programs are *Mathematica* (Unix and PC) and *Derive* (PC).

Text processing

Text or **word processing** may be your first use of a computer, but the computer can do far more than simply store and print typed material. There are complications caused by the historical development of computers being predominantly in the English-speaking community. It was for many years assumed that computing would be done in English regardless of the native tongue of the user. This is now changing, and leading software is being brought out in native-language or multilanguage versions. The use of English-biased computers to store and process non-English text is considered below.

Many research students nowadays type their own theses. This has benefits once the initial unfamiliarity of composing on screen has worn off. Apart from its use as a type-writer in which you can eliminate mistakes without retyping whole pages, the computer offers several tools to support the writing process.

When computers were scarce and users were competing for access, it was best to minimise thinking time at the keyboard. However, if you have free access, you should make use of features that make writing itself a creative part of the research process. In doing so, you need to know both the possibilities and limitations of each tool.

The **spell-checker** is a useful starting point, in that it works by looking up each word of your text in an extensive list. When the word has been found in the list there is no guarantee that it is the correct word. That is its main fault. Similarly a word that is not found there is not necessarily an error. All spell-checkers allow you to build up a personal addendum for word-forms not in the main list and for your technical vocabulary. The main dictionary file is usually in a machine-readable form to make look-ups fast and to keep the contents commercially confidential. This is irritating when you find a real spelling mistake within the dictionary, or if you have a fixed US-spelling dictionary but would prefer a UK version. You would also need a completely different dictionary for another language.

Characters are stored inside the computer as numbers. The commonest scheme in use is called ASCII, the American Standard Code for Information Interchange. This standard, established in 1962, assumed that no one would need letters other than the English A–Z, and it defines only the codes 0–127. (That's A–Z, a–z, 0–9, punctuation marks and a few others.) Thus, although computers work almost everywhere with eight-bit bytes (allowing 256 codes), the numbers 128–255 are *not standard* and it's essential you accept this if you are going to use any international networks or quote non-English text in your work. (No? Not even names of authors in citations?) In practice, each country or language has its own 'standard' to use the extra codes, leading to the joke: 'the great thing about standards is you can have so many of them'.

Style-checkers are less widely used and suffer more than spell-checkers from being mechanical. A style-checker can, however, be an aid to proofreading and you should at least run some samples of your text through one. The comments that come out should alert you to stylistic features of your writing that you should be able to defend, even if you will not change them. Typical points might include: long sentences, use of passive verbs, tautologies, vague qualifiers and cliché phrases. The style-checker may come within your word processor or may be run as a separate program but able to work directly on the word-processor document.

An **outliner** allows you to view your text at various levels. Each sentence and paragraph should make a point. It is good discipline to remind yourself of that point. Moving between an outline and the full text can help to avoid vacuous passages, ensure that points have not been overlooked, and that they flow in a logical order. People differ in how they work most productively so there is no requirement to plan the outline then expand it. It's equally valid to use stream-of-consciousness writing and then check or reorder the words. The outline may be purely used while writing, or may form part of the output, as a contents list.

A few companies and products dominate the word-processing market: WordPerfect, Word, WordStar and AmiPro. Despite the frequent claims that Macs are easier to use (which I personally find not proven) this is an area dominated by PCs. Unfortunately there are no standards for interfaces or file formats. As a result, products, and even versions of the same-named product, are often incompatible. This may be an inescapable effect of being in an immature and rapidly growing market.

You will find that any modern word processor provides numerous features that you will not need, and for those features you do need it provides several equivalent methods. You must be patient and be prepared to learn. Typing and text layout are real craft skills and while you may circumvent the need for employing a professional, you should not expect to ignore all the principles of the profession.

These sophisticated word processors now rival many programs marketed as **desktop publishing** (DTP). True DTP allows the visual manipulation of objects, lumps of text or graphics on the screen, to design a visually pleasing page. The word 'lump' was chosen deliberately, as one of the hallmarks of much amateur DTP is that the text gets corrupted and loses all sense while being moved about. Very commonly, the end of a sentence is [missing].

DTP is a productive tool in graphic design studios where it is ideal for producing brochures and advertising copy quickly. Less obviously, does it have much of a role in research? Once you have mastered its use, it might be ideal for designing eye-catching and informative conference posters.

There are some specialist areas in text formatting. All word processors claim to handle table layout but the examples and most published results do not bear scrutiny. See Ehrenberg (1982) or Chapman and Mahon (1986) for guidelines on the use and appearance of tables. Chemical structure diagrams are difficult because the conventions are less standardised. Mathematical formulae on the other hand contain numerous conventions built up by several generations of professional typesetters. What all these areas share is the need to place symbols in two dimensions rather than the simple linear sequence of text. While there are extensions or add-on products for the major word processors, in my experience the output is generally crude and inaccurate. This is partly caused by the inexperience of operators (it's very easy to see 'what you expect' rather than 'what there is' on a WYSIWYG screen) and partly by the software which underestimates the complexity of the task.

For top-quality text formatting in the academic area, a firm recommendation would be the $\LaTeX$ (pronounced *lah-tek*) system. This provides for all the areas listed in the last paragraph and provides unlimited access to symbols covering mathematics, logic, physics and many languages (not just Latin-based, but Greek, Arabic, Hebrew, Chinese, Sanskrit and many oriental and African scripts). $\LaTeX$ input is also acceptable (even preferred) by many academic publishers and journals. Most amazingly, the software is free and can be obtained for the price of a few computer disks.

Text as data

Text can be processed as data, though the operations are less tightly defined than the arithmetic performed on numbers. Text stored as data comprises not only the words (content) but any supporting or auxiliary information. If it is printed matter, this might include the fonts used, or the spacing, or any handwritten annotations; if it is transcribed interviews, it might include the identification of the speaker, the pauses between speech, even comments by the interviewer about body language.

The range of applications is wide and still growing. Many historic texts have been stored in great detail with much supporting material. Texts such as Shakespeare's *First Folio* are available for detailed literary or historical study. There is also a great deal of modern academic text on computers, either intentionally as a database or incidentally because computer networks are now a communication medium. Students can use this material for literary, psychological or sociological research.

Original text can be entered into a computer by direct typing or by devices such as **optical character readers** (OCR), though even the best of these still falls far below the accuracy (in reading) of an intelligent human. Recent marketing of pen-based hand-held computers has shown that software for reading handwriting is still rudimentary. Text that has been mechanically scanned may be stored as identified characters or as a graphic image. The supporting material may be inserted into the text, as markup, leading to tools such as SGML (Standard Generalised Markup Language). Alternatively, it may be stored as separate notes with cross-reference.

Text can be investigated in various ways. Transcribed answers to open questions on a questionnaire may need to be coded so that this information can be collated with the structured responses. This can be done by identification of keywords or phrases and a simple edit might suffice. More commonly, respondents will use synonyms, alternative word orders or alternative verb forms, so pattern matching is clumsy and error prone. The researcher must read and interpret each response.

Traditionally, researchers have used texts by making multiple copies, extracting quotations and cross-referencing. **Qualitative methods** based on computers support the same processes. As with numeric data, the computer is not doing magic, but it does the mechanical parts of the analysis far more quickly and accurately, enabling the researcher to handle more data and to consider more detailed analysis.

Graphic data

Computer graphics are popular and a strong selling point for many programs. They range from simple graphs that replace tables of output up to sophisticated visualisations of objects which could not be viewed otherwise. They may be generated for fixed display or you can get inside and navigate round **virtual reality**.

Graphics *as data* may be generated by one computer program and used by another or may be external images scanned into a file. They may be held in some page description meta-language (the best known is PostScript) or as a bit-map image (such as TIFF). The former can be transformed fairly readily and output at whatever resolution the display is capable of; the latter is already fixed at a certain size and any transformation or scaling will require some interpolation.

Images may be treated as data in the strict sense: for example satellite images of the earth need much processing and interpretation. At the other extreme the image may be saved so that it can be merged into another document. Images merged into word-processed text are best treated as **figures** so that commands can be applied to keep them associated with their captions. More sophisticated packages allow figures to be treated as **floats**, in which case the figure (and caption) can be moved to the top or bottom of the page and a cross-reference inserted in the text.

It's useful to distinguish between graphics for your own purpose, generated as part of the analysis of some data, and graphics for presentations. The former should be as neutral as possible, so that anything you read into the pattern is a reflection of the data, not of the method of presentation. The latter should emphasise clearly the message you intend. This should be information that is implicit in the data, but you should use graphics as precisely and effectively as you would use the written or spoken language.

References

Ehrenberg A S C (1982) *A Primer in Data Reduction*. John Wiley, London.
Chapman M and Mahon B (1986) *Plain Figures*. HMSO, London.

9
Buying your own computer

R Allan Reese

Hardware

Buying a computer is easy; choosing one is more difficult. By far the best method is to get experience of using computers so that you can make your own specification *before* talking to sales staff. The performance of computers has improved continuously and the price has dropped ever since the first; this process shows no signs of slowing, so whatever you buy will become obsolete in a couple of years, but provided you have been realistic about the type of work you expect to do, its working lifetime will be much longer. The first ever commercial computer (Leo) continued in use for 25 years!

Any advice from a printed book about buying a particular computer will be out of date. Since one of the great hidden costs of computing is support to *configure* (tune) the system and to link it with colleagues' machines, you should undoubtedly ask for local advice and follow it. Ask several people, and expect to receive contradictory answers. It's like asking car owners what make they recommend. In this section we try to alert you to some of the questions to ask. For example:

- Will you use your computer for note-taking, composing text or detailed page layout?
- Do you need to use particular programs, or would any program that provides equivalent functions be acceptable?
- Do you want to use your computer in libraries, in laboratories or in the field, or will it be based in the office or home?
- How can you transfer files between your system and other systems to which you have access? Do you want an identical system, a compatible system or merely a means for communication?
- What facilities do you need to connect your system to a communications network?
- (Most important) what will happen when (not if) your computer fails? What provision can you make to ensure you don't lose substantial amounts of work?

Lastly, how much do you want (can you afford) to spend? Do leave this until last. The worst thing to do is to go to a computer shop and say, 'I've got this much to spend. What can I buy?' Realistic figures at the time of writing would be: £300 for a simple machine for basic word processing and calculations; £800 for PC- or Mac-compatible system capable of running most modern software; £1500 for a current system that will run the latest software with acceptable response times.

Remember that while instant response may be desirable for interactive work, it is not strictly necessary and does mean that your computer is hardly used. If you know that a program will run even though it takes hours, you can leave the computer running overnight. This is quite safe and the electricity consumption is trivial. Other running costs include consumables (disks, printer ribbons or ink, paper), software and insurance.

Fast-response on-site maintenance contracts are probably not worthwhile; most machines show any defects well within the standard warranty period, or go on for years.

Parts of the computer

You interact with the computer through its display, so you must have a display that does not irritate you. Most software assumes the display can use colour, if only to mark out areas on the screen. On a monochrome (black and white) screen, colours have to be translated as tones. Desktop systems come with a VDU; monochrome systems are not much cheaper to buy.

You may also have a choice of screen size. The amount of detail is fixed within the computer by the choice of video card, so a larger screen will not show anything new, though it would help with fine-resolution work such as line drawing in computer-aided design (CAD). The commonest size screen is 14 inch. Larger screens (15, 17 and 20 inch) are more expensive and bulky. Laptop computers have to incorporate a flat display; colour adds a lot to the price. Most laptops will drive a VDU as an external monitor, so a versatile solution is a monochrome laptop for mobility, with a colour monitor to use at base.

Portable computers still suffer from limited battery life. You should consider whether or not you need a completely portable system (for use anywhere), or if you need it only to be portable between places where mains power is available.

Another consideration is the size of the hard (internal) disk. Current systems offer disks of at least 250 MB. While this is a huge capacity compared with just a few years ago, software writers seem ready to squander whatever you offer. However big the disk, you will quickly find it fills up and forces you to choose what programs and data to keep on it. The larger the disk, the more important it becomes to regularly use a disk maintenance program (such as Norton Utilities) and to keep backup copies of important files. Your computing service may help by providing backup onto cassette tape or via a network.

A recent trend is to supply software on CD-ROM. Adding a CD drive to your system makes it considerably more future proof. The current market in **multimedia** computers, incorporating CD and sound, is still changing rapidly; CD drives offer a range of speeds for data transfer, but so far the CDs are compatible with any available drives. A standard CD-drive can only read previously written CDs, which is why they are called ROMs (read-only memories). Devices for writing CDs are now appearing at prices around £2500 and they may become widely used for backup.

You will also probably want a printer. Despite the promise of the paperless office, most of us usually prefer to read and annotate on paper rather than on screen. Fortunately, a printer capable of giving acceptable quality for the throughput of an individual user can cost little over £100. For roughly £200 you can buy a bubble-jet printer that will print the same image as the highest quality printers except that it may be a bit fuzzier on the edges.

Except in a few special cases, the computer and printer are independent devices: you can plug any make of printer into the standard computer port. Different printers require different commands (control codes) but this is a software rather than hardware problem. When you need to use a higher quality printer, you should run the program to produce the output on your own computer with the appropriate device driver, save the output as a printer file, and take this file to the required printer. Note the difference between various types of system file (such as word-processor documents) and the printer file that is useful for only the one purpose.

You may be allowed to connect your personal system to a local network. One method is to use a modem which connects the computer to a standard telephone line. From your computer you can telephone to a central modem. This requires no changes to the computer, and the modem will be useful anywhere it is compatible with the telephone network. Watch out for firms who sell cheap modems within the UK that are *not* permitted for connection to the UK telephone system. As a bonus, most modems come with software that will turn a computer file into a fax message, so you do not need to print a document and rescan it to send a copy.

The second method is a connection to a dedicated computer network, and this generally requires an interface card to be installed into the computer. This is easily done, but the card may seem expensive if you will not be able to use it except on a particular campus.

A third method may soon be promoted as the best solution: this is the ISDN-standard card, which allows the computer to connect directly to a high-quality telephone line.

Software

Once you have your computer, you need to build a working portfolio of software. Choose a system that comes with preinstalled software. This will comprise the operating system, a graphical interface, and probably a bundle of basic but functional software tools. At one time you would have got this software as floppy disks packed with the machine. Now you do not have to install it, but companies usually do not provide the disks so you should make your own backup copy for security, to disks bought separately.

Undoubtedly you will get some software by copying from friends and colleagues. Please do not assume this is allowed. You must be aware that you may be stealing by copying and using. If you borrow a program, use it and like it, then buy a copy. Any student would be wise to check what software is available from the computing service for a reduced or nominal price. The exact licensing details may differ between programs and between sites.

A cheap way to expand your software is to buy computer magazines, many of which have disks or CDs of software attached. The usefulness of much of the software may be doubtful, but other items are priceless and you will get demo versions that allow you to try and understand programs before committing any real money.

Despite assurances from the publishers, you would be wise to invest in a certified and up-to-date virus-checking program if you are going to regularly insert any disks from elsewhere into your machine. Viruses are rogue programs, and a nuisance, but your computer can gain a virus only if you execute an infected program. If you never bring a new program onto your computer, you will be safe.

Commercial software is generally standardised and packaged. Most now comes shrink-wrapped as disk(s) or CD with a book and other documentation. Since most people (myself included) cannot bear to read the book first, there's generally a leaflet labelled 'read me first' or 'how to get started'. Follow a few simple instructions, and the program should be installed and running.

One problem is the attitude of software writers who assume their package will be the only one you install or use. They may demand that files be placed in particular directories, or that certain configuration parameters are set. If possible, any such features should be set in a shell or batch script file. The terminology varies, so seek advice from your local support staff. Such problems should decrease as applications become *plug 'n play* which is the cute buzzword for systems where the software can ask the computer about its special features.

In terms of price, the only advice is to shop around but be alert to minor differences in version numbers between apparently identical software. The computer magazines offer many mail-order suppliers, but sometimes the same product can be found in your high street. Whether software is the UK version or imported may be essential to its functioning or may affect the degree of support you can expect. Companies which sell assorted shrink-wrapped software do not offer much help in using it.

The final source of software is the **Internet**. It is estimated that over 30 million people now connect to the many individual networks that comprise Internet, so the variety of software is colossal. Join a discussion list or news group for a subject that interests you, and you will quickly get a feel for what programs are widely used and well regarded. Software can be picked up from many sites by a process called **anonymous ftp** (file transfer protocol). You make a direct call to the archive site, but since you are not a registered user of their machine you log on with the name 'anonymous' and type your e-mail address as a password. The password is used for their accounts (to see who's using their files) and so that their system can send you a message if the connection is prematurely lost. Ftp over international links can be slow and prolonged, so many sites round the world copy (mirror) each other with daily updates.

An example set-up

This is a reasonable specification for a medium-priced system at the time of writing. For reasons given, you may find that standards have changed, or that you can find a system that perfectly meets your needs at a far lower price.

Desktop PC

- Processors are rated by speed in several ways. The 486 series is the minimum for running Windows; 66 MHz (megahertz) is currently the best price/performance.
- **Pentium P75**. A Pentium is better for running Windows 95.
- **8 MB RAM** internal memory for running programs. Four megabytes may be offered, 8 MB is rapidly becoming the assumed minimum. You can increase this either at the time of purchase or later. You will also see references to *cache* and *video* (graphic) memory. These are separate memory chips included to further speed up the computer's running.
- **16 MB RAM** is the minimum for effective use of Windows 95.
- **PCI bus** is a technical point but it reflects the way the memory and processor interact. PCI is a new standard, VESA was the old one. VESA machines are therefore cheaper. The millions of VESA-based machines in use will ensure that repairs or upgrades will remain available.
- **500 MB hard disk** or larger! But take advice on how to keep a security backup of what is on that disk, especially your own data and documents.
- **3.5-inch floppy drive**. Changes in the physical size of disks have slowed down, but it's worth checking if the drive can use the 2.88 quad density.
- **CD-ROM drive** should be at least double-speed, though a quad-speed standard has been introduced. This is not the same as quad-density for floppy disks.
- **Sound card** is not strictly necessary, but the CD with sound card combination is bundled at a reduced price and does improve the computer for games or for playing music CDs.
- **Keyboard, monitor, mouse** are usually standard, but beware bargain-basement imports with a US keyboard.

Such a machine will typically come with preinstalled:

- **DOS** the fundamental and ancient command-based system for looking after the disk (file store) and running programs.
- **Windows** was designed as an interface between the user and DOS. The idea is that you point to names or pictures on the screen rather than type commands. The latest version, **Windows 95**, replaces DOS completely. The theory is great, but experience shows that Windows-based programs crash computers with depressing regularity and distressing loss of productivity. If you can get an expert to set up your machine and it works, use it and don't meddle!
- **OS2** was a minor rival to Windows for years, but the current release (OS2 version 3 WARP) has been well received and is now provided as well as, or instead of, Windows by several PC companies.

- **Works** or a similar integrated package. This provides simplified equivalents of most of the basic office programs: word processor, spreadsheet, presentation graphics. The word processor and spreadsheet may well meet all your needs, so use them. The graphics are usually less satisfactory.

One reason for looking for a higher performance machine than described would be to run desktop publishing and design software with advanced features. If that's your interest, you will need far greater depth of knowledge to make a wise purchase than is included here. In particular, you should seriously consider whether a PC or a Mac would be better.

Laptops

Should you decide that you need a portable computer, you could choose the same configuration as for a desktop, but would have to pay at least three times the price. If you will accept a smaller internal disk, monochrome screen and no multimedia capability, the price will be at the same level, and you will still have a machine that will run the same software, though slower. Laptops using the 386 processor are widely available, and excellent for running DOS versions of word processors and other standard programs.

For simple note-taking, you will not need the power or the software compatibility of a PC. The limiting factors may be weight and the delay after switching on. If you don't carry the machine, or if it takes two minutes to reach the point where you can enter notes, then you are unlikely to use it.

I use an Amstrad NC 200, advertised as 'learn to use it in five minutes or your money back'. It has its own word processor, but as it has a disk drive that uses PC-format disks, my main use is to type notes which are transferred to a desktop machine as plain text.

Macintosh

Macintosh is the only serious rival to PC-compatible computers for office use. It has remained a minority interest, though for a passionate minority. The strengths of Macs lie in graphic design areas. At one time Macs also had the advantage of uniformity, compared with the disparate designs of PCs. Nowadays, there are as many compatibility problems amongst Macs. The PowerMac was an attempt to bring together the best features of Macs and PCs. Whatever its merits, the political alliance underlying the idea did not last, and the outcome was a machine that is more expensive for running either type of software while requiring software vendors to produce a third version of their programs to take advantage of PowerMac's specific features.

Types of software

Depending on your subject and interests, you may want at least one program from each of the following groups:

- **word processing**
  Write, Works, MS Word, WordPerfect, AmiPro, WordStar
- **text layout/DTP**
  LaTeX, Timeworks, Ventura, PageMaker
- **spreadsheet**
  Works, Excel, Quattro, Supercalc
- **presentation graphics**
  Harvard Graphics, CorelDraw
- **data collection**
  EpiInfo

- **statistical analysis**
  SPSS, SAS, Stata, Splus, Minitab
- **database programming**
  Clipper, Paradox, Access
- **bibliographic database**
  Papyrus, Reference Manager, End Note
- **networking**
  Kermit, Procomm, Telcon
- **system maintenance**
  Norton Utilities, PC Tools, QEMM, Stacker

According to your choices, you might spend from almost nothing up to several hundred pounds. Be prepared to ask, and be flexible in what you will accept.

10
Who can help?

Shirley Coleman

Introduction

You have started your postgraduate studies – well done. Who can help you with ideas, support and encouragement? This chapter looks at supervisors and what to do if they are no use and the help available from other academic staff, librarians and technicians. It looks at a wide range of other sources of help including public services, government departments, trade unions and industrial companies. Even if you feel you have enough help, contact with these various sources can be very interesting and useful later on in your career. These experiences may add that little extra sparkle which will make your research excellent rather than just very good.

Supervisors

The potentially most useful resource is your personal supervisor. His or her task is to help you through your studies, getting the best from them and finishing successfully and ready to take up suitable, gainful employment. Some supervisors, however, are not too keen to dedicate sufficient time to you and this can be very frustrating.

If you look on the bright side, and assume your supervisor is interested in your studies and conscientious, what can you expect?

You can have regular meetings at which you summarise your work to date or since the last meeting. This means that you keep a good record of your activity throughout your period of study. Your supervisor can add his or her own thoughts and ideas to yours and offer suggestions which spark off further ideas in either of you. As well as you keeping an eye on published work and the advances in your field, your supervisor can be doing the same. This is useful as you can keep tabs on much more work and be more likely to stumble upon relevant information.

Besides providing technical support and inspiration, a good supervisor will deal with all the administration to do with your study. This could include the important question of choosing appropriate (and hopefully sympathetic) external examiners and applying for time extensions. The exact title of your thesis, if you are writing one, or of any publication, is very important; remember it will follow you around on your CV and job applications for ever more, so a good supervisor will help you choose the best wording which will allow you to adapt your past to suit your future.

A good supervisor can help you to meet academics and others who may be interested in you and your work. He or she may become a life mentor for you, helping you to find employment after you finish studying. If you have to defend your study in a *viva voce* a beneficent, supportive supervisor is a great asset as well as a comfort, especially if you are borderline.

Poor supervisors may be unwilling to see you or may not concentrate fully when you try to discuss your work with them.

If your supervisor is poor, what can you do about it? You can try to improve your availability and flexibility about when meetings are held; you can try to improve the quality of the presentation of your work. Your enthusiasm may ignite the interest of your supervisor, if you're lucky. If not, at least you will feel that you've done your best to improve the situation. In the final analysis if none of these efforts have any effect you just have to try to carry on regardless and look further afield for people to be involved with. It is worthwhile considering why your supervisor is underperforming. Sometimes supervisors are too busy; it is not always a good plan to go to study with someone whose fame is meteorically rising – they will not have much time for you. Perhaps you or your project bores your supervisor. Another possibility, however, is that your supervisor thinks you are highly competent, are doing a good job and don't need their help. Although this may be a flattering concept, it is also a nuisance as everyone can benefit from the input and interest of an eminent mentor, but again unless you go to great lengths to prove your incompetence there is not much you can do except get on as best you can on your own. Try to avoid becoming too self-centred. In a few years' time you yourself will have forgotten most of the vital details you are struggling with now.

In the rare (hopefully) cases where a supervisor is actually incompetent or maleficent, what can you do? You can try to change to another supervisor without becoming too involved in accusations and unpleasantness. Be careful how you choose another supervisor; someone who is keen, conscientious and compatible may be the best even if he or she is not the most renowned. Ask previous tutees how helpful they found their tutor. Try to check whether they are likely to move jobs or retire – racing them up the stairs or challenging them to tennis is not a reliable test! If it is not possible to change your supervisor or it's too late, the best you can do is keep a good record of your work. Don't rely on the supervisor at all but use the other resources (suggested in this chapter) in case you need to defend yourself later. Be positive and think of your traumatic period of study as good training for coping with later life. After all, if you're careful no one can stop you from doing your work, writing it up and passing successfully. This is more use than making a big fuss and trying to upset a structure that has so much inertia that it appealed to you as a stable place to study in the first place. If you do need to take matters further, you can always discuss your problems with the head of department, or dean of the faculty. If these people are unhelpful or you wish to go further, seek advice from the students' union and student affairs officers before you do anything rash – academics have long memories.

Even if you have a good supervisor, other sources of help can provide you with practical examples of where your study is relevant. This makes your work more interesting and allows unusual questions to be asked when you are interviewed for further study or jobs later on.

Attracting ideas

Try advertising yourself! If possible it's a good plan to give a talk or seminar about your work early on in your study. That way you get help in terms of ideas and suggestions from your audience and you can check what other work is being done in your field and by whom (this reinforces your own literature search).

Academic sources of help

Academic members of staff are usually happy to spend a few minutes discussing ideas which interest you arising from their study. It's often better just to drop in rather than be too formal. You can always ask their secretaries when is a good time to catch them.

The authors of relevant papers in journals that you have come across during your study are also likely to be enthusiastic about discussing their work with you. Use the 'address for correspondence' on the papers to contact them. This sort of proactivity not only provides you with state-of-the-art ideas but also gets you known and may be useful when you are looking for a job. You can combine a visit to more distant authors with the summer holidays but remember to take some presentable clothes away with you and a notebook!

Conferences often have attractively reduced rates for students and are usually well worth the money and effort of attending. You can cross-examine authors of relevant presentations and make some useful contacts for sharing ideas, help and inspiration. Occasionally there are conferences especially for postgraduate students. The subject matter will be wide ranging but they provide a good opportunity for you to make a presentation and for you to share your experiences with other postgraduates.

Universities and colleges usually run a series of seminars and lectures during term-time, again the subject areas covered will be wide but these are often well worth attending.

Other sources of help

Libraries

Help may be available from a surprising number of diverse sources as well as the more obvious places. University, college and public libraries are a standard starting point. Besides the books and journals available, try talking to the specialist libraries who have a vast pool of information at their fingertips. Government publications are surprisingly rich in all sorts of facts and figures; the monthly publication, *Social Trends*, for example, will tell you how many died from falling off ladders in the last month.

Have a look at the reports from government and other research institutes; they may give you some good ideas and contacts.

Government departments

The Department of Trade and Industry (DTI) has local offices which you can visit. You can find out which companies are involved in the latest initiatives and who you should contact to ask about them.

The DTI jointly sponsor the Teaching Company Scheme in which graduates are employed by a university as a research associate for two years but work full time in a company on a specific project. These projects may be relevant to your study and the list of current programmes could be very useful.

Other government departments may offer help and ideas, for example the Central Statistical Office. There is also the European Commission and the various ministries, for example the Ministry of Agriculture, Fisheries and Food (MAFF).

Local services

Town halls and civic centres have excellent information facilities and are able to supply data on many aspects of the local economic and social scene. Try looking up your city, country and local councils in the telephone book.

The local Chamber of Commerce is a good source of literature and contact names and addresses. They may be able to put you in touch with appropriate trades societies or trades unions. They are involved in research as well and may be willing to help you. The Chamber of Commerce will also have literature from the Institute of Directors, Confederation of British Industry and other institutions.

Miscellaneous

The financial pages in newspapers give useful information about companies. Keep a scrapbook of any articles which are relevant to your area of study.

Keep a check on the activities of local colleges and schools, they sometimes have project competitions in very interesting subject areas.

Groups of students may work with a local company and you could make use of the same contact.

Hospitals have personnel dedicated to disseminating information. They also have staff libraries which you may be allowed to browse through if your study is in an area associated with health.

Technical support

Laboratory staff know a lot more about the way things work than you can learn from a book, so make sure you are on good terms with them and listen carefully. Materials- and equipment-supply companies are very happy to furnish postgraduate students with literature and demonstrations. You may expect them not to be interested in you as you have no intention of buying anything, but on the contrary, they see you as the customers and people with influence of the future. They will be very keen to promote their products.

Companies

University or other careers services are probably the best places to learn about companies of interest to you. Usually the information is readily available for you to look at without having to make an appointment or talk to anyone. There is nothing more satisfying when you go to visit someone than knowing more about their company's business than they do! You can try to organise your own visit with a specific person in the company or you can contact their public relations department and try to join a factory tour. In either case remember to dress the part and take a notebook.

Making a visit

Arrange the visit when you are sure you are ready – don't risk having to cancel. Read up as much as possible about the place and the people who work there. Be clear in your mind what it is that you want to know or want to look at. Write out a checklist to be sure you don't forget anything. Double check your travel arrangements and take a map. Allow plenty of time to get there and plenty of time afterwards in case you're invited to stay for lunch, tea, evening meetings, and so on. Wear appropriate clothes. Afterwards, write a quick thank you note saying briefly what benefits you gained from the visit. That way they will be happy to have you back and will remember you if you apply for a job there!

Family and friends

Don't forget this valuable collection of committed supporters. Explain your ideas to them and try to cope with their common sense comments – they often say things which make you think very deeply about what you're trying to do. In most write-ups you need a layman's introduction, abstract or summary; try reading your attempt to a friend and see whether they can understand it. An outside view will help you to bring out the important points more clearly.

Summary

In summary, there are many sources of help if you have the energy to go out and seek them. The benefits will be widespread in terms of making your study more interesting and relevant and in the advantages it provides in your future career.

11
Managing your PhD

Stan Taylor

Introduction

While the knowledge and skills that you gained as an undergraduate and/or in studying for a master's degree have given you a basic background in your subject and perhaps some experience of and insight into the process of research, they may not necessarily have equipped you to undertake a PhD. As Salman (1992, p. 51) has put it,

> Unlike a certificate, a diploma, a bachelors or master's degree, a doctorate does not merely entail the consideration of already existing work within a prearranged structure but demands the creation of a personal project. To undertake a PhD is therefore to define oneself as having a contribution to make to the understanding of the area concerned.

Making such a contribution can, in itself, be a daunting and difficult experience, the more so because although you should have the advice, encouragement, and support of your supervisor, academic colleagues in the field, and fellow-graduate students, ultimately the responsibility is yours, i.e. you stand or fall largely by your own efforts. You have to create the project; undertake the research; write it up; complete it on time and submit; defend your thesis at your *viva*; and if you do all of these things successfully then you personally are awarded the title of 'Doctor'.

Many students start out on the road leading to a doctorate, but fall by the wayside and fail to complete their theses. While there are many reasons for this (see Phillips and Pugh, 1992, pp. 32–45), one of the most common is a failure to self-manage their research. Given that the latter is concerned with adding to knowledge, it obviously involves a creative process, and the argument can be made that this cannot, inherently, be managed, that is produced effectively and efficiently to order. This view was very much the orthodoxy a few years ago, and was a factor explaining why so many PhDs took a decade or more to complete or why, in some cases, students remained registered for the degree boy and man, girl and woman. However, particularly since the mid-1980s, research funding bodies led by the Economic and Social Research Council (Winfield, 1987; Young *et al*, 1987) have taken a tougher line on completion within three or four years including penalising departments which do not meet target rates of submission. So, if you are funded by the research councils, you are likely to be under pressure to complete, or if you are one of the increasing number of self-funded graduate students the pressures will be even greater. You cannot now rely upon the contention that creativity is a lumpy and lengthy process so that your thesis takes as long as it takes; you have to aim to complete within a given time frame and actively manage your thesis to, as far as possible, meet your targets.

The aim of this section of the book is to help you to think about how you are going to manage your PhD. With that in mind, the objectives of the chapter are to consider:

1. how to approach your PhD
2. how to start it
3. how to plan it
4. how to organise it
5. how to manage your thesis
6. how to manage your relationship with your supervisor
7. how to manage yourself
8. how to complete your thesis
9. how to prepare for examination.

These form the topics of this chapter and, hopefully, the points made will help you to manage your research better and to self-improve your chances of finishing and gaining the degree.

Approaching your PhD

If you are thinking about doing a PhD thesis, it makes sense to firstly enquire about precisely what you are letting yourself in for, that is: *what is a PhD thesis*?

All universities with PhD programmes have, in their rules and regulations, a definition of a PhD thesis for purposes of examination. While these vary considerably between institutions, they will usually include requirements that the thesis should be:

- on a specified and approved topic
- constitute a substantial piece of scholarship
- the work of the candidate
- make an original contribution to knowledge
- in principle worthy of publication.

On this basis, then, what you have to do is to specify a topic, have it approved by the relevant university body, produce a substantial piece of scholarship on your own which adds to knowledge and which is publishable.

This would be fine if there was some agreement as to the meanings of 'substantial' 'scholarship' 'own work' 'original contribution' or 'publishable'. But academics disagree strongly, both within and between disciplines about what these words mean, and there are divergent views among other stakeholders in graduate education such as the research funding councils, the government, and employers. As Partington *et al* (1993, pp. 67–68) have suggested, at one extreme, the PhD can be seen as a multivolume exhaustive account of a subject produced entirely by independent study, opening up new vistas in the discipline, and ready to go into proof as a major book; at the other it may be seen as the application of extensive research training to produce a narrow account of a detailed aspect of a phenomenon tackled as part of a research team and leading to an article or two in specialised journals. Between these extremes of the individually produced blockbuster and the narrow demonstration of ability to apply a research training lies a range of other conceptions of theses embodying a diversity of characteristics.

If, then, the nature of the beast remains elusive, what can you do? By far the best strategy is to go to the library and look at a few successful theses recently written in a cognate area to that which your are considering and ask yourself:

- Were they on big topics or relatively narrow ones?
- Did they involve a massive amount of detailed research or
- Was this limited and narrowly focused?
- Were they the product of a single scholar or was their considerable attribution of contributions by colleagues?

- Did they aim to open up broad new fields of knowledge or push the barriers out slightly in a tightly-defined area?
- What made them publishable?

By these means, you should hopefully have a reasonably clear idea of what, at the end of three or four years, you should expect to be able to achieve and you can embark on your own PhD.

Starting your PhD

Apart from those whose thesis topics are tied in to specific projects of their supervisor (which has been historically more common in the sciences), students beginning their PhDs usually have a broad idea about the field they want to work in, most frequently an area of interest identified in the course of undergraduate and/or earlier postgraduate study. But most are unsure about precisely what they want to do in their research, and the first few weeks or months are spent in deciding upon a topic. This usually involves looking at other work on or relating to the broad area, seeing where there are significant gaps/areas of disagreement/problems/conflicting interpretations, reflecting, refining, and coming up eventually with a detailed research proposal.

This can be a frustrating, difficult, and demoralising experience as your initial ideas and expectations often turn out to be too ambitious and you find yourself thrashing around in an intellectual vacuum and moving from the macrolevel of your subject through the meta- and so to the microlevel in the search for a topic. You should, however, receive strong support at this stage from your actual or potential supervisor who should act as a sounding board for your ideas and give you the benefit of his or her experience and that of appropriate colleagues. With such assistance you should eventually come up with a topic which, following Moses (1992, pp. 9–13) and Rudestram and Newton (1992, pp. 10–11), is:

- *viable* – can be tackled with available or specified obtainable resources, including the human one of a good supervisor with relevant experience and expertise in the area, library, computing facilities, and laboratories
- *do-able* – can be done by someone with your knowledge and skills or with specified additions to these and within a reasonable period of time
- *sustainable* – hold your interest and maintain your commitment over a period of several years
- *original* – has the potential to make an original contribution to knowledge
- *acceptable* – conforms otherwise to the requirements and standards that examiners may apply

Both you and your supervisor must be reasonably satisfied on all of these counts before you proceed further with the thesis.

Planning your PhD

When you were an undergraduate or a postgraduate taking a taught course, you worked to a plan, that of your degree programme which organised and directed your work over the duration of your studies. Postgraduate research is no different in so far as it needs to be conducted to a plan as well; one of the central causes of dropping out is failure to plan, resulting in loss of direction and time. But what is different is that it is now your responsibility, in conjunction with your supervisor, to draw up a plan and to direct your studies accordingly.

While PhDs vary hugely in their structures and format and in how they are tackled, most involve the nine components of:

1. review and evaluate general and specific literary and other sources in the area of the topic
2. identify the gap/problem/issue/conflict or other trigger for the research and formulate appropriate investigations/potential solutions questions/theories/alternative interpretations
3. decide upon an appropriate paradigm/conceptual framework/methodology within which to do the research; specify concepts, theories, hypotheses, statistical procedures and significance levels before you embark on survey or experimental work
4. tackle the substantive research, involving one or a combination of:

   - reading original documents
   - subject interviewing
   - elite interviewing
   - mass interviewing
   - scientific fieldwork
   - scientific experimentation
   - problem solving
   - computer modelling

5. sift, check, and analyse material
6. interpret findings/results and assess their implications for the initial focus of the inquiry
7. reformulate the field of knowledge in the light of your findings
8. summarise what your research has achieved, how it has been achieved, what conclusions have been reached and, if appropriate, directions for future research
9. write the drafts of your thesis.

These are the core components of most theses. There may be more, depending on the topic, which will define your tasks over the next few years.

While it is difficult to assign precise time values to each of these components and some, of course, may overlap, you should, in conjunction with your supervisor, try to plan a schedule for completing the stages of the research within a reasonable time limit.

So you may, for example, plan in outline to spend the first six months refining the topic, reviewing and evaluating the literature, and dealing with methodological questions, eighteen months undertaking the substantive research, three months analysing, interpreting and reformulating, and three months gluing the whole together, writing your summary and conclusions, and finally polishing the thesis for submission. Then, with these broad boundaries in place, you may be able to break the components up into subcomponents and try to set targets for their completion.

It is essential that, when you specify time for each of the components of the thesis, you should make a clear allocation for writing up that part of your work. This is critical for five reasons:

1. it gets you straight into the habit of writing rather than leaving this until the end when it can be difficult to acquire
2. it encourages you to reflect upon what you have done before you go on and may highlight problems/avenues of exploration
3. it provides a continuing record of your achievements so far
4. it enables your supervisor to see exactly what you have done and advise you how to proceed
5. if you write throughout you build up a portfolio of your work which you can then fashion into the shape of the final thesis rather than face beginning from scratch to assemble your research (and often finding that the sum of unconnected parts fails to add up to a thesis).

Your plan, then, should embody time for writing up each component. An example of such a plan, including writing time, is presented by Phillips and Pugh (1992, p. 85), and this can be adapted for your own use.

Organising your PhD

With an outline plan in place, you now have to decide how you are going to try and meet it in a organised way. This involves organising your time, working conditions, and materials.

Time

As a postgraduate student you are responsible for managing your own time and, without any sort of preparation for this, it is, as one student reported to Welsh (1979, p. 33), all 'too easy for the postgraduate to spend his [her] time pottering about' and fall behind.

Many universities now recognise that it is necessary to offer graduate students assistance with time management and embody a session in the appropriate induction programme, while others allow postgraduates access to courses run for the academic staff. If your institution doesn't offer assistance, you might consider:

- dividing your days into blocks of (say) two hours and keeping a diary of what you have done in that period
- reviewing this and calculating the time you spent on your thesis and the time you lost through distractions of one kind or another
- reorganising your day to ensure adequate working time and to minimise unwanted distractions.

Regular reviews of this kind can enable you to establish and maintain a stable pattern of work and help you to adjust psychologically to postgraduate study.

Working conditions

The demands of doctoral study are, or can be, very intense, and you need an appropriate working environment where you can read, reflect, think, evaluate and write. In line with this, the national body which represents postgraduates, the National Postgraduate Committee (NPC, 1995) has produced guidelines, including the provision of:

- office space preferably within the department where students are based with designated maximum numbers
- necessary equipment with a minimum of a desk, chair, lamp, bookcase, and lockable filing cabinet
- where appropriate, a carrel in the library or bench space in the laboratory
- access to computing facilities, preferably ones dedicated to postgraduate use
- private common space for postgraduates to study when necessary on their own
- access to the departmental staff common room and to the postgraduate common room for social purposes.

You should, preferably before signing up for a PhD, ensure that, as far as possible, these facilities are available to you. If there are deficiencies, you should bring these to the attention of your department as factors likely to delay your research and completion.

Materials

In the course of your PhD, you can expect to accumulate a vast amount of materials relating to your research, and it is important that you organise these properly. There is nothing worse than writing up and finding that you cannot trace the precise source of that key point or that your reference is inadequate and have to interrupt the flow while you trek to the library to sort it out. You should:

- find out which referencing system you will be expected to use in your thesis and note the exact requirements for footnotes or endnotes and for your bibliography
- assume anything that you read may eventually be cited and take full details of the reference. A master list of these should be stored in a card index or, even better, a database which can then later be manipulated and sorted to form your bibliography
- establish a filing system and file your materials under appropriate headings (this may be manual or computerised but in the latter case copies must be backed-up regularly);
- index your filing system so that it is easy to find any document.

Organisation of your material in this way should mean the minimum of delay while you are doing your research, particularly writing up, and hence the least interruption to your creative flow.

Managing your thesis

The creation of a plan for your PhD and organisation of your work is one thing, but it is hard to stick to it. While books and articles suggest that research consists of a seamless unrolling of the advancement of knowledge, in practice it may be two steps forward and one back, not to mention several sideways. Common problems with theses include the following.

Drifting from the topic

As the research progresses, highways and byways of new exploration open up which just have to be investigated because they could be vital. So every avenue is investigated until you become lost in the maze of possibilities and unable to find your way back to where you should be at that stage.

Difficulties with the methodology

The section on methodology can require you to grapple with a whole range of philosophical, theoretical, empirical and experimental problems, and it can require a major effort to try and identify, tackle, and resolve these, particularly when you are really itching to undertake the substantive research.

Frustration with the substantive research

You can expect a range of problems to occur as you undertake the substantive research: evidence that you can't obtain as easily as you hoped; experiments that don't work; apparently promising lines of enquiry which turn out to be dead ends; simulations which don't run properly – the list is endless.

Inconsistencies in findings

With the substantive research accomplished, you experience difficulties in analysing and interpreting: the evidence which is contradictory; the experiments which yield unexpected results; the cast-iron assumptions which are apparently falsified; the simulation results which defy predictions; variables which behave badly; and so on.

Writer's block

Finally, there is that dreadful experience when, after doing all the work for a chapter or more, you sit down at the word processor and find that you are unable to write it up in any acceptable form – the words won't come out or, if they do, they appear so trite that you reach instantly for the delete key and then repeat the process until you become completely fed up with yourself and go off in disgust and do something else.

Any or all of these experiences – and you are almost bound to meet them – can throw your planning out of gear, but there are several things that you can do to prevent problems getting out of hand. These include the following.

A regular review of your progress

Your supervisor is responsible from an institutional standpoint for reviewing your progress, but this can entail anything from weekly through to annual reviews. If your meetings with your supervisor are infrequent and/or irregular, you yourself should set aside time on a regular basis and ask yourself:

- What did I plan to achieve by this stage of my research?
- What have I actually achieved?

In this way you can identify shortfalls, hopefully at an early enough stage to correct them.

Acknowledge the existence of a problem

As an undergraduate or master's student, you may have sailed through with effortless brilliance and it can be an immense shock to encounter problems of the kinds outlined above, and acknowledging them can be seen as weakness or failure. But such problems are experienced virtually throughout the research community, and admission should not be conceived as a weakness, but as a strength. So, if you are falling behind, acknowledge that there is a problem, reflect upon the reasons why, and try to identify a solution.

Seek help

If you cannot quite define the problem or find a solution, don't simply pace your room waiting for inspiration and sink into the slough of despond – seek help. You can find this from:

- The literature – there are several books dealing generally with problems of researching a PhD (for example, Phillips and Pugh, 1992; Maunch and Birch, 1993; Preece, 1992; Rudestram and Newton, 1992; the present volume) or with PhDs in particular subject areas (for example, Allan and Skinner, 1991; Beynon, 1993; Frost and Stablein, 1992; Murrell, 1990) or with specific problems such as writer's block (Hall, 1993) which may offer help and assistance.
- Fellow-graduate students – they may well be experiencing similar problems or, if further on, have met them and sorted them out, so they may be able to assist and/or provide support and encouragement. Some departments now try to promote such mutual self-help by organising graduate-only workshops on a regular basis or by mentoring whereby a student near completion lends a sympathetic ear to those at earlier stages.
- A member of your department – junior members of your department will probably have fairly recently completed a PhD themselves, and will be able to point you in the direction of how to solve difficulties. Sometimes colleagues may take an interest in your research and

become, in effect, informal mentors, and there is some evidence (Lyons *et al*, 1990, pp. 277–85) that such support improves both the self-confidence of graduate students and their performance.

- Your supervisor – your ultimate source of assistance should be your supervisor who is charged with guiding you through your research on behalf of the university and who is, or should be, an expert in the broad field of your thesis and the processes and problems of research.

Your relationship with your supervisor

You should enjoy a supportive and productive relationship with your supervisor, one which starts as 'a master [or mistress] pupil relationship and ideally end(s) up as almost equal colleagues' (SERC 1983, quoted in Young *et al*, 1987, p. 28). These days, most supervisors receive at least some training in their responsibilities and often have to serve an apprenticeship period 'shadowing' an experienced supervisor before undertaking duties on their own (Elton *et al*, 1994, pp. 24–37). So, the chances are that you will have a good experience, provided of course that you yourself play your part and:

- turn up regularly for supervisory sessions
- take a constructive approach to criticisms and comments made (it can be painful to have pet theories criticised by your supervisor but not nearly as much as having them shot down in flames by your eventual examiners)
- treat supervisions in a business-like way as an opportunity to review your achievements so far and to set realistic objectives for progress (as an *aide memoire* it can be very useful immediately after the meeting to jot down summary notes of what was said, what you agreed you would do, and when you agreed to do it by).

But, what happens when your supervisor, your ultimate source of assistance, feels unable to help because of your subject matter, or is unable to find adequate time, or is disinclined to assist?

- *Inability to help* – it may be that, while your supervisor has a general expertise within the field, he or she is unable to assist with the specifics of your research, in other words your work goes beyond what he or she feels totally confident to supervise. In this case, he or she should at least have an extensive knowledge of the research network in the field and be able to point you towards sources of assistance outside your institution and, possibly, outside the country. Whereas, a few years ago, it was difficult to conduct a domestic or international dialogue about your research problems, this has of course been greatly facilitated by the Internet, which can be a very useful resource for graduate students.
- *Unable to find time* – all academics are under severe pressures to perform in research, teaching and administration, and time is at a premium, especially in institutions where postgraduate supervision is considered to be part of research activity and no allowance is made for this in allocating other duties. But, that said, it is no excuse for the occasional horror stories about postgraduates having most of their meetings with supervisors in railway stations or airports or receiving comments on their draft chapters months after they have been handed in for comments. If an academic agrees to supervise you then he or she takes on a commitment and should be prepared to fulfil it.

  So, if you are having problems, consult your university's guidelines for supervisors and, if available, postgraduate charter, and point these out to him or her. (If your institution has neither of these, the National Postgraduate Committee (1992) has produced *Guidelines for Codes of Practice* and you should consult these.) If you still experience problems, then a word with the head of the graduate school or department and/or a new supervisor may be the only way forward.
- *Unwilling to help* – supervisors may be unwilling to help because, in their judgement, it oversteps the boundaries between supervising your research and doing it for you. This may

reflect a traditional view of a PhD as a venture made worthwhile because it is alone and heroic with the weak falling off the mountain or alternatively a very real concern that, to continue the analogy, you are asking him or her to put in the supports and then carry you to the summit. If you are stuck and your supervisor proves unwilling to help or evasive, you should try and discuss the matter and resolve it. If that fails and you feel your feet giving way and are not thrown a rope, again consider taking the matter further.

Managing yourself

In addition to managing your thesis and your relationship with your supervisor over the course of your studies, you also have to manage yourself. That is, you must cope with the personal problems of postgraduate life. These can include initial adjustment, isolation and loneliness as you get into your research, and mid-thesis crisis.

Initial adjustment

At school and, to a lesser extent, as an undergraduate or a postgraduate on a taught degree, your programme of studies was mapped out for you and you had to meet deadlines set by your department for the delivery of work. As a postgraduate, you are, as noted in the quote at the start of this chapter, responsible for mapping out your own programme of study and for implementing it. This can be a liberating experience. Possibly for the first time in your life you are your own boss. But it can also be a frightening experience and it is not unusual for postgraduates in their first few months to feel as if they have been cast adrift on a stormy sea without a clear course to steer by. While, a few years ago, this period was regarded as one you had to survive to continue the degree, many institutions now offer help in the form of comprehensive induction programmes which are designed to help PhD students acquire appropriate skills, while training programmes for supervisors are now placing greater emphasis upon the need to support supervisees during this critical period. So, if you do feel lost and adrift, you are by no means alone and you should seek and expect support to help you to adjust to postgraduate life.

Isolation and loneliness

At school and as an undergraduate you studied in company. It can come as a shock to find yourself spending much of your time working on your own without human contact, and you can become isolated and depressed in consequence. Again, working alone was considered by many to be a necessary evil a few years ago but now some universities recognise the need to offer support to postgraduates. Academically, this may take the form of formal graduate supervisions involving not just the customary one, but two or three graduates, the use of graduate mentors, graduate-led workshops and seminars or, increasingly, through separate graduate schools. Socially, graduates are brought into contact with each other by such schools or by university or departmental postgraduate societies which organise a range of events, or through participation in student societies.

If these things are not done by your institution or they are not done enough and you still feel isolated and on your own, the answer may be a graduate self-support group. This usually consists of a few students, not necessarily from the same subject area, meeting once a week or once a fortnight to discuss their progress or the lack of it and helping each other in other ways. You should establish some appropriate ground rules, for example with regard to confidentiality and the making of constructive rather than destructive remarks, and try and help each other.

Mid-thesis crisis

One phenomenon which has been widely observed (see Phillips and Pugh, 1992, pp. 77–78) is the tendency towards a crisis in mid-thesis. You are now well into your research, churning it out day after day, and you become bored with the whole thing and ripe for distractions which will take your mind off the drudgery of your research and entertain you. While there is no simple solution to this problem, if you want to complete you have to continue the research, it can be beneficial to take a short break and then come back to it with a fresher mind. But, if you contemplate this, do stick to a defined break – there are many ex-PhD students who took a breather from their studies and then procrastinated and procrastinated about returning until it was far too late.

Completing your PhD

With the mid-thesis crisis overcome and the substantial research and the analysis complete, you still face what can be the momentous task of assembling all that you have done over a period of three or four years, perhaps more, into shape as a thesis. This task will be much easier if you have written draft chapters up as you go along, but these rarely fit together perfectly and work is still required to produce a coherent and cohesive account.

You have followed an intellectual journey across unknown territory. Think of yourself as the explorer producing a guidebook to where you have been and what you have seen and discovered in the process. As the author of the guidebook, you need to explain:

- your starting point and why you decided to embark on the journey (the literature and the deficiencies revealed by evaluation which led you to undertake the research)
- how you decided to undertake the journey (the methodology)
- the route you followed and the discoveries you made on the way (the substantive research chapters)
- how, in the light of the above, you redrew the route (analysis and interpretation)
- where you arrived at the end of your journey, how it differed from your starting point, and where you go from here (conclusions, knowledge added, and directions of future research in the subject).

You should start by, literally, drawing an outline map of these points. Ensuring that the various stages link together and are reasonably consistent with each other so that, in general, the route is clear and can be easily followed. Then, within each part of the route, each chapter, you need to decide what must be said to take your reader through that stage and lead him or her onto the next one, bearing in mind that the deviations up highways and byways which were so fascinating to you might be irrelevant to others. Concentrate on the essentials and leave extraneous materials to footnotes, endnotes or appendices.

If you do this you should have a master-map of the route as a whole and detailed guides to each of the sections of it: a template for your thesis. You should now try this out on your supervisor and ask for his or her comments before proceeding. Otherwise, when you have completed the draft, you may find major flaws in the structure which necessitate a major rewrite.

If your template is accepted, then you can begin to fill in the detail chapter by chapter; some feel the need to start at the beginning and work forwards, others in reverse, and still others begin in the middle and expand in either direction. But, whichever way you do it, the important thing is that you write, and write regularly so that you produce a complete first draft. Again, practice varies, but there is certainly one school of thought to the effect that the most important thing is simply to get it down rather than worry too much about the presentation, in other words you should be centrally concerned with substance and not worry too much about appearance at this stage.

If all goes well, then you should have a first draft of your thesis, one which may look rough and ready but at least it contains the essentials. You should add an introduction (explain what you aimed to do and why it was important) and show the draft to your supervisor. He or she should, at this stage, be able to advise you whether, in principle, your thesis has what it takes and, if not, what you need to do to improve it. Additionally, he or she may make recommendations as to presentation.

Often, in our haste to get the thesis out of the way or because we believe the substance stands alone, the presentation is hurried. But there are two reasons for taking some time to ensure that this is done properly. Firstly, while good presentation cannot rescue a poor thesis, it may help a marginal one, the examiners may be inclined to take a more charitable view if the thesis is easily readable and, as far as possible, error-free. Secondly, inadequacies in expression and errors in spelling and grammar are one of the most common reasons for the referral of theses, i.e. for these being accepted subject to minor corrections. It can be extremely galling to have to spend a month or two correcting elementary mistakes and errors, not just to you but to whichever examiner is landed with the task of listing your errors and checking that your have corrected them before the degree can be awarded. So it is important that you get this right before you go further.

You should:

- ensure that you have expressed yourself as clearly and concisely as possible (reading out loud can often help to identify sentences that are too long and unnecessary padding)
- check the grammar
- check that you have the right words (spell-checkers can tell you whether the word is spelled correctly but not if it is the right word in the first place)
- check the spelling
- check that you have followed appropriate conventions for footnotes, endnotes, quotations, citations, in the text and in the bibliography (a good guide is Turabian, 1982).

Given that many of us can be blind to our own deficiencies and errors, a friend may help with some expertise in the area to comment on the comprehensibility of the draft. Ask him or her or even someone from another discipline who will be less distracted by the content to check it for errors.

With this done, it is back to your supervisor for a final reread and, with luck, the green light to go ahead and submit the thesis for examination. If your supervisor still has reservations, you can still submit, ultimately it is your decision, but you should consider this carefully for fear of falling at the final fence.

If you do decide to submit, first you should check your university's regulations about submission. Usually, you must give notice of submission, and you may have to conform to a range of stylistic regulations, such as page margins and layout, as well as provide an abstract, a contents page, and a count of the number of words. Only when you are satisfied on this score can you print the required number of copies to be bound and submitted to your university.

Preparation for examination

When you have formally submitted and the university is satisfied that your thesis meets its regulations, the wheels are then set in motion for the process of examination, beginning with the appointment of examiners. In this respect, practice varies between universities, but there will normally be two, one from your department (but not usually your supervisor) and one external.

While there is not, and indeed should not, be any formal procedure for consulting candidates about the external, your head of department and/or supervisor will normally be asked for nominations and he or she may raise the matter with you and present some possible names. If you have reservations, such as that *Dr So and So* in your department has consistently been sceptical of the value of your research or that *Professor A N Other* outside might take unkindly to the fact that your thesis has refuted his or her life's work, you can and should raise these. There is, of course, no guarantee that your preferences will be taken into account and it may be that you end up with *Dr So* and *Prof Other*. If you do fear foul play, then in many universities it is possible to have your supervisor present at your *viva* (taking no part in the proceedings but able to observe) and in the unlikely event of perceived victimisation, there are usually appeal procedures which involve a fresh reading by a third examiner.

When examiners have been appointed, there is usually a hiatus of anything from a few weeks to several months while they read, digest, and form opinions about your work, following which they arrange the statutory oral examination, the *viva voce*.

While *viva*s play a part in most institutions at undergraduate level, this is usually limited, and for many postgraduates the *viva* on their thesis is a novel experience and it can be a daunting prospect. So it makes sense to, firstly, think about the possible purposes of your *viva* and, secondly, to prepare yourself for the experience.

You have already given your examiners written evidence of your abilities in the form of your thesis, and the purpose of the *viva* is to gain additional evidence relating to your suitability to be awarded the degree of PhD. Such evidence will normally be of two kinds, one relates specifically to issues arising from your thesis, the other, more generally, to your professional competence as a scholar.

As each PhD thesis is unique, there are no generic templates detailing what you are likely to be asked specifically about your work, but you can gain some insights into possible questions if you:

- check the publications of your examiners which may yield clues as to their likely interests
- looking at published guidelines for external examiners in your own or in adjacent subject areas. In the case of some disciplines there are highly detailed guidelines, for example in psychology (British Psychological Society/ UCosDA, 1995), and there are also suggested guidelines in Partington *et al* (1992, pp. 74–75) for assessing theses based on experiments and on documentary studies, respectively
- asking competent colleagues/friends to read your work and identify possible issues which could be taken up.

With regard to more general questions, as well as scrutinising your thesis, examiners will need (Partington *et al*, 1992, p. 77) to be satisfied that:

- the thesis is substantially your own work
- you have developed skills in research at this level
- you have an understanding not just of your thesis but of the general field to which it relates
- you have fully thought through the implications of your work.

You need, then, to be prepared for questions concerning these areas as well, in particular to show that you are fully familiar with what you have written, have a thorough mastery of the methodology to demonstrate your research skills, are aware of the wider literature pertaining to your field, and are clear about what your research means for your subject.

Thus prepared, a mock *viva* can be of help as a final step before the big day. Many departments now arrange these for their students by arranging for members of staff, preferably those who have been externals themselves, to read the thesis and ask you appropriate questions before giving you feedback on your performance. This can be invaluable in anticipating lines of inquiry and in improving your presentation skills. With these sharpened up, you should be ready for the real thing.

Before the *viva*, your examiners should have been in contact with each other to identify the strengths and weaknesses of your thesis and to reach a preliminary verdict and, upon this basis, to decide upon the objectives of the *viva*. They may:

- pass the thesis as it stands or with minor amendments and use the *viva* to discuss the latter
- pass the thesis as it stands or with minor amendments subject to satisfaction in the *viva* on your subject knowledge and professional competence
- regard the thesis as genuinely marginal and use the *viva* to probe problem areas of your thesis to determine whether you otherwise meet the standard for the award of the degree
- regard the thesis as unacceptable as it stands and use the *viva* to guide you towards the reasons.

Where there is no doubt that the thesis meets or exceeds the standards, humane examiners will put you out of your misery and tell you at the start of the *viva* that they intend to recommend that you be awarded the degree, perhaps subject to minor changes. But, even if your thesis is fine, other examiners may follow the letter of the law and give you no clue at the start as to the purpose of the *viva* and interrogate you at length on it to establish your subject and professional credentials. So, prolonged questioning is by no means necessarily an indication of failure, and you should not be disheartened by it.

Usually, if you have not been informed at the start, at the end of the *viva* you will be asked to leave while your examiners confer, and then brought back in and given an informal indication of the result prior to the formal recommendation to the university. This may be:

- the immediate award of the degree
- the award of the degree subject to minor revisions to be completed within a specified period
- the referral of the thesis for major revisions followed by resubmission and re-examination
- the award of a lower research degree or that no award be made.

If the outcome is either of the last two, you may wish to consider an appeal, and you should consult your university's procedures. If your thesis is referred, your examiners should give you a detailed list of the work required before resubmitting the thesis. If the degree is to be awarded subject to minor revisions, again you should receive a detailed list of the changes which have to be made and then the revised thesis is resubmitted to the internal examiner. Subject to his or her certification that the changes have been made within the time limit specified, the degree is then awarded and you join those who were lucky enough not to have to amend their theses with the title of 'Doctor'.

Conclusion

The road to a PhD is a long and hard one with many pitfalls on the way. But, as I hope this chapter has shown, you can improve your chances of gaining the degree by actively seeking to manage the processes of starting, planning, reviewing, undertaking, completing, and preparing for examination. While such management will not turn an inadequate thesis into a successful doctorate, its absence can, and in the past often has, meant that a promising topic has come to nothing, most likely the thesis has been abandoned. So, as you work for your PhD, remember that while you are being examined explicitly on your ability to make an original contribution to knowledge, you are also being examined implicitly upon your ability to manage your studies and yourself, and you need both to be awarded a doctorate.

References

Allan G and Skinner C (1991) *Handbook for Research Students in the Social Sciences*. Falmer, London.
Beynon R (1993) *Postgraduate Study in the Biological Sciences*. Portland, London.
British Psychological Society and UCosDA (1995) *Guidelines for Assessment of the PhD in Psychology and Related Disciplines*. CVCP, Sheffield.
Elton L and Task Force Three (1994) Staff development in relation to research. In Zuber-Skerritt O and Ryan Y (eds), *Quality in Postgraduate Education*. Kogan Page, London, pp. 24–37.
Frost P and Stablein R (1992) *Doing Exemplary Research*. Sage, Newbury Park, CA.
Hall C (1993) *Getting Down to Writing: A Students' Guide to Overcoming Writer's Block*. Centre for Research into Human Communication and Learning, Cambridge.
Lyons W, Scroggins P and Bonham Rule P (1990) The mentor in graduate education. *Studies in Higher Education*, Vol. 15(3), pp. 277–285.
Maunch J and Birch J (1993) *Guide to the Successful Thesis and Dissertation*. Marcel Dekker, New York.
Moses I (1992) *Supervising Postgraduates: Higher Education*. Research and Development Society of Australia, Campbelltown.
Murrell G (1990) *Research in Medicine: A Guide to Writing a Thesis in the Medical Sciences*. Cambridge University Press, Cambridge.
National Postgraduate Committee (1995) *Guidelines on Accommodation and Facilities for Postgraduate Research*. NPG, Troon; and (1995) *Guidelines on Codes of Practice for Postgraduate Research*. NPG, Troon.
Partington J, Brown G, and Gordon G (1993) *Handbook for External Examiners in Higher Education*. CVCP, Sheffield.
Phillips E and Pugh D (1992) *How to Get a PhD*. Open University Press, Buckingham.
Preece R (1992) *Starting Research: An Introduction to Academic Research and Dissertation Writing*. Sage, Newbury Park, CA.
Rudestram K and Newton R (1992) *Surviving Your Dissertation*. Sage, London.
Salman P (1992) *Achieving a PhD – Ten Students' Experience*. Trentham, London.
Turabian K (1982) *A Manual for Writers of Research Papers, Theses and Dissertations*. Heinemann, London.
Welsh J (1979) *The First Year of Postgraduate Research Study*. Society for Research into Higher Education, Guildford.
Winfield G (1987) *The Social Science PhD: The ESRC Enquiry on Submission Rates*. ESRC, London.
Young K, Fogarty S and McRae S (1987) *The Management of Doctoral Studies in the Social Sciences*. Policy Studies Institute, London.

Part Three

RESEARCH TYPES

12
Randomised trials

Douglas Altman

Introduction

A randomised trial is a planned experiment which is designed to compare two or more forms of treatment or behaviour. Randomised trials were originally developed in agriculture, as a means of getting valid comparisons between different ways of treating soil or plants where it was known that there was inherent variability in the land being used (such as drainage or exposure to wind). The same principles extend to numerous other disciplines where a comparison of treatments is needed, in particular when humans are being studied (such as in education or medicine).

The key idea of a controlled trial is that we wish to compare groups which differ only with respect to their treatment. The study must be prospective because biases are easily incurred when comparing groups treated at different times and possibly under different conditions. It must be comparative (controlled) because we cannot assume what will happen to the patients in the absence of any therapy.

Clinical trials merit special attention because of their medical importance, some particular problems in design, analysis and interpretation, and certain ethical problems. The methodology that is used was introduced into medical research 50 years or so ago, the most famous early example being a trial of streptomycin in the treatment of pulmonary tuberculosis (MRC, 1948). Randomised trials of medical treatments thus have all the general features of such studies in all fields but several additional problems. This chapter concentrates on clinical trials so that all aspects are covered. Some of the special problems (such as allocation bias, blinding, ethics) will not apply in other areas. It should be clear which these are. Nevertheless, even if not relevant to all circumstances these comments should serve to illustrate the way in which the design and analysis of research studies have to be tailored to individual circumstances.

Although clinical trials can be set up to compare more than two treatments I will concentrate on the two group case. Often one treatment is as an experimental treatment, perhaps a new drug, and the other is a control treatment, which may be a standard treatment, an ineffective 'placebo', or even no treatment at all, depending on circumstances. Trials may also compare different active treatments, or even different doses of the same treatment.

There are several books devoted to clinical trials, of which that by Pocock (1983) is particularly recommended. The papers by Peto *et al* (1976) and Pocock (1985) give a useful discussion of some of the trickier problems, see also Chapter 15 in Altman (1991).

Trial design

Random allocation

An essential issue in design is to ensure that, as far as is possible, the groups of patients receiving the different treatments are similar with regard to features that may affect how well they do. The usual way to achieve comparability is to use **random allocation** to determine which treatment each patient gets. This ensures that variation among subjects that might affect their response to treatment will on average be the same in the different treatment groups. There is no guarantee, however, that randomisation will lead to the groups being similar in a particular trial. Randomisation ensures that there is no bias in the way the treatments are allocated. Any differences that arise by chance can be at least inconvenient, and may lead to doubts being cast on the interpretation of the trial results.

While the results can be adjusted to take account of differences between the groups at the start of the trial, the problem should be controlled at the design stage. This can be done with **stratified randomisation.** In essence this involves randomisation of separate sub-groups of individuals, such as patients with mild or severe disease. If you know in advance that there are a few key variables that are strongly related to outcome they can be incorporated into a stratified randomisation scheme. It is essential that stratified randomisation uses **blocking**, otherwise there is no benefit over simple randomisation. There may be other important variables that we cannot measure or have not identified, and we rely on the randomisation to balance them out. With blocking, the patients are divided into relatively homogeneous subgroups, or **blocks**, and treatments are allocated randomly within blocks so that each block provides a small experiment. The precision of the overall comparisons between treatments is provided by the variability between blocks. Blocking is described more fully in chapter 14.

While randomisation is necessary it may not be a sufficient safeguard against bias (conscious or subconscious). The treatment allocation system should thus be set up so that the person entering patients does not know in advance which treatment the next person will get. A common way of doing this is to use a series of sealed opaque envelopes, each containing a treatment specification. For stratified randomisation, two or more sets of envelopes are needed. For drug trials the allocation is often done by the pharmacy, who produce numbered bottles which do not indicate the treatment contained.

Blindness

The key to a successful clinical trial is to avoid any biases in the comparison of the groups. Randomisation deals with possible bias at the treatment allocation, but bias can also creep in while the study is being run. Both the patients and the doctors may be affected in the way they respond and observe by knowledge of which treatment is given. For this reason, neither the patient nor the person who evaluates the patient should know which treatment is given. Such a trial is called **double-blind**. If only the patient is unaware, the trial is called **single-blind**. In a double-blind trial the treatments should be indistinguishable (see 'Placebos' section).

Alternative designs

The simplest and most frequently used design for a clinical trial is the **parallel group design**, in which two groups of patients receive different treatments concurrently. The most common alternative is the **cross-over design**, in which all the patients are given both treatments of interest in sequence. Here randomisation is used to determine the order in which the treatments are given. The cross-over design has some attractive features, in particular that the treatment comparison is **within-subject** rather than **between-subject**. Because within-subject variation is usually much smaller than between-subject variation the sample size needed is smaller. There are some important disadvantages, however. In particular, patients may drop out after the first treatment, and so not receive the second treatment. Withdrawal may be related to side-effects. Cross-over studies should not be used for conditions which can be cured.

Lastly, there may be a carry-over of treatment effect from one period to the next, so that the results obtained during the second period are affected. In other words, the observed difference between the treatments will depend upon the order in which they were received. Cross-over trials are covered well by Senn (1993).

Another type of design is the **sequential trial**, in which parallel groups are studied, but the trial continues to recruit patients until a clear benefit of one treatment is seen or it is unlikely that any difference will emerge. Sequential trials require that the data are analysed after each patient's results become available. There are problems with blinding, therefore, and also ethical difficulties. Their use is restricted to conditions where the outcome is known relatively quickly. Adjustment is made to significance tests to allow for multiple analyses of the data.

A useful variation on this principle is the **group sequential trial**, in which the data are analysed after each block of patients has been seen, typically three, four or five times in all. This design allows the trial to be stopped early if a clear treatment difference is seen, if side-effects are unacceptable, or if it is obvious that no difference will be found.

One further type of design is called the **factorial design**, in which two treatments, say A and B, are compared with each other and with a control. Patients are divided into four groups, who receive either the control treatment, A only, B only, or both A and B. This design allows the investigation of the interaction (or synergy) between A and B. It is rarely used in clinical trials, but its use is becoming more common. Such designs (and more complex ones) are standard practice in agricultural field trials and in industrial research (see Chapters 13 and 14).

Lastly, there are some situations in which groups of individuals are randomised, usually for logistic reasons. Such studies are common in health education, for example. In such trials the group is the unit of randomisation rather than the individual. Special considerations are necessary in the design and analysis of these trials.

Sample size

For parallel group studies the calculation of sample size is based on either a *t*-test or a comparison of proportions. For cross-over studies sample size calculations are based on the paired *t*-test. Sample size calculations are based on the idea of having a high probability (usually 80–90 percent) of getting a statistically significant difference if the true difference between the treatments is of a given size. Unfortunately, choice of this **effect size** is not easy.

Placebos

When we wish to evaluate a new treatment and there is no existing standard beneficial treatment, it is reasonable *not* to give the control group any active treatment. It is better to give the control group patients an inert or placebo treatment rather than nothing. Firstly, the act of taking a treatment may itself have some benefit to the patient, so that part of any benefit observed in the treatment group could be due to the knowledge/belief that they had taken a treatment. Secondly, for a study to be double-blind, the two treatments should be indistinguishable. Placebo tablets should therefore be identical in appearance and taste to the active treatment, but should be pharmacologically inactive. More generally, when the control treatment is an alternative active treatment rather than a placebo the different treatments should still be indistinguishable if possible. For surgical trials it is usually impossible to meet this requirement.

In many clinical trials we do find some apparent benefit of treatment in the placebo group, and there are often side-effects too. Without a placebo, we cannot know how specific any benefit (or harm) is to the active treatment.

Ethical issues

One of the main ethical issues concerns the amount of information given to the patient. In general the patient should be invited to be in the trial, and should be told what the alternative treatments are (although they will not know which they will get). The patient can decline, in which case the patient will receive the standard treatment. If the patient agrees to participate he may have to sign a form stating that he understood the trial. This **informed consent** is controversial, because it is likely that a lot of patients do not really understand what they are told, and that they may not always be told as much as they should be. There are some cases where it is not possible to get informed consent, for example when the patients are very young, very old, or unconscious.

No doctor should participate in a clinical trial if he or she believes that there is a clear preference overall for one of the treatments being investigated, and the doctor should not enter any patient for whom he thinks that a particular treatment is indicated. In other words, the ideal medical state to be in is one of ignorance: the trial is done because *we do not know which treatment is better*. This is sometimes called the **uncertainty principle**.

Outcome measures

In most clinical trials information about the effect of treatment is gathered about many variables, sometimes on more than one occasion. There is the temptation to analyse each of the variables and see which differences between treatment groups are significant. This approach leads to misleading results, because multiple testing will invalidate the results of significance tests. In particular, just presenting the most significant results as if these were the only analyses performed is fraudulent.

A preferable approach is to decide in advance of the analysis which outcome measure is of major interest and focus attention on this variable when the data are analysed. Other data can and should be analysed too, but these variables should be considered to be of secondary importance. Any interesting findings among the secondary variables should be interpreted rather cautiously, more as ideas for further research than as definitive results. Side-effects of treatment should be treated in this way.

Protocols

An important aspect of planning a clinical trial is to produce a protocol, which is a formal document outlining the proposed procedures for carrying out the trial. Pocock (1983) suggests the following main features of a study protocol:

1. background
2. specific objectives
3. patient selection criteria
4. treatment schedules
5. methods of patient evaluation (outcome measures)
6. trial design
7. registration and randomisation of patients
8. patient consent
9. required size of study

10. monitoring of trial progress
11. forms and data handling
12. protocol deviations
13. plans for statistical analysis
14. administrative responsibilities.

A protocol must accompany an application for a grant for a trial, and most of the above information will be required by the **local ethics committee**. Further, as well as aiding in the performance of a trial, a protocol makes the reporting of the results much easier as the introduction and methods section of the paper should be substantially the same as the description of features one to nine above.

For **multicentre studies** a detailed protocol is essential, and it is strongly recommended for any clinical trial. Indeed, a protocol is recommended for any research project.

Many difficulties can be avoided by having a pilot study, which is also valuable for assessing the quality of the data-collection forms, and for checking the logistics of the trial, such as the expected time to examine each patient (which affects the number that can be seen in a session).

Analysis

The general methods for the statistical analysis of data, which are described in chapters 22 and 23, apply to clinical trials and you should read those chapters. Several problems arise in the analysis of clinical trials, some of which are considered here. A fuller discussion of bias in analysis is given by May *et al* (1981).

Comparison of entry characteristics

Randomisation does not guarantee that the characteristics of the different groups are similar. The first analysis should be a summary of the baseline characteristics of the patients in the two groups. It is important to show that the groups were similar with respect to variables that may affect the patient's response. For example, we would usually wish to be happy that the age distribution was similar in the different groups, as many outcomes are age-related.

A common way to compare groups is with hypothesis tests (Altman and Doré, 1990), but a moment's thought should suffice to see that this is unhelpful. If the randomisation is performed fairly we know that any differences between the two treatment groups *must* be due to chance. A hypothesis test thus makes no sense, and in any case the question at issue is whether the groups differ in a way that might affect their response to treatment. This question is clearly a question of clinical importance rather than statistical significance. If we suspect that the observed differences (imbalance) between the groups may have affected the outcome we can take account of the imbalance in the analysis, as described below.

Incomplete data

Data may be incomplete for several reasons. For example, occasional laboratory measurements will be missing because of problems with the samples taken. It is important to use all the data available and to specify if any observations are missing. Also, some information may simply not have been recorded. It may seem reasonable to assume that a particular symptom was not present if it was not recorded, but such inferences are in general unsafe and should be made only after careful consideration of the circumstances.

The most important problem with missing information relates to patients who drop out of the study before the end. Withdrawal may be by the clinician, perhaps because of side-effects. Alternatively, the patient may move to another area or just fail to return without warning. Efforts should be made to obtain at least some information regarding the status of these patients at the end of the trial, but some data are still likely to be missing. One possible approach is to assign the most optimistic outcome to all these patients and analyse the data, and then repeat the analysis with the most pessimistic outcome. If the two analyses yield similar results, and results also similar to those from an analysis in which these patients are simply excluded, then we can be fairly confident in the findings. However, if there are many more withdrawals in one treatment group the results of the trial will be compromised, as it is likely that the withdrawals are treatment-related.

Protocol violations

A common problem relates to patients who have not followed the protocol, either deliberately or accidentally. Included here are patients who actually receive the wrong treatment (in other words, not the one allocated) and patients who do not take their treatment, known as **non-compliers**. Also, sometimes it may be discovered after the trial has begun that a patient was not eligible for the trial after all .

The only safe way to deal with all of these situations is to keep all randomised patients in the trial. The analysis is thus based on the groups as randomised, and is known as an **intention-to-treat analysis**. Any other policy towards protocol violations will involve subjective decisions and will create an opportunity for bias.

Adjusting for other variables

Most clinical trials are based on the simple idea of comparing two groups with respect to a main variable of interest, for which the statistical analysis is simple. We may, however, wish to take one or more other variables into consideration in the analysis. One reason might be that the two groups were not similar with respect to baseline variables. We can thus perform the analysis with and without adjustment. If the results are similar we can infer that the imbalance was not important, and can quote the simple comparison, but if the results are different we should use the adjusted analysis. Imbalance will only affect the results if the variable is related to the outcome measure. It will not matter if one group is much shorter than the other if height is unrelated to response to treatment. The use of some form of randomisation that is designed to give similar groups (such as stratified randomisation) is thus desirable as it simplifies the analysis of the data.

Outcome measures

One outcome measure should be treated as the main focus of attention in the analysis. There may be other outcome measures, and these can be analysed using the same methods. If there are genuinely several outcome measures of importance, then the P value considered statistically significant should be made smaller than the usual five percent to keep the risk of a **Type I error** small. One simple method is to use the **Bonferroni correction**, in which if there are k variables being analysed then results are considered significant only if P is less than $0.05/k$.

Sub-group analyses

Even with a single outcome variable there is often interest in identifying which patients do well on a treatment and which do badly. We can answer a question like this by analysing the data separately for sub-sets of the data. We may, for example, redo the analysis including only male patients and then only female patients. Sub-group analyses like these pose serious problems of interpretation similar to those resulting from multiple outcome measures. A preferable general approach is to compare the treatment effect observed in the sub-groups of interest, such as men and women. It is reasonable to do a small number of sub-group analyses if these were specified in the protocol, but on no account should the data be analysed in numerous different ways in the hope of discovering some 'significant' comparison, nor because some aspect of the data suggests a difference.

Interpretation of results

Single trials

In most cases the statistical analysis of a clinical trial will be simple, at least with respect to the main outcome measure, perhaps involving just a *t*-test or a *chi-squared* test. Interpretation seems straightforward, therefore, but for one difficulty.

Inference from a sample to a population relies on the assumption that the trial participants represent all such patients. However, in most trials, participants are selected to conform to certain inclusion criteria, so extrapolation of results to other types of patient may not be warranted. For example, most trials of antihypertensive agents, such as beta-blocking drugs, are on middle-aged men. Is it therefore reasonable to assume that the results apply to women too, or to very young or very old men? In the absence of any information to the contrary it is common to infer wider applicability of results, but the possibility that different groups would respond differently should be borne in mind. This issue is discussed by Ellenberg (1994).

All published trials

In many fields there have been several similar clinical trials, and it is natural to want to assess all the evidence at once. The first thing that becomes apparent when looking at the results of a series of clinical trials of the same treatment is that the results vary, sometimes markedly. We would expect to see some variation in treatment effect, because of random variation, and should not be worried by it. The confidence interval for the treatment benefit observed in a single trial gives an idea of the range of treatment benefit likely to be observed in a series of trials of the same size. A recent development has been to move towards the statistical analysis of all published trials to get an overall assessment of treatment effectiveness. The analysis is known either as an **overview** or a **meta-analysis**, although a better, more general term is **systematic review** (see Chalmers and Altman, 1995). Overviews have often shown a highly significant overall treatment benefit when most of the individual trials did not get a significant result. Again, this is to be expected, as many clinical trials are too small to detect anything other than an unrealistically huge treatment benefit.

Assessing the quality of clinical trials

Many studies have found that inadequate reporting of trials is widespread (such as Altman and Doré, 1990). For both writing up a clinical trial and assessing the quality of a published trial it is useful to have a checklist of the important issues (Gardner *et al*, 1986; SORT Group, 1995). Trials must be judged on the information that is included in the published report. We should not assume a satisfactory answer to any of the questions on the checklist if the information is not given.

References

Altman D G (1991) *Practical Statistics for Medical Research*. Chapman & Hall, London.

Altman D G and Doré C J (1990) Randomisation and baseline comparisons in clinical trials. *Lancet*, Vol. 335, pp. 149–153.

Chalmers I and Altman D G (eds) (1995) *Systematic Reviews*. British Medical Journal, London.

Ellenberg J H (1994) Selection bias in observational and experimental studies. *Statistics in Medicine*, Vol. 13, pp. 557–567.

Gardner M J, Machin D and Campbell M J (1986) Use of check lists in assessing the statistical content of medical studies. *British Medical Journal*, Vol. 292, pp. 810–812.

May G S, DeMets D L, Friedman L, Furberg C and Passamani E (1981) The randomized clinical trial: bias in analysis. *Circulation*, Vol. 64, pp. 669–673.

MRC (1948) Streptomycin treatment of pulmonary tuberculosis. *British Medical Journal*, Vol. 2, pp. 769–782.

Peto R, Pike M C, Armitage P *et al* (1976) Design and analysis of randomised trials requiring prolonged observation of each patient. I. Introduction and design. *British Journal of Cancer*, Vol. 34, pp. 585–612.

Pocock S (1983) *Clinical Trials: A Practical Approach*. John Wiley, Chichester.

Pocock S (1985) Current issues in the design and interpretation of clinical trials. *British Medical Journal*, Vol. 290, pp. 39–42.

Senn S (1993) *Cross-over Trials in Clinical Research*. John Wiley, Chichester.

Standards on Reporting of Trials Group (1995) A proposal for structured reporting of randomized controlled trials. *Journal of the American Medical Association*, Vol. 272, pp. 1926–1931.

13
Laboratory and industrial experiments

Tony Greenfield

Introduction

Experimental design and analysis is an essential part of scientific method.

Every experiment should be well designed, planned and managed to ensure that the results can be analysed, interpreted and presented. If you do not do this you will not understand properly what you are doing and you will face the hazard of failing to reach your research goals, of wasting great effort, time, money and other resources in fruitless pursuits.

Other chapters in this book (Chapters 22 to 24) are about the analysis of experimental data. This chapter is about the design of experiments.

An experimental design is:

the specification of the conditions at which experimental data will be observed.

Experimental design is a major part of applied statistics and there is an immense literature about it. In this chapter we present only those aspects of experimental design which have most to contribute to the physical sciences, specifically to laboratory and industrial experiments. Such experiments are aimed primarily at improving products and processes.

Descriptions are necessarily brief and selective. Please read this chapter as an introduction and a reference. There are many good books on the subject and a few of the best are listed at the end of this chapter.

There are many types of experimental design. In this chapter I describe a range of designs, selected for their usefulness to the physical sciences, including manufacturing and industries, and leading from the simplest to the more complex. These are given below.

Descriptive

Descriptive

A sample of several test pieces, all from a standard material, is tested to determine the elementary statistics of a characteristic of that material. You may, for example, wish to report the mean and the standard deviation of the tensile strength of a standard material.

Comparative

Comparison against a standard

You may wish to compare the characteristic of a new material against a specified industry standard. You would test a sample of several pieces and ask if there was sufficient evidence to conclude that the measured characteristic of this material was different from the standard specification.

Comparison of two materials with independent samples

You may have two materials, perhaps of different compositions, or made by slightly different processes, or, even if they are claimed to be of the same composition and made by exactly the same process, they are made at different places. You wish to determine if they have the same or different properties so you test a sample of several pieces from each material. These samples are independent of each other.

Comparison of two materials by paired samples

You may have two materials and you wish to determine if they have the same or different properties. However, you wish to ensure, in the presence of uncontrollable outside influences, to make a fair comparison. For example, you may wish to expose samples of structural steel to the weather and measure their corrosion. One approach would be to expose test pieces in pairs, each pair comprising one item piece of each material. The data to be analysed would be the difference in corrosion measured between each pair by weighing them separately.

Response

Factorial experiments

When new materials or manufacturing processes are being developed there are usually several variables, or factors, that can influence a material property. Experiments to investigate the effects of several variables should be designed in which all of those variables are set at several levels.

> **Warning**: There is a widespread belief, that is still taught in schools and to undergraduates, that the best approach is to experiment with one variable at a time and to fix all the others. That approach is inefficient, uneconomic, and will not provide information about interactions between variables.

Two-level factorial experiments, in which each factor is set at two levels, high and low, are widely used during development studies.

Response surface exploration with composite designs

In the final stage of a development study, when you are seeking the conditions (such as the values of composition and process variables) that will yield the best value of a material property (such as the highest value of tensile strength), additional points must be added to factorial experiments so that curvature of the response can be estimated. These designs are known as augmented or composite designs.

Interlaboratory trials and hierarchical designs

Another class of experiment used in industry is the interlaboratory trial for the purpose of estimating repeatability of test results within each of a set of laboratories and reproducibility of test results between laboratories. These are described more fully in several textbooks and in a British Standard (BS 5497).

Similar issues arise when comparing variability within and between batches of product.

Principles of experimental design

Statistical analysis of experimental results is necessary because of variation: *all test results vary*. This variation must therefore be considered when experiments are designed.

Descriptive experiments

In a descriptive experiment a characteristic of a standard material will be reported from the analysis of measurements on several test results. For example, the mean tensile strength of a sample of several test pieces will be calculated. This is unlikely to be the true value of the underlying population mean. If you calculated the mean tensile strength of another sample of several test pieces it would be different. The calculated sample mean is therefore only an estimate, a *point* estimate, of the underlying population mean. In reporting it you should report an interval in which you can confidently expect the population mean to lie: a confidence interval for the population mean.

This confidence interval depends on three things:

1. The variation of test results, expressed as the variance or standard deviation of the measured material property.
2. The number of pieces tested in the sample.
3. The degree of confidence of the interval, loosely interpreted as the probability that the population mean is truly in that interval, usually as a percentage (for example, a 95 percent confidence interval).

The variation will have to be determined from the experiment. The number of test pieces must be specified before the experiment is done. The degree of confidence is the choice of the experimenter and should be specified before the experiment is done. Ideally, the experimenter should specify how large a confidence interval, and with what confidence, he or she would like. For example he or she may specify: sample mean value ± 1.0 N/mm^2 with a confidence of 95 percent. The experiment would then proceed in four stages:

Stage one: a preliminary experiment to estimate the underlying population variance and/or a review of similar experiments reported in the literature
Stage two: a calculation of the sample size N needed to estimate the specified confidence interval using the variance estimated in stage one
Stage three: test measurements on a sample of N pieces
Stage four: calculation of the estimated mean, standard deviation and confidence interval.

These four stages are described fully in the referenced textbooks.

Comparative experiments

Statistical analysis of test results should never be regarded simply as a set of calculations leading to a clear-cut conclusion such as 'the effect is significant' or 'there is no significant effect'. The conclusion depends on the circumstances of the experiment and on the intentions of the experimenter which should be declared before the tests are done. For example, the circumstances of an experiment may ordain how likely an effect is to be declared as statistically significant *if it exists*; the intentions of the experimenter will include a statement of what he or she considers to be a *technically* significant effect.

Four major steps must be taken before starting an experiment:

Step one: state the alternative inferences that can be made from the experiment
Step two: specify the acceptable risks for making the wrong inference

Step three: specify the difference which must be demonstrated statistically so as to be of technical significance

Step four: compute the necessary sample size

These four steps are described more fully below and are described in detail in the referenced textbooks.

Step one: state the alternative inferences that can be made from the experiment. These should be stated as alternative prior hypotheses.

When two materials are to be compared according to some property, the most usual comparison is between the mean values of that property. Even though the sample mean values $\bar{x}_1$ and $\bar{x}_2$ will differ, is there sufficient evidence to infer that the mean values of the underlying populations(μ_1 and μ_2) differ? If there is not sufficient evidence you cannot refute the claim that the means of the underlying populations are the same. An assumption that they are the same is known as the null hypothesis (H_0). An assumption that they are different is known as the alternative hypothesis (H_a).

These may be stated symbolically as:

$$H_0\text{: } \mu_1 = \mu_2$$
$$H_a\text{: } \mu_1 <> \mu_2$$

In this case the experimenter is concerned about *any* difference between the population means. This will lead to a two-sided test.

If the experimenter is interested in a new material only if it has a greater mean strength than the standard material a one-sided test will be used and the alternative hypotheses will be:

$$H_0\text{: } \mu_1 = \mu_2$$
$$H_a\text{: } \mu_1 > \mu_2$$

The distinction must be made *before* the experiment is done. The calculation of the sample size depends on the distinction.

Step two: specify the acceptable risks for making the wrong inference. The wrong inferences are called the type one error and the type two error with probabilities α and β, respectively.

The possible inferences from a two-sided test may be understood from Table 13.1.

Table 13.1 Possible inferences from a two-sided test

		Inference	
		$\mu_1 = \mu_2$	$\mu_1 <> \mu_2$
Truth	$\mu_1 = \mu_2$	**correct** probability $= (1 - \alpha)$	**type one error** probability $= \alpha$
	$\mu_1 <> \mu_2$ or $\mu_1 - \mu_2 = \delta$ where δ is not zero	**type two error** probability $= \beta$ which depends on δ	**correct** probability $= (1 - \beta)$

A **type one error** occurs when the experimenter accepts the alternative hypothesis (H_a) although the null hypothesis (H_0) is true. The probability of this occurring is α. Usually α is specified as 0.05 (a five percent chance).

A **type two error** occurs when the experimenter accepts the null hypothesis (H_0) although the alternative hypothesis is true. The probability of this occurring is β which depends on the difference between the population means ($\mu_1 - \mu_2$). Usually β is specified as 0.05 or 0.10 for a specified difference of technical importance. The probability of detecting this difference is $(1 - \beta)$. Thus if β is chosen to be 0.05 and the experiment is designed accordingly there is a strong chance (a probability of 0.95) that the specified difference will be detected if it truly exists. More generally, a plot of the probability $(1 - \beta)$ of correctly detecting a true difference (Δ) against Δ is known as **the power curve of the test**.

Unfortunately, it is common for experiments to be done without consideration of β or the power and consequently important true effects may remain undetected. For example, consider the burst strengths (kPa) of two batches of filter membranes as shown in Table 13.2.

Table 13.2 Burst strengths (kPa) of filter membranes

Batch one	*Batch two*
267	284
262	279
261	274
261	271
259	268
258	265
258	263
258	262
257	260
256	259
251	246
250	241

Suppose that a purpose of the experiment was to detect any difference exceeding 10 kPa. A power calculation ($\alpha = 0.05$) shows that with only 12 results for each batch there is a probability of 0.75 of detecting that difference if it exists. Sample sizes of 23 would be needed to give a probability of 0.95 of detecting that difference.

Step three: specify the smallest difference (δ) of technical significance, which should have a specified probability $(1 - \delta)$ of being detected.

The purpose of many experiments is to discover an improvement in a material property, that which is being tested. In other experiments the purpose may be to show that, under different circumstances, there is no substantial difference in the material property.

In either case, the experimenter should be able to state the smallest difference which he or she would regard as likely to have a practical or technical significance. For example, how much stronger, in terms of tensile strength, should one material be over another to make its selection preferable for a particular application? One percent, or two, or 0.5? It may depend on the application.

The specification of this smallest difference (δ) is essential to the design of a comparative experiment.

Step four: compute the necessary sample size.

There are several formulae for calculating sample size. The correct choice of formula depends on the type of comparative experiment (comparison against a standard, comparison of two materials with independent samples, comparison of two materials by paired samples) and on whether the proposed test of comparison is one-tailed or two-tailed. The information needed in any of the calculations is the choice of α, β, δ and an estimate of the variance (σ^2) of the underlying population.

The choices of α, β, δ depend entirely on the opinions and purposes of the experimenter, as already explained. An estimate of the population variance (σ^2) may often be obtained from earlier experiments or from literature. Otherwise a preliminary experiment of at least five test pieces must be done to obtain that estimate.

Response experiments

Introduction to response experiments

Much research and development in the materials sciences is intended to establish relationships between materials properties and other factors which the technologist can control in the production of those materials. The properties are called **response variables** (also called *dependent variables*). The other factors which influence the response variables are called **control variables** (also called *independent variables*). The control variables are usully composition variables and process variables. All of these variables must be measurable.

- The variable we are most interested in studying is often called the *response* variable because it changes in response to changes in other variables. Also, it is usually a characteristic of a product by which we judge the usefulness of the product.
- For example, we may be doing an experiment about the production of an artificial silken thread. The process involves the stirring of a mixture and the tensile strength depends, at least partly, on the stirring speed. In this case tensile strength is an important property of artificial silk. It seems to respond to stirring speed so it is a response variable. We show it along the vertical, or Y axis (see Fig. 13.1).
- The variable which we think is influencing, or controlling, the response variable is, in this case, stirring speed. We call it a *control* variable and show it along the horizontal, or X axis.

All of the variables that may influence the response are collectively called **explanatory variables**. Sometimes there are other variables which may influence the response variables

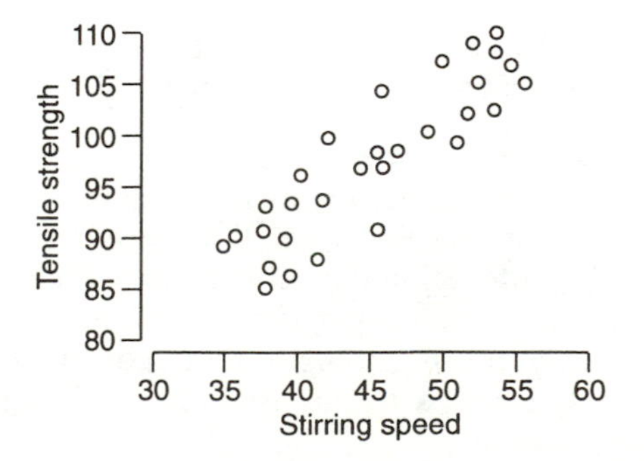

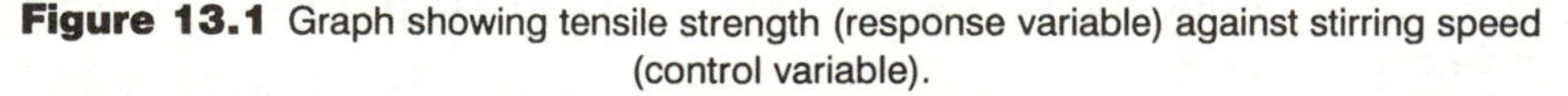
Figure 13.1 Graph showing tensile strength (response variable) against stirring speed (control variable).

but which cannot be controlled although they can be identified and measured. Common examples are temperature and humidity of a factory workshop atmosphere. These variables are called **concomitant variables** (also called *covariates*).

The experimental design is the specification of the values of the control variables at which the response variables will be measured. The experiment should be designed according to the expected relationship between the response variable and the control variables. The expected relationship is a hypothesis. The hypothesis should be formulated as an algebraic model that can be represented in terms of the measurable variables.

The experimental design would be very simple and few observations would be needed to fit the model if the expected relationship could predict the results exactly. However there are several reasons why this cannot be achieved:

- the exact relationship cannot be known; the model is only an approximation to reality
- all measurements may be subject to time dependent errors which we cannot identify but which show their presence by trends in the observed values of response variables
- all measurements are subject to random errors; these represent other unidentified variables which taken together show no pattern or trend.

The experiment must therefore be designed so as to reduce the influence of these unknowns.

The statistical objectives of designed response experiments are to specify:

1. the number of observations
2. the values of the control variables at every observation
3. the order of the observations

with a view to:

1. ensuring that all effects in the model can be estimated from the observed data
2. testing the reality of those effects by comparison with random variation
3. ensuring that all effects can be estimated with greatest possible precision (reducing the influence of random variation)
4. ensuring that all effects can be estimated with the least possible bias, or greatest accuracy (reducing the effects of time dependent errors)
5. suggesting improvements to the model
6. keeping within a budget of effort and cost.

Factorial experiments

Two-level factorial experiments. The two-level factorial design is fundamental to experimental design for the physical sciences.

When there is good reason to believe that over the range of values of a control variable the dependent variable is related to the control variable by a linear function of the form:

$$y = a + bx \tag{13.1}$$

where y is the response variable, x is the control variable, and a and b are coefficients to be estimated; then a and b can be estimated with the greatest precision if all observations are divided equally between the two ends of the range of x. A common fault among experimenters is to divide the range into $N - 1$ equal parts (where N is the number of planned observations) and to make one observation at each end and at each of the division points. A few intermediate points may be desirable as a check on the believed linearity if curvature is suspected.

In equation (13.1), the effect on y of a change in x of one unit is represented by the coefficient b, which is the slope of the line. Another way of representing the relationship between x and y is achieved by using a different notation. In this notation, the independent variables (the xs) are called factors and are represented by capital letters: A, B, C, ...

The range of a factor is specified by the two ends of the range: the high and the low values of the factor. These are represented by lower case letters with suffixes. For example, in the single factor experiment, the high and low values of factor A would be a_1 and a_0, respectively. This lower case notation is also used to represent the observed values of the dependent variable at the corresponding observation points.

The effect of factor A on the dependent variable (y) over the complete range of factor A is equal to:

$$(\text{mean value of } y \text{ at point } a_1) - (\text{mean value of } y \text{ at point } a_0)$$

which can be expressed as:

$$\text{effect of } A = \bar{y}\,(a_1) - \bar{y}\,(a_0)$$

or, more briefly,

$$A = a_1 - a_0 \tag{13.2}$$

Note a further abbreviation in that the capital letter A is used to denote the effect of factor A.

Similarly, the mean value of y is simply

$$M = (a_1 + a_0)/2 \tag{13.3}$$

Now consider two factors, A and B, which can be represented as two variables in a plane with the dependent variable y along a third dimension perpendicular to the plane. The design is shown in Fig. 13.2.

The high and low values of B are b_1 and b_0. If observations of y are made only at points defined by the extreme ranges of the two factors, there are four points which can be denoted by the combinations of letters as: a_0b_0, a_1b_0, a_0b_1, and a_1b_1. The notation can be abbreviated further to represent these four points as: (1), a, b, ab. The symbol (1) denotes the observation point at which all the factors are at their low levels. The point a is where factor A is at its high level but factor B is at its low level. The point ab

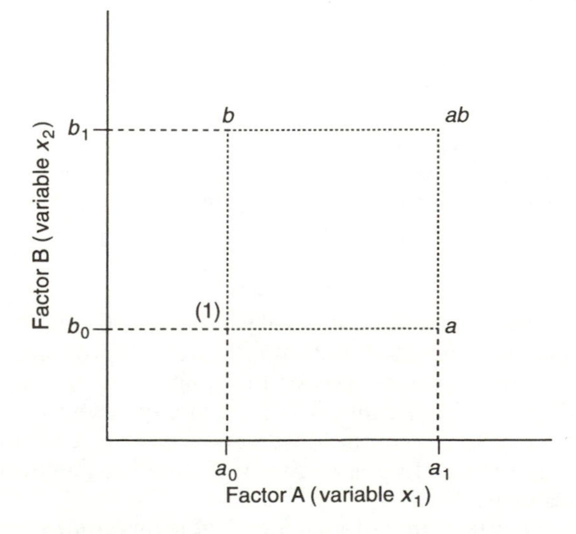

Figure 13.2 Graph showing two-factor design.

is where both factors are at their high levels. The rule is that the high and low levels of factors are represented by the presence or absence, respectively, of lower case letters.

Analysis is almost as easy as in the single factor case. Using the combinations of lower case letters to represent the values of y observed at the corresponding points, the average effect of factor A is:

$$A = \frac{a + ab}{2} - \frac{(1) + b}{2} \tag{13.4}$$

That is: the effect of A is the difference between (the mean value of y observed at all the points where A was at its high level) and (the mean value of y observed at all the points where A was at its low level).

Similarly:

$$B = \frac{b + ab}{2} - \frac{(1) + a}{2} \tag{13.5}$$

The interaction of factors A and B may be defined as the difference between (the effect of A at the high level of B) and (the effect of A at the low level of B). It is denoted by AB. Thus:

$$AB = (ab - b) - (a - (1)) \tag{13.6}$$

This is exactly the same as: the difference between (the effect of B at the high level of A) and (the effect of B at the low level of A).

The estimation of these effects is equivalent to fitting the algebraic model:

$$y = \beta_0 + \beta_1 x_1 + \beta_2 x_2 + \beta_{12} x_1 x_2 \tag{13.7}$$

where y is the response variable, x_1 and x_2 are two control variables and β_0, β_1, β_2 and β_{12} are algebraic coefficients.

Least squares regression analysis (see Chapter 23) is widely used for analysis of these and other experiments to be described. Computer software is available for this analysis which includes the estimation and testing of coefficients in equations such as (13.7).

Two-level fractional factorial experiments. These principles of design and analysis of two-level factorial experiments can be extended to experiments involving any number of factors. See Fig. 13.3 for an illustration of a three-factor situation. However the number of observations in such an experiment increases exponentially with the number of factors. If there are n factors, the number of observations is 2^n. Thus:

Number of factors	Number of observations
1	2
2	4
3	8
4	16
5	32
6	64
7	128
8	256
9	512
10	1024

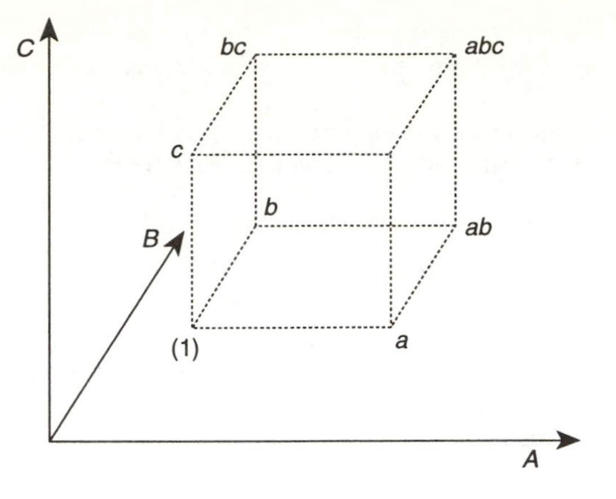

Figure 13.3 The three-factor situation.

It is not unusual to have experiments with seven or more factors (control variables). Thrift demands an experiment with only a fraction of the experiments in a full design. If a suitable fraction can be found the resulting experiment is called a two-level fractional factorial.

The theory and method of constructing these fractional experiments is described in textbooks. Also software is available for the automatic design and analysis of these experiments (see Bibliography).

Composite designs

Whereas two-level factorial experiments, and their fractions, are suitable for fitting models that are linear in the main effects and including interactions, they are not suitable for estimating curvature of response if it exists. For example, if there is a single control variable, equation (13.1) may be suitable either if the relationship is genuinely linear for all values of x, or on the rising or decreasing slope of a quadratic response.

However, if the experiment is to be done in a range of x which is close to the peak (or trough) of the quadratic response, curvature will have a major effect and must be estimated. This is particularly important if a purpose of the experiment is to estimate the value of x for which y is a maximum (or minimum).

Equation (13.1) must then be augmented as:

$$y = a + bx + cx^2 \qquad (13.8)$$

Similarly, equation (13.7) must be augmented as:

$$y = \beta_0 + \beta_1 x_1 + \beta_2 x_2 + \beta_{12} x_1 x_2 + \beta_{11} x_1^2 + \beta_{22} x_2^2 \qquad (13.9)$$

Designs for these augmented relationships are called **augmented** or **composite designs**. The theory and methodology of constructing them is described in several textbooks. Software is available for constructing and analysing them. Analysis is usually by least squares regression.

An example

Figure 13.4 shows how a catheter is fixed to a valve body. It enters through the A-channel and expands into the C-channel where it is held by a bush which is pressed into the end of the catheter. The purpose of the experiment was to discover the dimensions such that

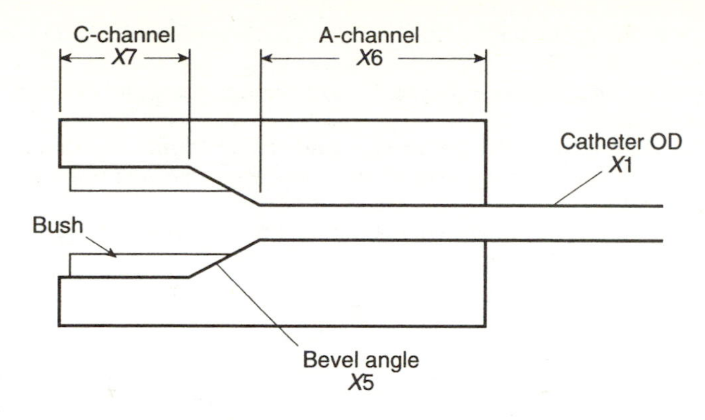

Figure 13.4 Catheter fixed to a valve body.

the catheter would be gripped with maximum security. The assembly is put into a tensile tester and force is gradually increased until the catheter is pulled out. The response variable (*Y*1) is the disassembly force measure in newtons.

There are some constraints in the dimensions. The C-channel inner diameter (ID) must be greater than the bush outer diameter (OD). The C-channel ID must be greater than the A-channel ID. These constraints are avoided by using differences in variables *X*3 and *X*4.

The control variables (measurements in mm, except for *X*5) are given in Table 13.3.

Table 13.3

		Low	*High*	*Increment*
*X*1	catheter OD	1.7	1.8	0.05
*X*2	bush OD	1.6	2.0	0.1
*X*3	C-channel ID – bush OD	0.20	0.45	0.05
*X*4	C-channel ID – A channel ID	0.00	0.10	0.05
*X*5	bevel angle	30°	90°	5°
*X*6	A-channel length	3.0	5.5	0.5
*X*7	C-channel length	0.5	4.0	0.5

The experiment was designed, using DEX, to fit a model which included all quadratic effects and the first order interactions: *X*1.*X*2, *X*1.*X*3, *X*1.*X*4, *X*1.*X*6, *X*1.*X*7, *X*3.*X*5, *X*4.*X*5. The choice of interactions was based on experience, mechanical judgement, and analysis of earlier trials.

The designed experiment has 47 observations of which the first 32 comprise a quarter of a 2^7 factorial. The following 15 observations are axial and centre points which are added to estimate curvature as quadratic terms in the model. Thus, this is a composite experiment constructed by augmenting a fractional two-level experiment. It achieved its purpose of discovering the conditions which would yield a maximum disassembly force for the catheter valve. You can see a further explanation of the designed experiment on p. 420 of Metcalfe (1994).

Bibliography

Atkinson A C and Donev A N (1992) *Optimum Experimental Designs*. Oxford University Press, Oxford.

Box G E P and Draper N R (1987) *Empirical Model-building and Response Surfaces*. John Wiley, New York.

Box G E P, Hunter W G and Hunter J S (1978) *Statistics for Experimenters. An Introduction to Design*. John Wiley, New York.
Diamond W J (1981) *Practical Experimental Designs for Engineers and Scientists*. Van Nostrand Reinhold, New York.
Greenfield T *DEX: A Program for the Design and Analysis of Experiments*. Greenfield, Derbyshire.
Grove D M and Davis T P (1992) *Engineering Quality and Experimental Design*. Longman Scientific and Technical, Harlow.
Metcalfe A V (1994) *Statistics in Engineering*. Chapman & Hall, London.

14
Agricultural experiments

Roger Payne

Introduction

Experiments are a fundamental part of agricultural research. Indeed much of the theory of design and analysis of experiments was originally developed by statisticians at agricultural institutes. The methods are useful, however, in many other application areas. So this chapter should be of interest to all researchers whether or not their chosen area of study is agriculture.

The aim of the chapter is to cover the main principles of the design and analysis of experiments. Further information can be obtained either from the books listed at the end of the chapter, or from your local statistician.

First we consider the basic experimental *units*. In a field experiment these may be different *plots* of land within a field. In animal experiments they may be individual *animals* or perhaps *pens* each containing several animals. There is a great deal of practical expertise involved in setting out, managing, sampling and harvesting field plots. The accuracy of these aspects can be vital to the success of an investigation, and potential field experimenters are encouraged to read Dyke (1988, Chapters 2–6) for further advice.

Blocking

The units frequently have either an intrinsic or an imposed underlying structure. Usually this consists of a grouping of the units into sets known as **blocks**. For example, the animals in an experiment may be of several different breeds (this would be an intrinsic grouping). Alternatively, in a field experiment, you might decide to partition up the field into several different blocks of plots. The aim when allocating plots to blocks is to make the plots in each block as similar as possible: to try to ensure that pairs of plots in the same block are more similar than pairs in different blocks. Often this can be achieved by taking contiguous areas of the field as blocks, chosen so that the blocks change along a suspected trend in fertility. However, blocks need not be contiguous. For example, in an experiment on fruit trees you might wish to put trees of similar heights in the same block, irrespective of where they occur on the field.

Blocks need not all have the same shape. In a glasshouse you might find that the main variability is between the pots at the side and those in the middle. So you could have two L-shaped blocks around the edge and other rectangular blocks down the middle. The important principle is to think about the inherent variability of your experimental units, and then block them accordingly.

Further advice can be found, for example, in Dyke (1988, Chapter 2) or Mead (1988, Chapter 2). An important final point is that you should remember the blocking when you are planning the husbandry of the trial. For example if you cannot harvest all the units of the experiment in one day, you should aim to harvest one complete set of blocks on the first day, another set on the second, and so on. Then any differences arising from the different days of harvest will be removed in the statistical analysis along with the differences between blocks, and will not bias your estimates of treatments.

More complicated blocking arrangements can also occur, and are described later in this chapter.

Treatment structure

The purpose of the experiment will be to investigate how various treatments affect the experimental units. In the simplest situation there is a single set of treatments – perhaps different varieties of a crop; or different amounts of a fertiliser; or different dietary régimes for animals. In Fig. 14.1 we show the field plan of a simple example in which fungicidal seed treatments to control the disease take-all were examined in a **randomised complete block design**.

In this design the units (in this case field plots) are grouped into blocks as described in the section on blocking. Each treatment occurs on one plot in every block, and the allocation of treatments to plots is *randomised* independently within each block; so here, a random permutation of the numbers one to three was selected for each block and used to determine which treatment was applied to each of its plots. Designed experiments are usually analysed by analysis of variance. Figure 14.2 shows output from the statistical system Genstat (Payne *et al,* 1993). This is developed by statisticians at an agricultural research institute, and so it has especially comprehensive facilities for the design and analysis of experiments.

First there is an **analysis of variance** table which allows the effects of the treatments to be assessed against the inherent variability of the units. The variability between the blocks is contained in the **Block stratum**, and the variability of the plots within the blocks is in the **Block.Plot stratum**. Within the *Block.Plot stratum*, the *Seed_trt* line contains the variability that can be explained by the assumption that the seed treatments do have differing effects, and the residual line contains the variability between units that cannot be explained by either block or treatment differences. The *mean square* (m.s.) for blocks is over nine times that for the residual, indicating that the choice of blocks has been successful in increasing the precision of the experiment. Similarly the variance ratio (v.r.) of 12.05 for *Seed_trt* suggests that the treatments are different. The analysis will also

|← Block →|← Plot →|

150	0	100	150	100	0	100	150	0	0	150	100

Figure 14.1 An example of a randomised complete block design.

```
***** Analysis of variance *****

Source of variation      d.f.        s.s.        m.s.      v.r.     F pr.

Block stratum               3     2.39990     0.79997      9.12

Block.Plot stratum
Seed_trt                    2     2.11449     1.05724     12.05     0.008
Residual                    6     0.52630     0.08772

Total                      11     5.04069

***** Tables of means *****

Grand mean 7.149

 Seed_trt           0.         100.         150.
                 6.590        7.256        7.601

*** Standard errors of differences of means ***

Table          Seed_trt
rep.                  4
s.e.d.           0.2094
```

Figure 14.2 Analysis of variance.

provide tables showing the estimated mean of the units with each treatment, usually (as here) accompanied by a standard error to use in assessing differences between the means.

It is important to realise that the analysis is actually fitting a *linear model* in which the yield y_{ij} on the plot of block i to receive treatment j is represented as follows:

$$y_{ij} = \mu + \beta_i + t_j + \varepsilon_{ij}$$

where μ is the overall mean of the experimental units, β_i is the *effect* of block i, t_j is the *effect* of treatment j, and ε_{ij} is the *residual* for that plot (representing the unexplained variability after allowing for differences between treatments and between blocks).

The residuals are assumed to have independent normal distributions with zero mean and equal variances. The variance ratio of 12.05 for treatments in the analysis of variance table can then be assumed to have an F distribution on two and six degrees of freedom (the degrees of freedom from the treatment and residual lines, respectively), leading to the probability value of 0.008 in the right-hand column, and the differences between the means divided by the standard error of difference can then be assumed to have a t distribution with six degrees of freedom (the degrees of freedom for the residual).

The effect t_j of treatment j represents the difference between the mean for treatment j and the overall mean. So when we are assessing whether the treatments are identical, we are actually seeing whether there is evidence that their effects are different from zero.

This way of representing the analysis becomes more useful in experiments when the treatments given to the units may differ in several different ways. For example, we may have several different fungicides to study and we may also want to try a range of different amounts; or we may want to investigate the effect of varying the amounts of several different types of fertiliser; or we may want to see how well different varieties of wheat are protected by different makes of fungicide. Each of these types of treatment is then represented by a different treatment *factor*, with *levels* defined to represent the various possibilities. One of the great advantages of a designed experiment is that it allows you to examine several different treatment factors at once. Suppose that we have two treatment factors N (nitrogen at levels 0, 180 and 230) and S (sulphur at levels 0, 10, 20 and 40) and we wish to examine all their combinations, again in a randomised complete block design. The factors N and S are said to have a *crossed* or *factorial* structure, and we can represent the yield y by the model

$$y_{ijk} = \mu + \beta_i + n_j + s_k + ns_{jk} + \varepsilon_{ijk}$$

We now have three *terms* to represent the effects of the treatments: the parameters n_j represent the *main effect* of nitrogen, s_k represent the *main effect* of sulphur, and ns_{jk} represent the *interaction* between nitrogen and sulphur.

The analysis of variance table shown in Fig. 14.3 now contains a line for each of these, to allow you to decide how complicated a model is required to describe the results of the experiment.

```
***** Analysis of variance *****

Source of variation      d.f.         s.s.         m.s.      v.r.     F pr.

Block stratum               2      0.03793      0.01896      0.33

Block.Plot stratum
N                           2      4.59223      2.29611     40.19     <.001
S                           3      0.97720      0.32573      5.70     0.005
N.S                         6      0.64851      0.10808      1.89     0.127
Residual                   22      1.25683      0.05713

Total                      35      7.51269

***** Tables of means *****

Grand mean 1.104

          N           0         180         230
                  0.601       1.313       1.398

          S           0          10          20         40
                  0.829       1.155       1.167      1.266

          N           S           0          10         20          40
          0                   0.560       0.770      0.524       0.552
        180                   0.894       1.289      1.525       1.545
        230                   1.032       1.404      1.454       1.700

*** Standard errors of differences of means ***

Table             N          S          N
                                        S
rep.             12          9          3
s.e.d.       0.0976     0.1127     0.1952
```

Figure 14.3 Analysis of variance.

When analysing a factorial experiment we would like to find a simple model to explain the situation. The full model above will estimate the means for the sulphur and nitrogen treatments as

$S \times N$ means	$N0$	$N180$	$N230$
$S0$	0.560	0.894	1.032
$S10$	0.770	1.289	1.404
$S20$	0.524	1.525	1.454
$S40$	0.552	1.545	1.700

=

μ
1.104

+

n_j: $N0$	$N180$	$N230$
−0.503	0.209	0.294

+

s_k	
$S0$	−0.276
$S10$	0.051
$S20$	0.063
$S40$	0.162

+

ns_{jk}	$N0$	$N180$	$N230$
$S0$	0.234	−0.144	−0.090
$S10$	0.118	−0.075	−0.044
$S20$	−0.141	0.148	−0.007
$S40$	−0.211	0.071	0.141

It will be much easier to describe what is happening if there is no interaction. The model will then be

$$y_{ijk} = \mu + \beta_i + n_j + s_k + \varepsilon_{ijk}$$

leading to a table of means:

$S \times N$ means	$N0$	$N180$	$N230$
$S0$	0.326	1.038	1.122
$S10$	0.652	1.364	1.448
$S20$	0.665	1.377	1.461
$S40$	0.763	1.457	1.559

=

μ
1.104

+

n_j: $N0$	$N180$	$N230$
−0.503	0.209	0.294

+

s_k	
$S0$	−0.276
$S10$	0.051
$S20$	0.063
$S40$	0.162

and you will notice that we can decide on the best level of nitrogen without needing to consider how much sulphur is to be applied, and on the best level of sulphur without needing to think about the level of nitrogen on the plot. This is what we mean by saying that the two factors do not interact: the *interaction* assesses the way in which the changes in yield caused by the various levels of nitrogen differ according to the amount of sulphur or, equivalently, the way in which the response to amount of sulphur differs according to the level of nitrogen.

This idea can be extended similarly to three or more factors. These **factorial** arrangements thus have the advantage that we can examine several different types of treatment at once. If there are no interactions we can present one-way tables of means and these will each have the same replication (number of units for each level of the treatment) as they would have had if we had performed individual experiments of this size for each treatment factor in turn; the only difference is that the factorial experiment has fewer residual degrees of freedom, but 22 (above) is ample! More important, by examining the interaction we can assess how valid our conclusions for each factor are over a range of values of other factors.

Treatment factors need not always have a crossed structure. They can also be *nested* one within another. For example, we may have several strains of two different species of aphid, *Mp* and *Rp*, and wish to examine their esterase levels. Suppose that we have strains Mp_1 ... Mp_4 and Rp_1 ... Rp_3 with several individuals of each strain. We will certainly be interested in assessing differences between *Mp* and *Rp*. We may also be interested in how much variation there is amongst {Mp_1, Mp_2, Mp_3 and Mp_4} and amongst {Rp_1, Rp_2 and Rp_3}; that is whether there is variability of the strains beyond the variability of the individual aphids. The model of interest (assuming that there is no blocking) would then be

$$y_{ijk} = \mu + s_i + st_{ij} + \varepsilon_{ijk}$$

where the parameters s_i represent the effects of the species ($i = 1, 2$), and st_{ij} represent the strain *within* species effects ($j = 1, ..., 4$, for $i = 1$; $j = 1, ..., 3$, for $i = 2$).

Notice that the model does not contain a strain main effect. The actual number allocated to each stain is only a labelling; it does not imply any special similarity for example between the strain numbered 2 for *Mp* and the strain numbered 2 for *Rp*.

Other types of block structure

Sometimes more complicated blocking structures may be required. For example, there may be more than one way of forming the units into groups. Perhaps we need to cater for large fertility trends running both along and across the field; or we may have insufficient pens in an animal experiment to complete the experiment all at once and need to sub-divide the animals, from several different breeds, into batches to be examined in successive weeks. One possibility, if we have two blocking factors each with the same number of levels as the number of treatments that we wish to examine, is to use a **Latin square**. An example, for five dietary treatments, is shown in Table 14.1.

Table 14.1

5 × 5 Latin square (before randomisation)	Breed 1	Breed 2	Breed 3	Breed 4	Breed 5
Week 1	Diet A	Diet B	Diet C	Diet D	Diet E
Week 2	Diet B	Diet C	Diet D	Diet E	Diet A
Week 3	Diet C	Diet D	Diet E	Diet A	Diet B
Week 4	Diet D	Diet E	Diet A	Diet B	Diet C
Week 5	Diet E	Diet A	Diet B	Diet C	Diet D

Notice that each diet occurs once in each row and once in each column. So we have simultaneously blocked the units by rows (week), and by columns (breed). In the analysis we will be able to estimate and eliminate both the row differences and the column differences, leading to a smaller residual mean square – and thus smaller standard errors for differences between the treatment means. The table shows the plan before randomisation. To randomise we need to select a random permutation for the rows and then another one for the columns.

We have now seen two types of block structure. The randomised complete block design has a *nested* structure with the individual units nested within the blocks, while the Latin square has a *crossed* structure of rows crossed with columns. These operations of nesting and crossing provide the basis for the more complicated arrangements that are sometimes needed for sophisticated trials. For example, the *split-plot design* extends the ideas of the randomised block design to have a further nesting of *sub-plots* within plots. Alternatively, we may have replicated Latin squares (rows crossed with columns, all nested within replicates). Further examples can be found in Cochran and Cox (1957), and Payne *et al* (1993).

Assumptions of the analysis

Analysis of variance assumes firstly that the model is additive: that is, differences between treatment effects must remain the same however large or small the underlying size of the variable measured. So, for example, in a randomised-block design, we are assuming that the theoretical value of the difference between two treatments remains the same within a block where the recorded values are generally low, as in one where the values are generally high. If your design has more than one replicate of each treatment within each block you can check this by fitting block × treatment interactions, but usually this is not possible. An alternative method, that checks for the common form of non-additivity where treatment effects are proportional, is to fit Tukey's single degree of freedom for non-additivity. Non-additivity can also cause interactions between treatment factors but of course, these may also occur for genuine reasons, for example caused by one treatment modifying the mode of action of another.

Secondly, the variance must be homogeneous: the variability of the residuals should be the same at high values of the response variable as at low values. If this assumption does not hold, the standard errors presented will be too large for differences between treatments with low means and too small for differences between larger means, causing incorrect conclusions to be drawn. Homogeneity of variance can easily be assessed by plotting the residuals against the fitted values: if the variance is homogeneous, the residuals should lie within a uniform band.

Thirdly, the residuals are assumed to have normal distributions. Non-normality can be assessed by plotting the residuals as a histogram or by plotting the residuals, sorted into ascending order, against values that would be expected from a normal distribution (a **normal plot**). Non-normality is usually also associated with non-homogeneity of variances.

Transformations

Failures of the assumptions can often be corrected by transforming the data. The transformations described in most textbooks are designed to stabilise the variance (assumption 2): for example, the square root transformation for counts or the angular transformation for percentages. A frequent mistake is to use the angular transformation blindly, without regard to the way in which the percentages have been obtained; it is important to realise that it is appropriate only when they are based on binomial data (for example a number r diseased out of n examined).

However, it is equally, if not more important, to consider the additivity of the model. Otherwise, as mentioned above, the resulting interactions can make the results difficult if not impossible to interpret.

In some situations a transformation can be chosen both to provide additivity and to stabilise the variance. With data where the treatments take the effect of a proportionate increase (or decrease), the standard errors will often be proportional to the means; a logarithmic transformation will then correct both aspects. With percentages representing proportions of diseased material, treatment effects are often found to be approximately proportional to the amount infected for low percentages, while for percentages near to 100 percent they tend to be proportional to the amount uninfected. If the percentages are obtained by visual assessment of areas such as infected parts of leaves, the standard errors tend to show the same pattern: for low percentages the eye tends to examine the amount infected, while nearer to 100 percent it is the amount uninfected that is assessed. In this situation, a logit transformation, $\log(p/(100-p))$, would be appropriate.

Further information about model assumptions and transformations can be found in Mead (1988, Chapter 11), Mead and Curnow (1983, Chapter 7) or John and Quenouille (1977, Chapter 14).

Generalised linear models

If additivity and homogeneity of variance cannot both be corrected simultaneously by transformation, a generalised linear model should be used (McCullagh and Nelder, 1989). However, although these are readily available in statistical systems such as Genstat and GLIM, they do require rather more statistical expertise to specify the models and to interpret the results than in ordinary analysis of variance.

Repeated measurements

Special care is needed with experiments where the same units are observed at successive times. **Repeated measurements** like these can show complicated correlation patterns, and you may not be able to assume that the necessary **distributional** assumptions hold for analysis of variance (see the section on **Assumptions of the Analysis**). A statistician should be able to advise on alternative methods like analysis of summary statistics, multivariate analysis of variance or the use of antedependence structure.

Conclusion

We have illustrated above the three main principles of experimentation (known as the three R's): *replication* – the need to have more than one unit for each treatment so that you can ascertain the intrinsic level of variability of the units and so decide whether the differences between the treatments go beyond what we might expect to occur by chance; *randomisation* – the need to allocate units to treatments at random, to avoid any biases; and *blocking* – ways of grouping the units in order to improve the precision of the experiment. These concepts, together with the ideas of factorial treatment structure, are fundamental to a successful experiment – whether in agriculture or in any other research area.

References

Cochran W G and Cox G M (1957) *Experimental Designs* (2nd edn). John Wiley, New York.
Dyke, G V (1988) *Comparative Experiments with Field Crops* (2nd edn). Griffin, London.
John J A and Quenouille M H (1977) *Experiments: Design and Analysis*. Griffin, London.
McCullagh P and Nelder J A (1989) *Generalized Linear Models* (2nd edn). Chapman & Hall, London.
Mead R (1988) *The Design of Experiments: Statistical Principles for Practical Applications*. Cambridge University Press, Cambridge.
Mead R and Curnow R N (1983) *Statistical Methods in Agriculture and Experimental Biology*. Chapman & Hall, London.
Payne R W, Lane P W, Digby P G N, Harding S A, Leech P K, Morgan G W, Todd A D, Thompson R, Tunnicliffe Wilson G, Welham S J and White R P (1993) *Genstat 5 Reference Manual, Release 3*. Oxford University Press, Oxford.
Pearce S C (1983) *The Agricultural Field Experiment*. John Wiley, Chichester.

15 Surveys

Roger Thomas

What is a survey?

A survey is a procedure in which information is collected systematically about a set of **cases** (such as people, organisations, objects). The cases (or **sample units**) are selected from a defined population and the aim is to construct a data set from which estimates can be made and conclusions reached about this population.

Research surveys resemble laboratory experiments in that they aim to collect data in a systematic way and to make inferences from the results. However, surveys are usually done under natural conditions. For example, you may ask people to give interviews in their homes or to respond to postal questionnaires. In sociology or economics it is only rarely possible to arrange matched or randomised experimental and control groups.

Some surveys are mainly for description: for example, to estimate how much butter is purchased annually by private households in the UK. However, most research surveys, while describing detail as a preliminary, aim to **test hypotheses** or to **estimate relationships between variables**. For example, you may wish to test if it is true that people in one group consume more fat than people in another group.

Some studies described as surveys use qualitative methods. These are important in research but in this chapter we deal only with **standardised quantitative surveys**.

Standardised quantitative surveys

The quantitative survey has four basic, consecutive operations. These are:

- draw a sample of units from some population
- develop and test standardised ways to measure these units
- apply them to the sample units
- make inferences to the population from which the sample was drawn.

Quantitative surveys use explicit, standardised and objective methods of sampling, data collection and data analysis. Comparability and reproducibility are critical goals to which flexibility and depth may have to be sacrificed to some extent. The skill of the quantitative survey designer is to devise measures which are relevant and revealing, but at the same time objective and standardised. The strength of the quantitative survey method is its ability to produce results for which the precision and reliability can be estimated, using methods which can be replicated, or which themselves replicate methods used in previous studies.

Survey measurement can take many forms. For example, we might measure the height or weight of individuals; or we might ask standard attitude questions and regard the answers as indicating the position of the respondents on some dimension of attitudes, say 'authoritarianism'. The output of the measurement stage consists of counts and proportions of sample units with particular measured attributes. For example: being male, being so many centimetres tall, or scoring at a particular level on 'authoritarianism'.

Sampling and sampling error

Most surveys are based on samples from some population. The approach to the design must:

- define in both conceptual and practical terms the population(s) to which you wish the results of the survey to apply (*conceptual definition*: 'all adults in Doncaster'; *practical* or *operational definition*: 'all adults whose names appear in the 1994 electoral register for Doncaster')
- define the estimates and comparisons that you wish to make (such as the proportion of adults in Doncaster who played tennis in the last four weeks; proportions of males and females separately who played tennis; mean number of tennis games played in the last month)
- define the degree of precision needed (for example, to provide useful estimates of level or change, to detect hypothesised differences between groups).

Sample based estimates which refer to the population from which the sample is drawn are subject to **sampling error**. This falls into two categories: **random error** and **bias**. Random sampling error is always present, but with correct sampling methods the degree of error or variability can be estimated. The most important factor in determining the size of random sampling error is usually sample size. Sampling bias arises where the nature of the sampling frame or the sampling method is such that they systematically fail to represent the population in particular ways. For example, a sample drawn from the electoral register will not represent foreigners, or adults who fail to register.

If you select the sample properly, you can draw conclusions about the population, such as that males have a mean height of 176.5 centimetres plus or minus 1.3 centimetres or that the mean score for males on a scale measuring 'authoritarianism' differs from the mean score for females. Such conclusions will always be subject to margins of sampling error and may also be affected by error or bias in the measurement process.

Sampling and error are discussed more fully in Chapters 16–18.

Survey measurement

Survey measurement can take many forms. For example, we might measure the height or weight of individuals, or we might ask standard attitude questions and regard the answers as indicating the position of the respondent on some dimension of attitudes such as 'authoritarianism'. Survey measures need to be as **valid**, **reliable** and **unbiased** as possible. By 'valid' we mean that they are pure and adequate measures of what they are intended to measure. By 'reliable' we mean that they are not subject to large amounts of random error. By 'unbiased' we mean that the mean value of the measures will, for a large sample, be the same as the true mean value for the population.

Development of measures which satisfy these criteria requires careful thought and empirical pretesting. This applies to intuitively straightforward measures, such as answers to single survey questions, as well as to more complex measurement procedures such as formal tests.

The risk of measurement error constantly lurks, as in the following examples.

A single question such as 'have you ever committed a criminal offence?' is unlikely to be a *valid* and *adequate* measure of the attribute of 'criminality'. The people who answer 'yes' are as likely to be very honest as very dishonest. Also, some people will not know the difference between criminal and non-criminal offences.

Whether or not a child can calculate the right answer to a single arithmetic problem will not be a *reliable* or *precise* measure of his or her level of achievement at arithmetic.

A whole battery of problems chosen to test different aspects of understanding and performance is needed.

The answer to the question 'would you agree that Ace Electronics is the foremost computer manufacturer in this country?' is likely to provide a *biased* measure of where Ace Electronics actually stands in relation to its competitors. The question leads respondents who know little about the subject to agree.

Planning and conducting a survey

Planning and conducting a quantitative survey requires a large investment of time and effort. You may well be attracted by the idea of basing your research on original data, but the alternatives of secondary data analysis or other forms of original research which are less demanding should be considered seriously.

To conduct a survey successfully it is necessary, but not sufficient, to get the theoretical and statistical aspects of the design right. In practice, project design and project planning interact: a perfect design which cannot be implemented practically is of no use; but neither is a practically convenient scheme which will not support the scientific conclusions that you want to draw.

You must estimate, plan, obtain and manage the resources needed to draw the sample, collect and process the data and analyse and write the results. For most research students, the prime resource is their own time. However, you will also need to draw on the time and cooperation of survey respondents, and possibly that of people who have control over key resources such as sampling lists and computing or clerical resources.

You are likely to need several planning iterations, and probably a pilot test, to arrive at the best set of practical and theoretical compromises and a workable and robust project plan. Survey planning is often sufficiently complex to justify the drafting of a time chart, which will help you to identify systematically:

- lead times and planned beginning and end dates of separate phases of the operation (such as sampling, fieldwork)
- interdependencies and deadlines for completing operations (for example: sampling must be completed, clearances obtained and documents printed before fieldwork can begin)
- time profiles of demand for particular resources, particularly your own time.

Apart from people's time, most surveys require some office and technical support such as stationery, printing, data entry and computing.

The following are some key areas of survey planning and implementation.

Access, cooperation and confidentiality

Obtaining a sample

To define in conceptual terms the population you wish to sample is one thing; to find and gain access to some means of enumerating and sampling the population may be quite another. You cannot do a survey of, say teenage unmarried mothers, unless there is some practical and ethical way to obtain a suitable list of teenage mothers' addresses, or information on where they may be found, sampled and interviewed. Publicly available sources exist for some populations, such as trade directories and the electoral registers, but drawing a sample from them can be laborious.

If you have privileged access to some suitable subjects or respondents (such as members, employees or clients of an organisation) you must still consider if this group can be regarded as a representative and unbiased sample of the population about which you want to draw conclusions. Thus a sample of teenage mothers whom the researcher knows may not represent teenage mothers generally.

Data collection

Surveys generally involve contacting prospective respondents and asking them to help. Contacting can often be difficult. Many people can be contacted face to face in their own homes, but only in the evenings and at weekends. Young and socially active people may be elusive even then. Postal surveys avoid the legwork problem but then another problem is to obtain an adequate level of response.

If the population to be studied is a sub-group (say, individuals who are economically active), you may need to screen a much larger group of individuals or addresses to find eligible cases. It is important to estimate in advance the effective yield of such an operation, estimating non-contact and cooperation rates as well as expected eligibility rate.

You can seldom be sure in advance that a given number of respondents will be available and cooperative. The outcome depends on the accuracy of guesses and assumptions about eligibility, contact and refusal rates. You need to calculate carefully how many sample selections you will need to produce the required sample and must budget for the time and effort indicated.

Do not guess blindly. Draw on evidence from similar surveys or, better still, from a pilot study. You must then work through the full implications for data collections and for other stages of the survey. The accuracies of estimates and assumption are crucial. If you are uncertain about them, a pilot study is needed.

Gain cooperation

When you make direct contact with potential survey respondents you should be prepared to make clear to them what cooperation will involve. This is further discussed below.

In other cases, the practical and ethical issued may be more complex. An example is where you need samples or other information from institutional sources such as employers and agencies. Access to samples of the required population may be controlled by **gatekeepers** whose cooperation may be needed either during sampling, or data collection, or both. You need to plan and timetable your access to the gatekeeper and ensure his cooperation. Gatekeepers have their own criteria for agreeing or refusing cooperation. Demands on busy people at short notice invite refusal so you must define carefully the minimum cooperation you require before making your approach. At the same time you need to stay flexible, in terms of dates and ways and means, so as to adapt to their priorities.

Data collection and confidentiality

The 1989 Data Protection Act, which prevents abuses of confidential information, has made things more difficult for survey researchers. Gatekeepers, even if personally sympathetic to the research, may be bound by rules which say that, if people have supplied information without being explicitly told that it may be used for research, they will have to be reapproached for permission before their identities or details are released to you. This can easily mean that a lot of non-response is incurred before you even start your approach; that you may be expected to pay the costs of the preliminary approach; and that your plans must allow for the delays this extra work will entail.

During fieldwork, you may need permission from a parent or carer where children or sick or elderly people are involved. However, you will still need to ask the sample members for their personal agreement to cooperate and give them appropriate assurances of confidentiality. In some surveys (concerning say, alcohol or drug use or sexual behaviour by minors) there may be difficult issues of confidentiality within households.

Mode of data collection

The **mode of data collection** is the medium through which information is elicited and recorded. Examples are:

- face to face interview
- telephone interview
- postal or other self-completion questionnaire, including diaries
- observation of behaviour
- physical weighing, measuring or testing.

These different modes have different implications for costs and resources and for the detail and complexity of data which can be collected successfully. Research students with low budgets must keep unit costs low so that they can achieve worthwhile sample sizes. They often use postal and other self-completion modes. Field interviews of a dispersed sample may be too time consuming and expensive, but interviews at a central place visited by members of the population may be possible. Telephone interviews may be more cost-effective provided all members of the sample are accessible by telephone. A lot of time and money can be spent on calls that do not lead immediately to interviews.

With good interviewing technique, face-to-face interviews can cover complex topics and may achieve high rates of response. However, that is not certain and the researcher with little interviewing experience should not be too ambitious. For postal and other self-completion modes, the design of questionnaires and questions should allow for the lowest expected levels of literacy and motivation of the intended respondents. For most populations, the layout and internal structure of self-completion questionnaires must be clear and simple. Failure to observe these cautions may lead to disastrously low levels of response and data quality. To increase rates of response to postal questionnaires you may need to send reminders, or even visit non-responders, so adding to your time and expenses.

A good survey design often combines several modes, for example a short and simple self-completion questionnaire followed by in-depth interviews in depth with individuals selected by reference to their self-completed responses.

Where information will be collected by questioning, you must decide who the respondent should be. In many surveys of individuals the respondent will be the sample member but in others it may be a well-placed observer such as a carer or parent. In other cases the sample units may be households or organisations rather than individuals. The choice of informant may then be wider but still difficult.

Design of data collection instruments and procedures

Design of data collection instruments and procedures is one of the most crucial aspects of practical survey design. Here are some general principles, mostly obvious but often breached in practice because of other pressures on the survey designer.

There are several categories of data which survey documents are designed to obtain:

- answers to substantive questions focusing on the topic of research

- answers to questions designed to provide means of classifying cases into sub-groups which will be used in analysis (such as residents of different areas, age groups, people with and without dependent children)
- information carried over from the sampling frame (such as place of residence, case identity) or recorded by observation (sex of respondent, time and place of interview).

Researchers focus most readily on the substantive concepts of their research and the questions they need to ask to measure them. They may also identify classificatory variables that they need to measure. They often fail to anticipate and collect all the information needed, say, to classify people according to socio-economic group. More often they neglect some of the mechanics of document design which should be thought about first.

With small samples, ancillary details, such as household data and time and place of interview, may not seem worth bothering with, but things which were clear at the time are hard to remember a year later. A systematic, well documented approach pays dividends.

Some examples of questionnaire mechanics are:

- devise case numbering systems which are foolproof at data collection and convenient at data processing
- devise a complete system for accounting for every sample unit, including not-eligibles, non-contacts, refusals, partial responses, full responses
- lay out questionnaires and questions so that they are clear to the interviewer or respondent, and that responses are recorded so as to be convenient for sorting, coding and keying the data for entry to a computer
- identify and enumerate the ultimate units of analysis (persons, households, trips, career episodes)
- capture information which is needed in analysis but is not obtained as responses to questions.

It is a good exercise to start by roughing out a questionnaire which has blanks for all the substantive questions (apart from an indication of the response format), but sets out all the mechanics just mentioned.

Postal and self-completion packages

A data collection instrument in the form of a package designed for self-completion needs to satisfy several requirements. You should:

- make clear who is asking for information, for what purpose and who will have access to the information
- make clear to each respondent exactly what he is expected to do by way of completing and returning the document
- provide a painless way to return the document (such as an addressed pre-paid envelope)
- aim to attract, not frighten off the prospective respondent
- use instructions and questions which are clearly and unambiguously expressed in simple language
- provide appropriate precoding boxes or spaces for verbatim replies
- make data processing easy, for example by arranging that codes to be keyed are in a coding column and by pre-assigning a data file location for each data item.

A questionnaire administered remotely must have a covering letter. This should be short: one side of an A4 sheet should be enough. Describe your purpose in a single sentence. Most recipients will want to discover quickly what they are being asked to do, how much effort is required, if the subject is interesting, and if the researcher and the research are respectable and trustworthy. You should assure recipients of confidentiality. Long explanations of how confidentiality will be safeguarded may be counterproductive. An official address (such as a university) and a telephone number where the researcher can be contacted are useful indicators of authenticity.

Design of questionnaires

A questionnaire should not be long and complicated. More pages with a clear and user-friendly layout are better than fewer pages with a cramped and forbidding layout. Word processors have made it easy to create an attractive appearance.

It may not be easy to keep the questionnaire simple. Consider, for example, a questionnaire about smoking behaviour and giving up smoking. Should you design separate question sequences for smokers and occasional smokers? If so, can occasionals and regulars be distinguished reliably by their answers to a simple question? If not, is it possible to design subsequent questions which will be meaningful to both occasional and regular smokers? How should you handle the fact that some smokers make several attempts to give up smoking? Should you ask about the most recent attempt or about all attempts in the past year, or about the most long-lasting attempt?

You can signal a change of topic of questions with a short preamble (such as 'That was about journeys you make as part of your work. Now can we talk a bit about journeys you make in your own time?').

Key aims in the design of questions

- Do not lead respondents who may not have strong or clear views of their own towards a particular answer. (Don't use questions like: 'Would you agree that Ace Electronics is Britain's foremost computer manufacturer?'.)
- Use simple grammar and limit the number of different things the respondents has to keep in mind to understand the question. (Don't use questions like: 'Apart from visits to places which charge entrance fees, on how many days in an average month between May and September do you make visits to the countryside?'.)
- Limit the number of different mental tasks the respondent has to perform to answer the question. (Don't ask: 'Please list for me all the alcoholic drinks you have had in the last week?'. Ask first: 'Have you had any alcoholic drinks in the seven days ending yesterday?' and then ask for details day by day.)
- Use vocabulary of the level used in the tabloid press. It is often better to use four simple words than one abstract latinate word.
- Write in concrete and specific terms rather than abstract ones. (For example: '... learn to read and write better' rather than '... improve literacy skills'.)
- Avoid slang and fashionable language even when you associate it with the population you are surveying (for example many young people will not know what 'street credibility' is).
- Do not ask double-barrelled questions (such as 'Do you approve of children getting up and going to bed when they like?').
- Make it easy for the respondent to understand how to answer (such as 'Tick one box only', tick all that apply', 'Please write in below').

Skilled interviewing permits instruments to greater length and complexity, so the design of interview documents can be more complex than that of self-completion documents. Even if you are doing all the interviewing yourself, you will have many other things to think about. It is worth making the mechanics of asking questions, recording information and deciding where to go next, as simple and foolproof as possible.

Pretests and pilots

Even long experience in the design of survey documents and procedures provides no guarantee against misjudgements, particularly when surveying new populations or new topics. Instruments and procedures must therefore be pretested. The effort and time are invariably justified. You can pretest piecemeal but a pilot survey is usually best. Many pilots focus on data collection, because that is where the researcher is most vulnerable to uncontrollable factors, but sampling, data processing and other procedures also need to be tested.

A pilot has two main functions. The first is *development* of instruments and procedures, where the pilot is a step on the way towards the final design. Here you need to make specific checks on reliability and validity. The developmental pilot may include features which will not be part of the main survey, such as asking respondents if they understood the questions in the way intended. Sometimes similar questions may be asked near the beginning and near the end so as to check reliability.

The second function is *rehearsal* of instruments and procedures, where the aim is to fine tune a design. The researcher often wants the pilot to be part development and part rehearsal. It is possible to combine them but there is a danger that the two functions will interfere with each other.

Much of the benefit of both types of pilot springs from seeing how real respondents drawn from the target population respond and react to the survey approach. The typical questionnaire has different question sequences addressed to respondents in different circumstances and it may be best to choose a non-random pilot sample which includes a wide range of circumstances. However, simply going through the procedures with a respondent often alerts the question designer to mistakes and oversights, so it is better to pilot with two or three colleagues as respondents than not to pilot at all.

Most student researchers are anxious to have as large a main sample as their resources allow. This raises the question: can pilot cases be added into the main survey sample? The answer depends on how close the procedures, including sampling, at the pilot were to the main survey procedures. If the questions were changed between pilot and main survey you may have problems.

Data preparation

There are several stages between collecting data and analysing data. These are: coding, editing, and preparing the data for analysis. These must be planned in advance and included in the design of documents and procedures.

Coding

Coding is the conversion of verbatim answers to categorised data. A simplifying frame must be imposed on the large variety of real-life situations. One way is to provide precoded answers for the respondents to choose from. Another is to record verbatim answers and then to fit them into categories devised after the event (a coding frame). Each method has advantages and drawbacks.

Precoded answers are easier to process so most large-scale surveys rely on them. It is common to prompt the respondent with a list of precoded answers, either orally or with a prompt card, so that there is no doubt about which answer has been selected. In other cases the interviewer, but not the respondent may be provided with answer categories and the interviewer decides into which category the respondents reply fits. This can be unreliable if different interviewers interpret the same response differently.

Some questions can be precoded easily because the possible answers are few and determinate. This is not true of more probing questions such as those beginning 'Why ...'. It is common to add a category 'Other answers, please specify' in case the proffered list is not exhaustive, but this does not prevent bias since respondents may accept prompted answers as delimiting the possible range.

Answers to be coded after the event must first be recorded verbatim. The process of devising a suitable coding frame and then coding individual responses may be time-consuming. Variability between coders may seriously increase the variability and bias of the survey results.

Data editing

Even the best run surveys provide data which do not exactly conform to the specification implicit in the logical structure of the questionnaire and the response categories for individual questions. Defects in the formal structure of the data, or **edit errors**, are marginal indicators of data quality. Some common ways in which editable errors arise are:

- interviewer fails to ask question
- interviewer fails to record response
- interviewer codes wrong response
- respondent overlooks the question
- respondent is unable to answer a question
- respondent is unwilling to answer a question
- respondent misunderstands the question or the response categories
- mistakes in transcription of data
- data are miskeyed.

The effect in every case is either a gap in the data or the recording of an incorrect value.

Treatments are:

- infer or impute what a missing or suspect item should have been
- record a 'no answer' response which will appear as such in the data analyses
- change data items which appear inconsistent.

Some editing can be done automatically when the data are entered into the computer, but some requires case-by-case treatment which is time-consuming. One method is for a computer program to check for errors which are then referred to the source documents so that corrections can be made manually. Editing is usually worthwhile since uncorrected edit errors can cause trouble and confusion, but dealing with them probably does not greatly increase the quality of the substantive data.

Case non-response and item non-response

Case non-response, the failure of some sample units to respond at all, may be more damaging than item non-response, specific data items missing from an otherwise complete case record or other edit errors. All the originally selected cases which were believed to be eligible should therefore be carried through to analysis even if they have failed to respond. There will still be some information, such as place of residence, available on both responders and non-responders and this may help to identify non-response bias even though there is no questionnaire data for the non-responders.

Preparation for analysis

When the survey data have been coded and edited manually, the data must be entered to computer for analysis. Few researchers need to be convinced that computer analysis is worthwhile. Any form of hand analysis is laborious. Even with samples as small as 25 the time and effort of data preparation for computer analysis is profitable.

Professional data entry clerks are able to read paper documents and key data into a computer file at high speed provided the documents are well designed and there is a suitable data layout. If you, the researcher, have to do this job it may be tedious but it will make you familiar with the data set and its blemishes. Keying, like other survey operations, is prone to error and the cost of verification (a check on each keystroke) may be worthwhile.

Part Four

DATA: MEASUREMENT

16
Principles of sampling

Peter Lynn

Introduction

In much research it is necessary to **sample** units for study. Examples are:

- A geographer who wishes to estimate the prevalence of a certain plant in a field will not study the entire field but will divide the field into small areas and sample some of those areas.
- A sociologist who wants to ascertain the proportion of the population who have experienced a certain event will not interview the whole population but will select a sample.
- In everyday life you may:
  — taste your cooking to see if a dish is ready
  — sample cheese before you buy it.

In all of these situations it is important *how* the sample is selected. If the sample is not typical of the total set of units in which you are interested it will fail to serve its purpose.

Why sample?

It is usually possible, at least in principle, to study all of the units which form the population of interest to a study. Reasons why this is rarely done include cost feasibility and quality.

Cost

There is often a real marginal cost associated with the inclusion of each unit of study: the cost of the time of the researcher, experimenter, interviewer, fieldworker and the cost of equipment and materials. So the budget may constrain the sample size. Even without that constraint a smaller sample may leave more money for other stages of the project or for other projects.

Feasibility

If a study of the quality of an industrial output involves destruction of the samples there is no point in studying the entire output. If results are needed by a particular deadline there may be insufficient time to study all units.

Quality

Concentration of effort on a sample can increase the quality of the research which may then lead to more accurate results (see the section on bias, variance and accuracy). For example, in an interview survey with a modest sample size it should be possible to recruit a team of highly capable interviewers and provide them with personal training and briefing. With a much larger sample it may be more difficult to find enough high quality interviewers and infeasible to brief them personally. This could adversely affect the quality of data. The main reasons for sampling are summarised in Fig. 16.1.

Sample design

The procedures and mechanisms that collectively constitute the method of sample selection are known as the **sample design**. At its simplest, a sample design is a **sampling frame** (the list of units from which the sample will be selected), with a specification of the procedures to sample from the frame. In scientific study, the procedures are usually **objective**, with units chosen by a chance mechanism rather than by subjective selection. Thus, if the same sample design is applied repeatedly, it is likely that different units will be sampled each time due to the play of chance.

Inference

Scientific sampling is to provide a means of making **inferences** about the population of interest using observations made on the sample. The observations are collectively called **sample statistics**, whereas the unknown descriptors of the population are called **population parameters**. So, sample statistics are used to make inferences about population parameters. For example, if we observe a sample mean, $\bar{x}$, we might then make an inference about the probability of the corresponding population mean, μ, being within a certain distance of $\bar{x}$. We might decide on the appropriate probability and distance, using information about how the sample was selected, or information about the composition of the sample. A particular sample statistic may be called an **estimator** of a particular population parameter. For example, the sample mean might be an estimator of the population mean. The particular value of that sample statistic observed by the study will be called the study **estimate**. So, if the sample mean is 12.6, then 12.6 is the estimate of the population mean.

There are two basic approaches to inference. One is the **design-based** approach and the other is the **model-based** approach. Chapter 10 of Thompson (1992) provides an introduction to similarities and differences between the two approaches which are described and discussed in Cassel *et al* (1977) and Särndal (1978).

The design-based approach relies on the randomisation present in the design to provide the theoretical basis for estimation. This approach produces estimation methods which are unbiased, and the precision of which can be estimated, without requiring knowledge of the population structure.

With the model-based approach, the researcher must develop a statistical model which adequately describes the distribution of variables. With this method, the sampling mechanism is irrelevant, and the accuracy of estimators depends on the adequacy of the

Why sample?

Cost
Feasibility
Time
Quality

Figure 16.1.

specification of the model. Great care must be taken with this approach. An adequately specified model can prove an impossible goal.

The remainder of this chapter assumes application of the design-based approach to inference.

Bias, variance, accuracy

Any method of selecting a sample for scientific study should be objective and should result in a **representative** sample (Kiaer, 1895). Objectivity is usually interpreted as meaning that the selection method should not permit any subjective influence, and should be **unbiased**. If a sample design is unbiased, then the average value of a sample statistic, across a large number of repetitions of the study, will equal the corresponding population parameter (Sections 4.3 and 5.1 of Moser and Kalton, 1971). This does not mean that the statistic based on the one sample actually selected will necessarily equal the population value. There could be a lot of variation across the different samples that could be selected. This variation is measured by the **sampling variance** (Section 4.3 of Moser and Kalton, 1971, and Section 3 of Mohr, 1990). The smaller the sampling variance, the greater the chance that the sample statistic based on the one sample actually selected will be close to the corresponding population parameter.

Bias and variance are properties of the sample design, not of the particular sample selected. The complete set of all estimates that could be obtained, corresponding to all the samples that could be selected using the chosen sample design, is known as the **sampling distribution**. A sampling distribution may be summarised in tabular form, or as a histogram or graph. For example, Fig. 16.2 shows three sampling distributions produced by three different sampling designs, all proposed to measure the same thing. Design A produces a symmetrical distribution, centred on the population parameter. So design A is unbiased. Design B has an identically shaped sampling distribution to design A but is not centred on the population parameter. So design B is biased and is inferior to design A if other factors, such as cost and ease of implementation, are equal. Design C also has a symmetrical sampling distribution, centred on the same value as design B, so it is equally biased. But the distribution is more compact. C has smaller sampling variance than B. So C is superior to B, other things being equal.

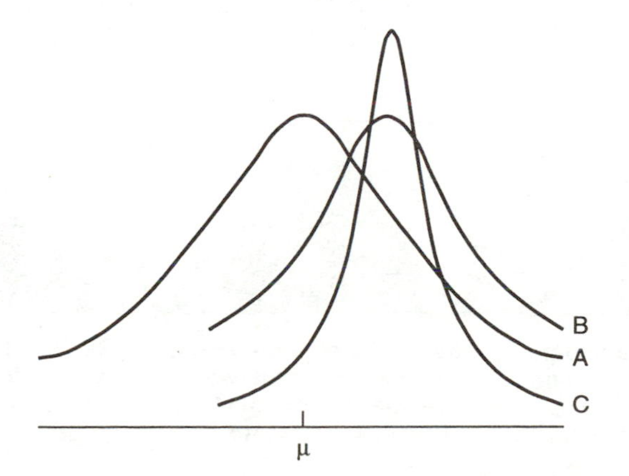

Figure 16.2 Three different sampling distributions produced by three different sample designs A, B and C, where μ is the population parameter.

One question is left unanswered by Fig. 16.2: is design A superior to C? If we consider bias, A is preferable, but if we consider variance, C performs better. The concept of **error** provides a means of answering the question. Error is a property of a particular selected sample not of the sample design. It is simply the difference between the sample estimate and the population parameter. A sensible objective is to minimise the expected magnitude of error. A design can never guarantee an error-free estimate unless it is unbiased and has zero variance. This can be achieved, using the concept of **statistical accuracy**, a performance criterion that encompasses bias and variance simultaneously.

Accuracy can be measured by a quantity known as **root mean square error** (RMSE). The RMSE of an estimator can be thought of as a measure of the average magnitude, across the sampling distribution, of the difference between the estimate and the population value: the average error (see Kish, 1965, pp. 13 and 60 for a precise definition). RMSE might not always be the most appropriate criterion for comparing the desirability of different research designs, but in general it is a very useful concept, and has the advantage of ease of interpretation. A biased design with low variance can produce a more accurate estimator than an unbiased design with higher variance.

Another term sometimes used when discussing estimators is **precision**. Precision is simply the converse of variance. A precise estimator is one with low sampling variance, and vice versa. If unbiased designs are being considered, then precision is synonymous with accuracy. In general, precision is only one component of accuracy; bias is the other.

Precision is usually measured by a quantity known as the **standard error** (see Sections 4.3 and 4.4 of Moser and Kalton, 1971, for a definition and discussion). For many sorts of unbiased estimators, the standard error has a known and fixed relationship to the area under the curve of the sampling distribution. For example, in most situations the population parameter plus or minus two standard errors will encompass approximately 95 percent of the area under the curve (see Fig. 16.3). In other words, 95 percent of the possible samples under the specified design will produce an estimate within plus or minus two standard errors of the true population value. It is this relationship which allows the computation of confidence intervals and the testing of hypotheses (Mohr, 1990).

Precision (sampling variance) is determined by three factors: the inherent variation among the population, the sample size, and other aspects of the sample design such as clustering and stratification. The sample size, n, is an important determinant of precision (larger samples yield greater precision) but it is not the only one. In most situations, variance is approximately inversely proportional to n, so doubling the sample size will halve the sampling variance. The number of units in the population, N, is virtually irrelevant. But it is possible to obtain precise estimators with small n. And it is equally

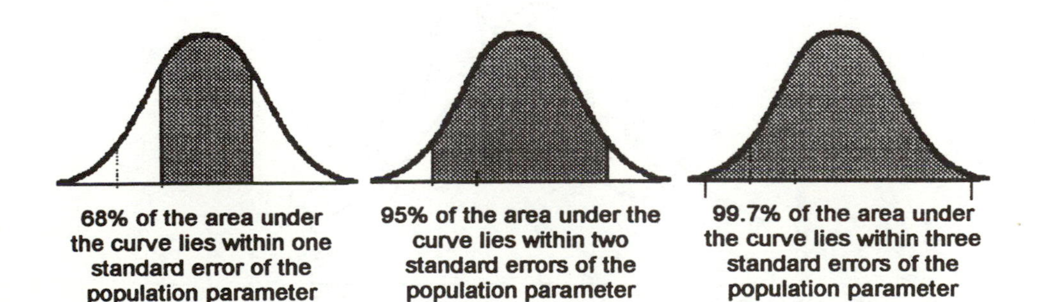

Figure 16.3 Ninety-five percent of the area under the curve lies within two standard errors of the population parameter.

possible to obtain imprecise estimators with large *n*, if the sample is not appropriately designed. It is the **design** of a sample that should give us faith in a study's results. Sample size alone is a fairly meaningless indicator.

Cost is important in the design of a study. It has a valid influence on study design. A main aim of the researcher, though not often stated explicitly, should be to maximise accuracy per unit cost. It is nearly always possible to improve the accuracy of a procedure, but this usually incurs a cost. The task of the research designer is to effect an appropriate balance between cost and accuracy.

Non-probability sampling

There are sometimes compelling reasons for a researcher to consider sample designs which involve selecting the more easily accessible units in the population. **Accessibility sampling** is the term for designs where this is the main consideration. Such designs can have cost and administrative benefits and are common in some fields of research. But there is a drawback. There is a risk of bias and the design provides no means of assessing this bias. For example if the health of fish in a lake was assessed by measuring the first ten fish caught from the most convenient place on the shoreline, the design would be biased if the propensity to be in that part of the lake and to be caught was in any way related to health.

To reduce the risk of such bias, the research may turn to **purposive sampling**. With purposive sampling, the researcher recognises that there may be inherent variation in the population of interest. He attempts to control this by using subjective judgement to select a sample which he believes to be 'representative' of the population. Purposive sampling *can* lead to very good samples, but there is no guarantee that it will. Its success depends on two assumptions:

- the researcher can identify in advance the characteristics which collectively capture all variation
- the chosen sample will correctly reflect the distributions of these characteristics.

Two factors likely to cause contravention of these assumptions are imperfect knowledge of the population structure, and prejudiced selection. Yates (1935) showed that even experts are generally unable to purposively select an error-free sample.

Accessibility and purposive sampling are often combined, as with quota sampling methods used in some interview surveys (see Chapter 17). However, any design which involves an element of purposive selection is open to criticism. Such designs do not necessarily lead to 'unrepresentative' samples but there is no way to measure the likely quality of the samples. Probability sampling provides the means to do this.

Probability sampling

Probability sampling or **random sampling** (they mean the same thing) is often thought the only defensible selection method for serious scientific study unless it is simply not feasible. This is a natural corollary of the design-based approach to inference (see the earlier section on inference).

Probability sampling refers to sample designs where units are selected by some probability mechanism, allowing no scope for the influence of subjectivity. Every unit in the population must have a known, non-zero, selection probability. Note that the probabilities need not be equal. The advantages of probability sampling are that it enables the avoidance of selection biases and that it permits the precision of estimators to be assessed, using only information that is collected from the selected sample. Furthermore, because precision can be estimated, this provides a tool for making informed estimates of the likely effect of changes to aspects of the design. Thus you can choose between competing potential designs, according to precision and cost.

The theory of probability sampling is well developed. Thompson (1992) provides a good introduction with references to more specialised texts. There are practical strategies for implementing probability sampling designs in various fields. For example, see Cochran (1977) or Kish (1965) on surveys; Cormack *et al* (1979) on biological studies; Hohn (1988) on geological studies; Keith (1988) on environmental sampling; Ralph and Scott (1981) on sampling birds; Seber (1982, 1986) on sampling animals; Deming (1960) on business research; Metcalfe (1994) on industrial and engineering research. See also Chapters 12–15 of this book.

The simplest form of **probability sampling**, and the one with which other forms can be compared, is **simple random sampling**. This is the design under which every possible combination of n units from the population of N units is equally likely to be selected. Note that this implies that each individual unit has an equal selection probability. Simple random sampling is therefore **design-unbiased**.

Systematic sampling

Systematic sampling is another design that gives each unit an equal selection probability. The population units are listed and a sample is taken by selecting units at fixed intervals down the list. For example, to sample 100 from a list of 2000 you would generate a random start between one and 20 and select that unit and every 20th thereafter until the end of the list was reached. Each unit has an equal chance of selection because each of the 20 possible samples, corresponding to the 20 possible random start points, has an equal probability of being selected and each unit belongs to one, and only one, of those 20 samples. But it is not simple random sampling because there are only 20 samples that could be drawn. Any other combination of units cannot be selected.

There are two main reasons for sampling systematically. One is administrative convenience. It may be easier for the sampler to count through a list or a set of files than to sample randomly. The other is that systematic sampling incorporates an element of stratification if the list is ordered in a way that is related to the subject of the study.

Stratification

The aim of stratification is to guarantee that the sample reflects the structure of the population, at least in terms of one or more important variables. For example, if the sample design for a medical study is to select patients systematically from a list ordered, or stratified, by age, the sample will be certain to reflect the age distribution of the population. If age is related to the variable(s) of interest, this is beneficial. Simple random sampling would not guarantee reflection of the age structure. You might, just by chance, select all the older patients. Stratification can be incorporated within a probability sample design. It does not violate the principle of random sampling.

Stratification can be either implicit or explicit. **Explicit stratification** requires the creation of distinct strata, the determination of a sample size and the selection of a sample from each stratum. For example, the geographical region of interest to a study may be divided into grid squares each of which would be assigned to one of three strata: densely inhabited, sparsely inhabited, uninhabited. Within each stratum, a simple random sample of grid squares could be selected for study. One advantage of explicit stratification is that it enables the use of unequal sampling fractions which are described below. Stratification ususally brings modest gains in precision (see earlier section on bias, variance and accuracy) and is worth doing unless it is prohibitively time consuming or expensive.

With **implicit stratification** the sample frame is sorted before it is sampled systematically. The sample size in each stratum is not fixed in advance although it cannot vary much. One advantage of implicit stratification is that you can stratify using continuous variables without losing information by having to create broad bands as strata. Stratification usually brings modest gains in precision (see earlier section on **Bias, variance and accuracy**) and is worth doing unless it is prohibitively time consuming and expensive.

Unequal sampling fractions

Sometimes different proportions of units are required in different strata – **unequal sampling fractions.** Explicit stratification is necessary. One reason for unequal sampling fractions is to provide sufficiently large samples for separate analysis in each stratum. For example, in a study of the population of England and Wales a sample of 2000 people might provide sufficient precision for many purposes. But if separate estimates were required for each of the two countries the sample in Wales may be too small for estimates to be useful. Simple random sampling may select 1900 in England and only 100 in Wales. An unequal sampling fraction could be imposed to ensure a sample of 1725 for England and 275 for Wales.

Unequal sampling fractions may be used if it is impractical to give all units an equal selection probability. For example, in line-intercept sampling of vegetation cover, the size of a patch of vegetation is measured whenever a randomly selected line intersects it. This results in large patches having higher probabilities of inclusion in the sample.

Unequal fractions may be used to increase the precision of estimates based on the total sample. This can be achieved by using larger sampling fractions in strata which are inherently more variable (Moser and Kalton, 1971, pp. 93–99; Thompson, 1992, pp. 107–108). That this is sensible can be seen by considering an extreme situation where the population falls into two strata within one of which there is no variation at all. There would be no point in selecting more than one unit from that stratum because knowledge of any one unit confers knowledge of the complete stratum. So, the sampling fraction should be much higher in the other stratum.

In the examples above, unequal sampling fractions were obtained by specifying an exact sample size for each stratum. Another method is **weighted sampling** in which units in one stratum would have a different probability of selection than units in another. In the study of England and Wales, each person may have a selection probability three times that of each person in England. To produce unbiased estimates for England and Wales as a whole it is necessary to *weight* the data to restore the correct distribution across the strata and this can cause a loss in precision.

Multistage sampling

It is sometimes desirable to select the sample in two or more stages, to reduce cost or effort. For example, in a survey where people are interviewed in their own homes, a sample of small geographical areas such as postcode sectors or electoral wards may be selected first followed by a sample of addresses drawn from each sampled area. This provides efficient workloads for interviewers, and is much more cost-effective than a sample spread thinly over the whole country. Similarly a survey collecting data from patients' medical records might first select a sample of GP practices, and then a sample of patients registered at each practice. This reduces the number of practices that need to be contacted, thus easing study administration.

Multistage sample designs are efficient solutions in many contexts but they may result in less precise estimators than single-stage samples of the same size. This is because study units within each first stage entity are often less variable than units in the whole population. This increases sampling variance. For example, two people living in the same postal sector, or registered at the same GP practice, may be more similar than two people sampled at random from the whole population, in terms of many variables. The heterogeneity of first stage entities can be measured by the **intraclass correlation coefficient**, ρ (Kish, 1965, Section 5.4). The higher the intraclass correlation, the more detrimental it will be, in precision, to cluster the sample.

In multistage sampling, the selection probability of each unit will be the product of the probability of selecting the first-stage entity to which the unit belongs and the conditional probability of selecting the unit given that the first-stage entity has been selected. Thus the relationship between these probabilities needs to be considered carefully when the sample is designed and the probabilities must be recorded carefully to allow appropriate weighting (see the section on weighting below).

Capture–recapture sampling

The classic use of **capture–recapture sampling** is to estimate the total number of units in a population. There are variants of the technique, but the basic idea is to select a sample, attach an identifying mark to each sample unit, return these units to the population, select another independent sample, and observe what proportion of the second sample has already been included in the first sample. If both samples are truly independent, this proportion should be an unbiased estimator of the population proportion included in the first sample. Thus population size can be estimated as the size of the first sample divided by this proportion.

Capture–recapture sampling has been used on many animal populations and on elusive human populations such as the homeless. See Cormack (1979) and Chapter 18 of Thompson (1992) for further discussion.

Adaptive sampling

Adaptive sampling is frequently used for sampling rare or elusive populations. It requires an initial selection and study of a probability sample. Then, whenever a unit exhibits a high or interesting value of the variable of interest, a further sample of units is taken close, usually geographically, to the interesting unit. For example, in a survey of a rare mineral resource a probability sample of small areas might be studied. Most areas would reveal a zero occurrence of the mineral but where the mineral is found there may be an increased probability that adjacent areas will also host the mineral. This intuitive idea of increasing the study coverage often brings increases in precision.

Adaptive sampling is used in the study of rare animal and vegetation populations, as well as in geology, and in epidemiological studies of human populations. Thompson (1988, 1990, and 1992 Chapter 24) provide further discussion.

Sample size

An important element of sample design is the determination of the sample size. As described in the section on bias, variance and accuracy, sample size affects the precision of estimators, although it is not the only element of sample design to do so. This means that if the research can specify in advance the required level of precision, it is possible to determine, at least approximately, the sample size that would be required to deliver that precision. See Barnett (1976), Sections 2.5 and 2.9.

Weighting

Weighting is when, in order to assume a desired degree of importance, an observation is attached to a numerical coefficient (frequently by multiplication).

There are several reasons why it may be desirable to *weight* data before analysis. The most common are:

To correct for unequal selection probabilities

If different units in the population had different selection probabilities, sample estimates would be biased unless each sampled unit is given a weight proportion to the reciprocal of its selection probability. For example, in the unequal sampling fractions discussed in the section on probability sampling, units sampled in Wales should be given a weight one-third that of units in England.

To adjust for non-response

If there are some selected units for which no data could be collected (this is common in interview surveys and in medical studies) then sample estimates will be biased if the propensity to respond is related to any of the variables of interest. Non-response bias may be compensated by weighting. This requires some knowledge about the relationship between responding and non-responding units, or between responding units and the total population in terms of some characteristics which may be related to propensity to respond and to variables of interest. There are various ways to develop such weighting. See, for example, Elliott (1991).

Post-stratification

The proportion of units in a random sample that have a certain characteristic may happen, by chance, to differ greatly from the corresponding population proportion. If the characteristic is believed to be related to variables of interest to the study, the data could be weighted to match the population profile.

Weighting to correct for unequal selection probabilities or to adjust for non-response is to eliminate bias. Weighting by post-stratification is to improve sampling variance. Holt and Smith (1979) discuss post-stratification and its likely effects.

Summary

Sampling is a complex discipline, yet it is of primary importance in many studies. It is the foundation on which much study is built. For many purposes it is not necessary to be closely familiar with even the few sampling techniques mentioned in this chapter but it is always important to consider how a sample is to be drawn and what effect that sampling method might have on the data. Sampling methods for scientific study should be objective and should maximise accuracy of estimation, per unit cost, as far as possible. This will require the strict application of appropriate probability selection methods which will then allow estimation of the accuracy obtained.

The particular sample design issues that are of primary importance vary between disciplines and are discussed in other chapters of this book.

References

Barnett V (1991) *Sample Survey Principles and Methods* (2nd edn). Edward Arnold, London.
Cassel C M, Särndal C E and Wretman J H (1977) *Foundations of Inference in Survey Sampling*. John Wiley, New York.
Cochran W G (1977) *Sampling Techniques* (3rd edn). John Wiley, New York.
Cormack R M (1979) Models for capture–recapture. In Cormack R M, Patil G P and Robson D S (eds), *Sampling Biological Populations.* International Co-operative Publishing House, Fairland, MD.
Deming W E (1960) *Sample Design in Business Research*. John Wiley, New York.
Elliot D (1991) *Weighting for Non-response*. OPCS, London.
Fisher R A (1966) *The Design of Experiments* (8th edn). Hafner, New York.
Hohn M E (1988) *Geostatistics and Petroleum Geology*. Van Nostrand Reinhold, New York.
Holt D and Smith T M F (1979) Post-stratification. *JRSS (A)* Vol. 142, pp. 33–46.
Keith L (ed.) (1988) *Principles of Environmental Sampling*. American Chemical Society.
Kiaer (1895) Observations et experiences concernment des denombrements représentifs. *Bull Int Stat Inst*, Vol. 8(2), pp. 176–183.
Kish L (1965) *Survey Sampling*. John Wiley, New York.
Metcalfe A V (1994) *Statistics in Engineering*. Chapman & Hall, London.
Mohr L B (1990) Understanding significance testing. Paper 73 in the series *Quantitative Applications in the Social Sciences*. Sage, Newbury Park.
Moser C A and Kalton G (1971) *Survey Methods in Social Investigation* (2nd edn). Gower, Aldershot.
Ralph C J and Scott J M (eds) (1981) *Estimating Numbers of Terrestrial Birds: Studies in Avian Biology no 6*. Pergamon, Oxford.
Särndal C E (1978) Design-based and model-based inference in survey sampling. *Scandinavian Journal of Statistics*, Vol. 5, pp. 27–52.
Seber G A F (1982) *The Estimation of Animal Abundance* (2nd edn). Griffin, London.
Seber G A F (1986) A review of estimating animal abundance. *Biometrics*, Vol. 42, pp. 267–292.
Thompson S K (1988) Adaptive sampling. *Proceedings of the Section on Survey Research Methods of the American Statistical Association*, pp. 784–786.
Thompson S K (1990) Adaptive cluster sampling. *JASA*, Vol. 85, pp. 1050–1059.
Thompson S K (1992) *Sampling*. John Wiley, New York.
Yates F (1935) Some examples of biased sampling. *Annals of Eugenics*, Vol. 6, pp. 202–213.

17
Sampling in human studies

Peter Lynn

Introduction

This chapter describes the main issues to be considered when a sample of people is designed for research. Units may include groups of people such as households or families. The most common research method used with human subjects is the interview survey and much of the discussion is about sampling for interview surveys. The central issues are similar for any research which takes the collection of data directly from the selected sample. Important sampling issues for other research methods will be mentioned. These have much in common with interview surveys.

Much of this chapter refers specifically to studies in the UK, particularly the section on sampling general populations. Some of the issues may be similar for research in other countries but you should seek advice of a local expert.

Important considerations

Target population

Before you embark on the design of a sample, think carefully about the definition of the population about which you intend to make inferences. This is the **target population** (Chapter 16 describes the concept of inference). For example, consider a study about the attitudes of parents towards schools. Is the study concerned only with parents who currently have a child in school? Or are the views of those whose children will soon be starting school also relevant? And what about those whose children have recently left school? How should the study treat foster parents and other carers? Are you interested only in schools that cater for a particular age range? Is the study about all schools? Or only state schools? What about grant-maintained schools? Are you attempting to represent the whole of the UK? Or Great Britain? Or England and Wales?

Only when you have defined the population of interest *precisely* will you be able to work out the best way to sample it. Then, if you realise that it is not possible to give all members of the population a chance of selection, you must restrict the **survey population** more than the target population. You should make this distinction explicit and acknowledge it in your thesis. Consideration of the nature of the excluded part of the target population may lead you to conclude that you cannot reasonably make inferences about the target population but that instead you should make statements only about the survey population. See Section 3.1 of Moser and Kalton (1971) for further discussion of the problem of defining the population.

Efficient fieldwork

Much research involves visits to members of the sample to collect information from them. For example, you might interview each person, or measure them, or observe certain behaviour. You therefore need to consider, and perhaps control, the geographical location of the sample.

For example, for large scale interview surveys it is common to select a sample of small areas, such as postcode sectors, and then a sample of addresses within each. It is possible to sample the areas so that, by taking an equal number of addresses in each, you will have an equal-probability sample (see section on 'clustering' below). Then the number to sample within each area can be set to provide an efficient work load for one interviewer, baring in mind the amount of work that one interviewer may want to do.

Even if you are doing all the fieldwork yourself, you will want as large a geographical spread of the population of interest as possible. This may be achieved by sampling some small areas and then concentrating your sample of people within those areas. It is better to use probability sampling, rather than purposive selection for the reasons given in Chapter 16.

Fieldwork efficiency for some research methods is not related to the geographical distribution of the sample. An example is where the data are to be collected at a central point using a self-administered postal questionnaire or the telephone.

Clustering

The samples from sample designs where the study subjects are concentrated in several small areas are called **clustered** samples. Samples may be clustered in ways other than geographical, though geographical clustering is the most common form. Some degree of geographical clustering can be achieved indirectly, by sampling first-stage entities which provide some implicit clustering of study subjects, even though the geographical boundaries of the entities may overlap.

For example, in a study of hospital outpatients, it is likely that most outpatients will live close to the hospital where they were treated. If several hospitals are sampled at the first stage, the resultant sample will be clustered even though it might contain a few people who live a long distance from the hospital they attended.

In the rest of this chapter, for simplicity, first-stage entities in a sample design will be called *areas*, though the arguments apply equally to other sorts of first-stage entities.

Multistage sampling (see p. 133) is a necessary, but not sufficient, prerequisite for clustered sampling. There is usually a trade-off between the reduced costs achieved with a clustered design and reduced precision (see p. 133).

Multistage designs should result in equal probability samples. The two simplest ways to achieve this are:

1. select areas with equal probabilities and then select individuals with equal probabilities
2. select areas with probability proportional to size (**PPS sampling**: see Lynn and Lievesley, 1991, pp. 16–17) and then select individuals with probability proportional to the reciprocal of the selection probability of the area to which they belong.

The first method uses the same sampling *fraction* in each area so the sample size will vary across areas in proportion to the population sizes. In the second method, the same *number* of individuals is selected in each area so the fraction is inversely proportional to the area population size. The second method is preferable, for fieldwork planning and for statistical efficiency, but it is not always possible as a suitable measure of size for each area must be known in advance.

Samples of the general population in the UK are typically clustered within administrative areas such as electoral wards or polling districts, or areas defined by postcodes such as postal sectors (Lynn and Lievesley, 1991; Bond and Lievesley, 1993). The choice of clustering unit is often influenced by the choice of sampling frame (see the section on sampling frames below).

Stratification

If you need to oversample some sub-groups of the population so that you can analyse them with reasonable precision, then the sub-groups must be identified before sample selection. There may be other reasons to stratify (see p. 132) so you must consider carefully what stratification factors are available and how they should be used.

Sampling frames

The list of population members from which a sample is drawn is known as the **sampling frame**. The frame may have a physical existence, such as a printed list or a computer file. Alternatively, it may result from the application of a sampling method. In this section we discuss the important properties of a sampling frame or sampling method. These properties are summarised in Fig. 17.1 and are discussed in more detail by Lynn and Lievesley (1991, Section 2.3). These characteristics of a frame will not always be attainable, as lists and records used as frames may have been compiled for purposes different from the purpose of the study.

The frame should completely cover the target population. There may be omissions because the frame was not designed to include all of your study or because some people have been excluded for other reasons. Reasons for omissions should be investigated and careful consideration given to the extent that they might bias the results of your study (see pp. 129 and 137).

The frame should not include individuals who are not part of your target population. Otherwise your results may be distorted or, at best, you may have to expend some time identifying and excluding ineligible sampled cases. Similarly, duplicate entries may introduce bias unless the number of entries for each individual is known.

Ideally, the units listed on the frame should correspond exactly with the study units. If this is not the case, there should be a known linkage which enables calculation of selection probabilities. For example, to sample households, addresses are usually selected from a frame such as the Post Office's file of addresses (see the following section on sampling general populations). There is not always a unique one-to-one link between an address and a household, but fieldworkers can establish the number of households at each sampled address and whether each sampled household could have been selected via any other address.

The information on the frame should be sufficient to unambiguously identify each sampled individual and allow easy location on the ground. If this is not the case it may be worth looking for supplementary information from other sources, such as better quality address details, before starting fieldwork.

Ideal properties of a sampling frame

- No omissions (up-to-date)
- No ineligibles listed
- No duplicates
- Frame units correspond to study units
- Units are uniquely and fully identified
- Frame permits stratification and clustering
- Easy and inexpensive access
- Familiarity

Figure 17.1 Ideal properties of a sampling frame.

For multistage sampling, the frame must identify suitable areas. Preferably these should be areas for which there is also information available for stratification. The frame should be inexpensive and easy to use and there are advantages in using a frame which has already been used as a sampling frame on previous occasions. You can learn from previous experiences and minimise the possibility of unexpected nasty surprises.

Sampling general populations

Populations which might be of interest to particular studies can broadly be classified as either **general** or **special**. There is no precise definition of the difference between the two but a general population can be thought of as one which includes a large proportion of the total population in the geographical area of interest to the study, perhaps a quarter or more. This would include 'all adults', 'all households with a telephone', and 'all married women'. A special population would be a smaller proportion of the total. It may be called a **minority** population.

Methods commonly used to sample general populations are quite distinct from those used to sample special populations. In this section we discuss the former, and in the following section, the latter.

General population sample designs in the UK usually involve the selection of addresses followed by the identification of relevant individuals associated with each selected address, using some clear definition of association. This is because there are no comprehensive lists of individuals available from which to sample directly in this country. The one exception is the electoral register which is used to sample electors but it has a rather biased coverage of adults as a whole.

The two lists most commonly used as sampling frames of addresses in the UK are the Postcode Address File (PAF) and the Electoral Register (ER). An outline follows of these two frames and how they might be applied. See Lynn and Lievesley (1991, Chapter 3) for further details. The council tax lists are also used sometimes but generally just for local studies as they can only be studied through local authorities. Area sampling (Kish, 1965, pp. 301–358) is another way to sample residential addresses and is commonly used in the US. It has no advantages in the UK where good lists of addresses exist. A general population sample can also be selected by non-probability methods such as quota sampling. These are generally easier and cheaper to implement than probability methods but have the serious defects described in Chapter 16.

The ER is a list of all people eligible to vote and includes the name and address of each person. It is compiled annually by each local authority district (see Hickman, 1993). Because it includes addresses it can be used as a frame of addresses rather than individuals. This is its most common use as its coverage of addresses is better than its coverage of individuals (Smith, 1993; Foster, 1993). However, about five percent of inhabited residential addresses are not listed on the ER and this can introduce bias. Coverage is particularly poor for some ethnic minorities, people who have moved recently, people aged 20–24, and inner London. The main advantage of the ER is that addresses can be selected with probability proportional to the number of registered electors. This can be efficient if you want a sample of individuals, rather than households, and do not want to select more than one individual at an address.

The PAF is a computerised list of every address to which the Post Office delivers mail in the UK. It is split into two files: **large users** and **small users**. The latter is the one used to sample residential address, though it includes some non-residential property too. The PAF contains no information about the occupants at each address so the only way to sample addresses is with equal probability. The main advantage of the PAF over the ER is greater coverage of residential addresses: only one or two percent are missing. It is also generally more up-to-date than the ER and, because it is computerised, complicated and large samples can be drawn easily. This can also be a disadvantage as samples must be commissioned from computer agencies whose charges may be high. ER samples can be drawn by hand from public documents at either OPCS or the British Library.

Sampling special populations

Broadly there are three methods to sample a special, or minority, population:

1. screening a general population sample
2. sampling from an existing administrative list
3. constructing a sampling frame.

A brief description of each follows. There is fuller treatment in Hedges (1978), and Kalton and Anderson (1986).

Screening a general population sample

Screening involves the selection of a sample of addresses, as you would for a general population sample, and then checking each sampled address (called a **screen interview**) to identify members of the population of interest. Those identified can then be interviewed. This process can involve a large amount of work. Its viability depends largely on the incidence of the special population in the general population. For example, if each of only one percent of addresses contains a member of the special population, then to find 300, say, you would need to screen 30,000 address. If the incidence were ten percent you would need to screen only 3000.

The amount of detail that needs to be collected in the screen interview is important. In some situations it may be possible to screen by methods other than personal interviewing, such as with a postal questionnaire or by telephone. **Screening errors** could also be important, particularly if the definition of membership of the special population is complicated. If the screening questions are designed to be conservative, in the sense that some people identified as eligible turn out to be ineligible when the detailed data are collected by the main study instrument, then some effort will have been wasted in interviewing these **false positives.** On the other hand, if the questions are tightened to minimise the risk of false positives, some eligibles might be missed. These are **false negatives**. This could introduce bias. A balance has to be struck.

Sampling from an existing administrative list

It is sometimes possible to identify an existing list which is adequate as a frame. But there may be problems of access which have to be negotiated carefully. The main problems to realise when you use an existing list are those of coverage (omissions and ineligibles) and selection probabilities (duplicates). A clustered sample for fieldwork efficiency may be difficult to design as administrative files are often inflexible. Sometimes a choice must be made between the use of an imperfect existing list and the screening of a general population sample. For example, it might be possible to use the DVLA files of registered keepers of motorcycles to sample motorcycle riders. But a keeper is not the same as a rider. Some people will keep more than one motorcycle each and will be listed more than once. On the other hand, there are around one million motorcycles in the UK so it might be feasible to screen a general population sample, though it would be expensive.

Constructing a sampling frame

It may be possible to construct a frame where none exists. Perhaps several existing lists, each of which covers part of the population of interest, could be combined. **Snowballing** techniques (Goodman, 1952) may be used to expand the frame. The main problem could be to assess the completeness of the constructed frame. When you combine lists, you should pay special attention to overlap, with duplicate entries on the combined list.

It may not be necessary to construct a frame of the individuals of interest if a frame can be constructed of first-stage units which can be sampled before enumeration of individuals within each sampled unit. For example, it would be an enormous task to identify all families who live in bed and breakfast accommodation in a particular area and to list them. It might be possible to identify and list all bed and breakfast *establishments*, a sample of which could be drawn and eligible families identified in each.

Weighting

As described on pp. 134–135, study data should usually be weighted, for several reasons. With general population samples, different units will usually have different selection probabilities. This will certainly be true if the design is to select addresses and then to sample one person at each address. In this situation, people who live alone will have a greater chance of selection than those in multiperson households. The data should be weighted to correct the imbalance (see Lynn and Lievesley, 1991, Chapter 4).

In studies of human subjects, there will always be some non-response. If the sampling frame provides some information about each individual, this information could be used to develop weighting to adjust for non-response. Even if the frame does not contain any useful information, as with the general population frames of addresses, it may be possible to obtain some basic details about non-respondents as well as respondents, by observation in the field. For example, variables related to housing type and area characteristics could be observed. If variables observed for all sample members are related to propensity to respond, and also to survey measures, then the information can be used to reduce non-response bias (Elliott, 1991).

General population samples can also be compared with external population data, such as the decennial census of population. If discrepancies are apparent, weighting can be used to adjust the sample profile. If discrepancies can be assumed to be caused by non-response bias or by sampling variance then weighting is an appropriate remedy. However, you should be careful when you compare your data with external data. Differences, even quite subtle ones, in the definitions used, question wording, and data collection mode, can produce an artificial outcome to your analysis. In this situation, weighting might do more harm than good by introducing biases.

References

Bond D and Lievesley D (1993) Address-based sampling in Northern Ireland. *JRSS (Series D: The Statistician)*, Vol. 42, pp. 297–304.

Elliot D (1991) *Weighting for Non-response*. OPCS, London.

Foster K (1993) The Electoral register as a sampling frame. *Survey Methodology Bulletin number 33*. OPCS, London.

Goodman L A (1952) On the analysis of samples from **k** lists. *Ann Math Statist*, Vol. 23, p. 632.

Hedges B (1978). Sampling minority populations. In Wilson M (ed.) *Social and Educational Research in Action*. Longman, London.

Hickman M (1993) *Compiling the Electoral Register 1992*. HMSO, London.

Kalton G and Anderson D (1986) Sampling rare populations. *JRSS (series A)*, Vol. 149, pp. 65–82.
Kish L (1965) *Survey Sampling*. John Wiley, New York.
Lynn P and Lievesley D (1991) *Drawing General Population Samples in Great Britain.* SCPR, London.
Moser C A, Kalton G (1971) *Survey Methods in Social Investigation* (2nd edn). Gower, Aldershot.
Smith S (1993) *Electoral Registration in 1991*. HMSO, London.

18
Problems of measurement

Jim Rowlands

Looking for trouble

In any programme of research involving measurement it is essential to consider critically at the outset the likely quality of the data that the system of measurement will provide. No amount of data analysis will save the situation if the data are unsound or insufficiently informative.

Always try to identify and counteract sources of bias that may distort the picture. For example, in a study of a certain disease in which measurements on a group of diseased individuals are to be compared with measurements on a control group of individuals who are free of that disease, if it is possible that the control group has its own peculiarity and so is not representative of the general disease-free population then an observed difference between the groups may not mean what it appears to mean. If nothing is done about this beforehand, such as selecting the control group from the disease-free population at random, then the conclusions of the study may be unsound and will be open to challenge.

If the output of a system is measured over a period of time and two measured variables move together during that period, such as one being high when the other is high and low when the other is low, then it will be impossible to determine which variable has the greater effect on the output. The two variables are said to be **confounded**. It is possible for the effect of a measured variable to be confounded with the effect of some unconsidered variable. For example, if the level of a stimulus is increased gradually over time, then the changes in the response may not be due to the changes in the stimulus. They could be due solely to the action of some other variable that happens to be changing with time. One counter measure is to randomise the time order in which the different levels of the stimulus are applied. The question of whether an unconsidered variable is in play can then be investigated by comparing the graph of the response plotted against time with the graph of the response plotted against the levels of the stimulus.

Try to ensure that the information content of the data set will be sufficiently high for its intended purpose. It is important to appreciate that merely increasing the number of measurements will not necessarily achieve this. For example, in a study of the effects of a number of variables on the thickness of an extruded material, eight strips of material were produced. Each strip was produced under a different combination of settings of the variables. Subsequently, thickness was measured with negligible error at frequent intervals along each strip, so creating a sizeable data set. The investigator was disappointed to find that the results shed little light on the question of which variables influence thickness. Doubling or tripling the number of thickness measurements made on each strip could not help; quite enough was known about the thickness profiles of the eight strips already. What was needed was more strips. The root of the problem was that setting up the extruder to produce material under a given set of conditions was an error-prone exercise, so that each set-up would produce a strip with a somewhat different average thickness even under the same nominal conditions. A pilot study in which a small number of strips were produced under the same set of conditions, with the extruder being set up afresh each time, would have revealed this and prevented puzzlement later.

It is easier to guard against problems that complicate the interpretation of results, such as bias and confounding, for some types of investigation than for others. At one end of the spectrum are experimental studies in which the system under study is set up and controlled by the investigator. At the other end of the spectrum are pure observational studies (Cochran, 1983) where the investigator has no control over the collection of the data. This type of study can arise in an industrial context. Experimenting on an industrial plant in full scale production is usually out of the question. On the other hand, massive amounts of data may be provided on many variables by automatic measuring devices. Unfortunately the information content of such data can be very low even when every measuring device is operating satisfactorily. Continual compensatory adjustments made by operators or by automatic controllers can cause confounding between key variables. Also, by contrast, the data may have nothing to say about the effects of some other key variables because the operating policy is to hold these constant throughout.

Another consideration is the sampling rate. This can be set too high. Values of measurements on the same variable taken closely together in time can be highly correlated. If this is so then the amount of data to be analysed can be reduced without serious loss of information by working with a sub-sequence of the stream of measurements. A simple graphical method can be used to examine the correlation between successive measurements. This is the construction of a scatter plot of the time-ordered sequence of measurements against the same sequence off-set by one time unit. For example, seven successive time-ordered measurements such as 6.2, 6.5, 8.4, 5.0, 9.0, 9.9, 9.7 yield the six points (6.2,6.5), (6.5,8.4), (8.4,5.0), (5.0,9.0), (9.0,9.9), (9.9,9.7) on the scatter plot. Figure 18.1(a) shows such a scatter plot in a case where there is little or no correlation between successive measurements, while Fig. 18.1(b) shows a case where successive measurements are highly positively correlated.

It is best not to take measurements provided by other people entirely on trust. Always investigate how they were really obtained. In one investigation it was discovered that values of pressure recorded on a data sheet were in fact eyeballed averages obtained by viewing six separate pressure gauges simultaneously. In another investigation, measurements that should have been taken at three different times of the day were found to have

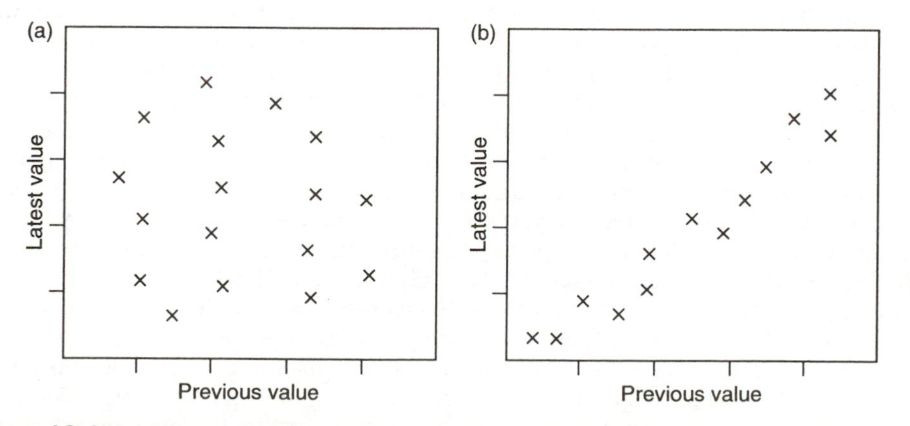

Figure 18.1 Scatterplots showing (a) no correlation, and (b) high positive correlation, between successive time-ordered measurements.

been taken all at the same time. Also, important information may not be made available because no one thought to give it and the researcher did not think to ask for it. In one case, flow rate, which turned out to be a key variable, was not included on the data sheet. The rate had to be estimated using measurements of pressure and viscosity and Poiseuille's formula. Only later was it discovered that flow rate was being measured directly and routinely with comparatively little error. In another case, an investigation of the strength of a material made little progress until it was discovered that the recorded strength was actually the average of longitudinal strength and transverse strength. The material had no obvious grain but it was found that these two strengths behaved quite differently. Consequently their average had little meaning.

What should I measure, exactly?

The choice of what to measure is not always straightforward. Ideally the measurement should:

1. characterise the property or phenomenon of interest
2. be easy to obtain
3. be largely error-free.

The final choice usually represents what is felt to be the best compromise between these three requirements.

Special subject matter knowledge is often needed in order to satisfy the requirement that the measurement should characterise the phenomenon of interest. Always try to speak to an expert about what it is best to measure before making an irrevocable decision. The aim of an experimental programme described by Grove and Davis (1992) was to study *submarining* in car crashes where the lap portion of the seat-belt rides up and crushes the abdomen of the wearer. As Grove and Davis remark, it would have been highly inefficient merely to record at each trial whether submarining occurred or not. Instead, the investigators measured a number of characteristics known to be associated with the tendency to submarine. One such characteristic was the angle (in radians) that the wearer's back makes with the vertical at the moment of maximum restraint. It is unlikely that someone unfamiliar with the subject would have thought to measure this variable.

The crashes in the seat-belt study were in fact simulated on a computer. Therefore, to ensure the correctness of the results it was necessary to calibrate the simulation model using data obtained from a limited number of real crash tests carried out on dummies under various conditions. The use of calibration curves (or surfaces) is a feature of many measurement processes. For example, the concentration of a chemical that is difficult to measure directly might be obtained indirectly by using a calibration curve that relates the value of an easily measured surrogate response, such as electrical resistance, to the concentration of the chemical. The calibration curve is constructed in the first place by measuring the surrogate response at standard concentrations of the chemical. The scientific challenge is to find a surrogate response that is specific to the quantity of interest. For a discussion of the statistical aspects of calibration curves, such as the construction of confidence intervals, see Mandel (1984) or Carroll *et al* (1988).

It can happen that measurements cannot be expressed on a well-defined quantitative scale or that such measurements are unavailable. For example, a defect that sometimes occurred in a type of transparent acrylic sheet was the appearance of parallel dark bands when the sheet was lit at an angle. The number of dark bands per unit length was easily measured but it would not matter how many bands there were if they were not dark, that is if they were invisible. Consequently the characteristic of most interest was the darkness of the bands. In principle the degree of darkness could have been measured on a continuous scale by referring it, directly or indirectly, to some physically meaningful standard but such a measure was not available at the time of the study. Instead darkness was assessed on an ordered classificatory scale. The scale values chosen were zero (no bands visible), one (bands just visible), two (bands quite dark) and three (bands very dark). Sheets were produced under a number of different conditions. Each one was then inspected visually and assigned to one of the categories zero, one, two or three. Once all the sheets had been classified in this way, the complete data set consisted of the counts for the four categories for each set of conditions.

Classificatory data should not be analysed as if they consisted of measurements made on a well defined scale even if, as was the case in the above example, the categories can be arranged in a meaningful order. Special methods of analysis are required. For further reading see Siegel (1956); Maxwell (1961); Dobson (1983); McCullagh and Nelder (1989).

In subject areas such as education or the behavioural sciences, researchers are on the safest ground when attention is confined to classificatory data. Much progress has been made since the 1960s but early attempts at meaningful quantification in these areas were often unsatisfactory. A familiar example is IQ testing. The testing procedure produces a partial ordering of the subject population but it is not clear what it actually measures. For further reading see Roberts (1979) and Michell (1990).

It may be necessary to measure more than one property or response in order to characterise the feature of interest, that is, the complete measurement may be a vector. This was the case in the seat-belt study discussed above. Another example is provided by the measurement of colour. Any particular colour is characterised by the values of three quantities, conventionally denoted by L, A and B. These are related to the electromagnetic spectrum. A polar transformation of L, A, B gives a characterisation of the colour in terms of its hue, chroma and intensity.

For some kinds of measuring device the output is not a vector but a curve on a graph. The graph can be difficult to interpret but the method of principal components can be used to analyse the curve and describe the results. See Church (1966).

Falling into error

A series of measurements of a fixed quantity T made under constant conditions will usually vary to some extent. In fact, if the measurements do not vary this may well be an indication that the system of measurement lacks sensitivity. If the series is extended indefinitely then the average or *mean* $\bar{x}$ of the measurements x can be expected to settle down to some constant value μ. The measuring device or method of measurement is said to be *unbiased* if $\mu = T$, so that the measurements cluster around the correct value. Otherwise, the device is said to have a *bias* equal to $\mu - T$.

Comparisons of different quantities involving the *differences* of measurements will not be affected by the existence of a constant measurement bias, but in general, it is important to detect and estimate bias when it exists. If the variation in a run of measurements of a fixed quantity T is not too great then the value of $\bar{x}$ will quickly stabilise. Then, if the value of T is known, the value of $\bar{x} - T$ will give a good indication of the bias of the system of measurement in the region of T.

Youden (1962) discusses how measurement bias can arise as the aggregate of small non-random errors associated with the individual components of a measuring device and describes efficient strategies, known as **weighing designs**, for identifying the most important sources of these errors.

It is good practice to monitor the stability of the measurement system by carrying out short runs of repeat measurements of a reference quantity at regular intervals. Instability may appear over the long term, such as a trend in the results indicating an increasingly serious bias and the need for recalibration, or it may appear within a single run of repeats. At one manufacturing company, the same item was passed repeatedly through an on-line colour-monitoring device. The device measured the item's colour in terms of L, A and B at each pass. The values of L, A and B are displayed simultaneously on the time plot shown in Fig. 18.2. The result of each pass is presented as a triangle in which the A-vertex is off-set horizontally to the right of the L-vertex by a fixed amount and likewise for the B-vertex. This graphical presentation of the results shows clearly that the device was unreliable.

A measure of the variation in a set of *n* measurements is the *variance* whose formula is

$$s^2 = \frac{1}{n}\sum(x - \bar{x})^2$$

Unlike the **range**, which is the difference of the largest and smallest measurements in the set, the variance does not tend to increase in size if more measurements are added to the set. The **standard deviation** is the square root of the variance. It is preferable to the variance as a measure of variability in that it has the same units as x.

If the standard deviation is too large then the **precision**, that is the resolution or limit of detection, of the system of measurement will be adversely affected. For example, if there is a need to measure the diameters of fine fibres to the nearest micron but the standard deviation is 0.5 microns then the desired precision will not be truly attainable. With this standard deviation and assuming no bias, a single measurement whose value is ten microns, for instance, could easily have arisen either from a fibre whose diameter is nine microns or from one whose diameter is 11 microns.

This begs the question of what data should be used in the calculation of the standard deviation. The guiding principle is that the standard deviation should be calculated from

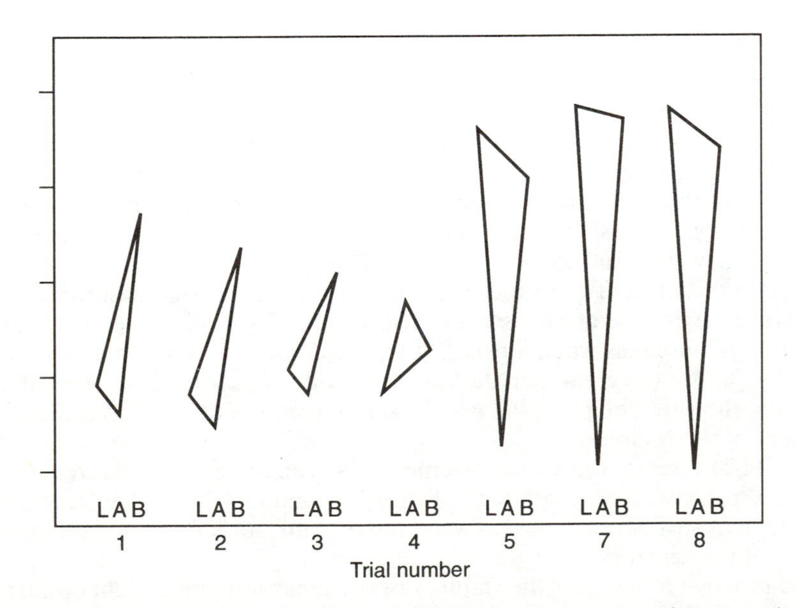

Figure 18.2 Time plot of successive L, A, B colour measurements of the same item.

measurements of the same nominal quantity made in circumstances where the usual sources of error are allowed to operate. For example, the measurement system should be set up afresh between successive measurements, so that set-up error can make its contribution to the calculated standard deviation. If this is not done then the calculated value is likely to be far too small and will give a false impression of experimental error.

Getting the picture

Usually much of the information contained in a data set is obscured by the detail of the individual values. Data summaries are produced to extract this information and present it in a digestible and useable form. The calculation of the mean $\bar{x}$ of a homogeneous data set, that is one obtained under constant conditions, may be viewed as a way of arriving at a value that typifies the set. Calculating the standard deviation s then gives some indication of the degree of variation around this typical value.

One method of summarisation that sacrifices less information than merely representing the data set by its values of $\bar{x}$ and s is the construction of a **frequency table**. Table 18.1 shows a frequency table that summarises the results of 2608 experimental trials reported by Rutherford and Geiger (1910). Each trial consisted of counting the number of alpha particles emitted from a bar of Polonium during an eight-minute interval. The table shows, for example, that exactly two particles were emitted during 383 intervals while 11 or more were emitted during only six intervals. With the data summarised in this way, a *chi-square* test could be used to check the theory that radioactive emissions occur purely at random (Cramer, 1951).

A frequency table may be constructed for measurements made on a continuous scale by counting the number of results that fall in each of a number of non-overlapping arbitrarily chosen **class-intervals**. Table 18.2 shows a frequency table constructed from 550 roughness measurements of a ship's hull as given by Metcalfe (1994). A pictorial representation of this frequency table is the **histogram** shown in Fig. 18.3. It is important to note that the essential feature of a histogram is that the area of each rectangle is proportional to the corresponding class-frequency. It is this that enables the histogram to convey the shape of the distribution of the measurements.

Often, there is not enough data to allow a histogram to be constructed or a less detailed picture of the distribution is all that is required. In either case, a **box and whisker plot** may be used. Here, the central 50 percent of the ordered data set is represented by a rectangular box and the whiskers are lines drawn from the ends of the box to the largest and smallest results in the set. Finally, the box is divided in two by the **median**, that is, by the middle value of the ordered data set.

Figure 18.4 compares three sets of results obtained under three different conditions A, B and C, by comparing their box and whisker plots. The results are degrees of contamination expressed as percentages and there are 16 results in each set. Despite the wide spread of results, it appears that condition A is preferable to B and C.

It should be noted that means should not be calculated if the data are purely classificatory because addition is not properly defined in that case. In the food industry, for example, taste is usually assessed by using a panel of tasters. Each member of the

Table 18.1 Frequencies of intervals containing k alpha-particles

k	0	1	2	3	4	5	6	7	8	9	10	$\geqslant$11
Frequency	57	203	383	525	532	408	273	139	45	27	10	6

Table 18.2 Frequency table of 550 roughness measurements (lowest trough to highest peak) of a ship's hull

Roughness (*µm*)	*Frequency*
47.5–57.4	7
57.5–67.4	24
67.5–77.4	63
77.5–82.4	42
82.5–87.4	57
87.5–92.4	85
92.5–97.4	78
97.5–107.4	70
107.5–117.4	46
117.5–127.4	26
127.5–137.4	14
137.5–207.4	25
207.5–277.4	13

panel awards each recipe a score on an ordered classificatory scale. The results should be summarised by quoting the median score of each recipe or, better, by constructing a box plot of the scores for each recipe.

The conclusions of a programme of research, and the evidence that supports those conclusions, must be presented clearly. Graphs, such as time plots, scatter plots and histograms, are often the best way of communicating results but this will not be the case if the quality of the graphs is poor. The principles of effective graphical presentation should always be observed. These principles are discussed and illustrated by Cleveland (1985). For example, if the reader is expected to make a visual comparison of two graphs then the scales should be the same for both. Superposed graphs must be easily

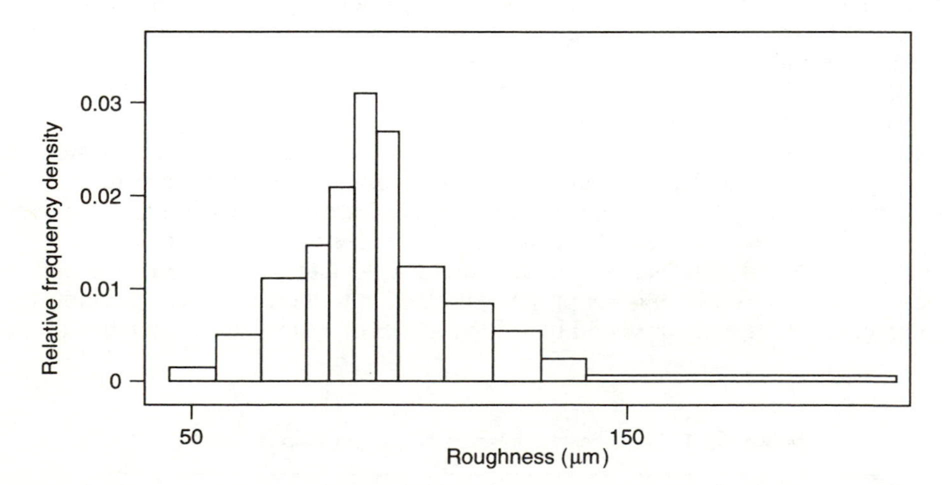

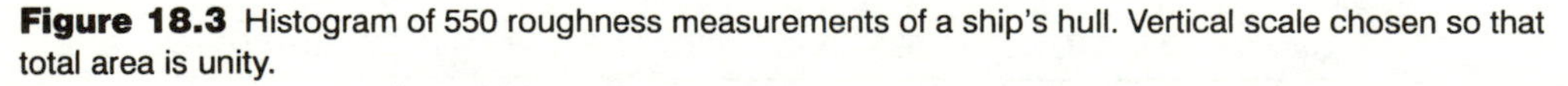

Figure 18.3 Histogram of 550 roughness measurements of a ship's hull. Vertical scale chosen so that total area is unity.

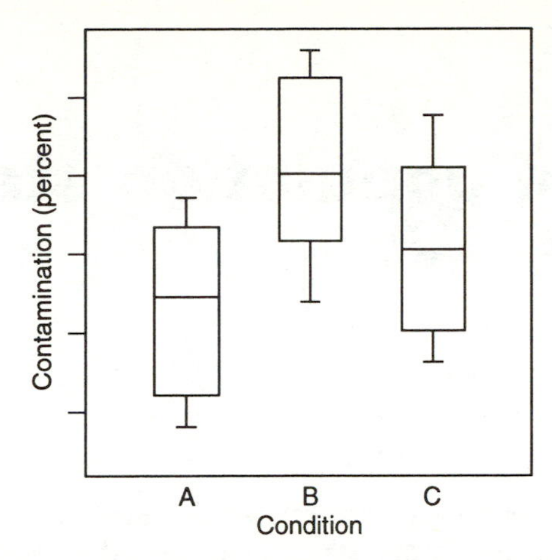

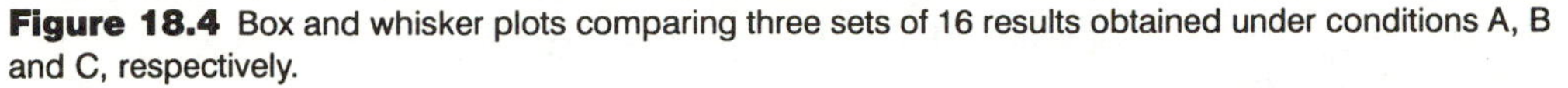

Figure 18.4 Box and whisker plots comparing three sets of 16 results obtained under conditions A, B and C, respectively.

distinguishable. Zero should not be included in a scale if this will destroy the resolution of the data on the graph. Percentages should not be graphed without indicating the number of results on which they are based. Finally, graphs should be proofread and their visual clarity should be good enough to survive reduction and reproduction; it is not unknown for key features of a graph to disappear at publication stage when it is too late to do anything about it.

References

Carroll R J, Spiegelman J and Sacks J (1988) A quick and easy multiple-use calibration curve procedure. *Technometrics*, Vol. 30, pp. 137–141.
Church A (1966) Analysis of data when the response is curve. *Technometrics*, Vol. 8, pp. 229–246.
Cleveland WS (1985) *The Elements of Graphing Data*. Wadsworth.
Cochran W G (1983) *Planning and Analysis of Observational Studies*. John Wiley.
Cramer H (1951) *Mathematical Methods in Statistics*. Princeton University Press.
Dobson A J (1983) *An Introduction to Statistical Modelling*. Chapman & Hall.
Grove D M and Davis T P (1992) *Engineering Quality and Experimental Design*. Longman.
Mandel J (1984) *The Statistical Analysis of Experimental Data*. John Wiley.
Maxwell A E (1961) *Analysing Qualitative Data*. Methuen.
McCullagh P and Nelder J A (1989) *Generalised Linear Models* (2nd edn). Chapman & Hall.
Metcalfe A V (1994) *Statistics in Engineering*. Chapman & Hall.
Michell J (1990) *An Introduction to the Logic of Psychological Measurement*. Lawrence Erlbaum.
Roberts F S (1979) *Measurement Theory*. Addison-Wesley.
Rutherford E and Geiger H (1910) The probability variation in the distribution of α particles. *Philosophical Magazine*, Vol. 20, pp. 698–707.
Siegel S (1956) *Non-parametric Statistics*. McGraw-Hill.
Youden W (1962) Systematic errors in physical constants. *Technometrics*, Vol. 4, pp. 111–123

19
Sources of population statistics

Tony Rowntree

Introduction

By **population statistics** here is meant statistics of the number of people and their characteristics (such as age, sex, marital status) living in defined areas. The focus of attention is population statistics relating to the UK, although passing reference is made to international statistics.

The interest of the research worker in population statistics is likely to be the need to relate the results of a particular research project to a more general population. Thus, there may be a need to make use of population figures as denominators in the calculation of rates. Again, there may be a need to compare the characteristics of the sub-group or sample on which the research is based with the corresponding characteristics of the general population. Such comparisons will need to take account of the date or period to which the data concerned relate, the geographic areas covered by the research and the definitions used in the research project. There will often be a need to assess the available evidence regarding the possible presence of bias in the research data when compared with similar statistical data for the general population.

It is desirable to avoid differences between the coverage or definitions used in a research project and those which have been used in the production of statistics about the general population. This makes it prudent whenever possible to examine the relevant population statistics before starting on data collection or assembly for the research project. If that is not possible such a scrutiny should be carried out before starting to analyse the data from the research project.

The census of population

The primary source of statistics about the population of the country and its constituent areas is the official Census of Population. The census is usually taken every ten years, the latest being held in 1991. It involves a comprehensive enumeration and is designed to give a statistical description of the population at the census date in each area of the country. Its value lies in both from the information produced directly and in the benchmark which the census count provides for estimates made between successive censuses.

The census includes questions about individuals, the households to which they belong and the accommodation in which those households live. The questions asked vary somewhat between censuses. Those included in the 1991 census in Great Britain are listed below.

Questions relating to individuals

- Sex, date of birth
- marital status

- relationship to head of household
- whereabouts on census night
- usual address
- term-time address (for students and schoolchildren)
- usual address a year ago
- country of birth
- ethnic group
- whether suffering from long-term illness
- whether working/retired/looking after home etc.
- hours worked (for those working)
- occupation
- name and address of employer
- address of workplace
- means of daily journey to work
- degree/professional/vocational qualification.

Questions regarding dwelling

- number of rooms
- type of accommodation (house/flat, etc.).

Questions relating to household

- tenure of accommodation
- possession of amenities (bath, shower, WC, central heating)
- possession of car or van.

The results of the 1991 census were disseminated in a series of publications, some relating to areas, some to individual census topics. The main published volumes were as follows.

Local base reports

- county/region, parts 1 and 2 (67 volumes)
- Greater London parts 1 and 2
- Great Britain, parts 1 and 2
- England, regional health authorities, parts 1 and 2
- Wales, parts 1 and 2
- Scotland, parts 1 and 2
- new towns (Scotland).

Topics volumes

- historical tables
- sex, age and marital status
- persons aged 60 and over
- usual residence
- housing and availability of cars
- communal establishments
- Gaelic language
- Welsh language
- limiting long-term illness

- ethnic group and country of birth
- migration, parts 1 and 2
- topic report for health areas
- economic activity, Great Britain
- economic activity, Scotland
- workplace and transport to work, Great Britain
- workplace and transport to work, Scotland
- children and young adults
- qualified manpower
- household composition
- household and family composition (10 percent)
- key statistics for local authorities
- key statistics for urban areas, Great Britain
- key statistics for urban areas, regions (6).

Monitors

- county monitors (55)
- regional monitors (12)
- national monitor, Great Britain
- national monitor, Scotland
- national monitor, Wales
- health authorities, Great Britain
- health authorities, Wales
- health authorities, Regions (14)
- postcode sectors, England and Wales (19)
- postcode sectors, Scotland (12)
- parliamentary constituencies (12)
- European constituencies
- wards and civil parishes, England (46)
- wards and civil parishes, Scotland (12)
- communities, Wales (8)
- summary and review monitor
- topics monitors (15).

In addition to the above published tables, the census result dissemination programme includes a number of **products**. These are available in paper output, magnetic tape or floppy disk media. These include:

- Local base statistics (99 tables) and/or small area statistics (86 tables) for each enumeration district in England and Wales and for each output area in Scotland and for aggregates of these basic areas up to the national level.
- Special migration statistics, identifying combinations of area of usual residence at census date with area of usual residence a year earlier.
- Special workplace statistics, analysing journey to work and means of transport for combinations of area of residence and area of workplace.

In addition, **samples of anonymised records** can be purchased. These are samples of records which have been anonymised so that they do not contain information which would lead to the disclosure of information about an individual or household. There are two such files, one being a two percent file of individuals living in private and communal establishments and the other a one percent hierarchical file of private households and the individuals within them. Both files are held by the census Microdata Unit at the University of Manchester.

A number of explanatory volumes are available which enable the user to use census information as effectively as possible.

The above description relates to the 1991 census. Broadly similar material was produced from earlier censuses, though for each census the questions asked, some of the definitions and classifications used, and the detailed pattern of publication differed to some extent. If you are working on a project which involves drawing on more than one census, you will need to take care that comparisons between censuses take account of any such changes.

Further information on all census results may be obtained from the customer services sections at the census offices:

(*For England and Wales*)
Office of Population Censuses and Surveys
Segensworth Road
Titchfield
Fareham
Hants PO15 5RR
Tel: 01329 813800
Fax: 01329 813532

(*For Scotland*)
General Register Office (Scotland)
Ladywell House
Ladywell Road
Edinburgh EH12 7TF
Tel: 0131-314 4254
Fax: 0131-314 4344

Problems which arose at the enumeration stage of the 1991 census have raised questions about the quality of census-based statistics. While these problems were not negligible, the census remains an unrivalled statistical data source for the population as a whole. Individual census volumes give details of points which need to be taken into account when using the statistics produced from the census.

Intercensal estimates

Given that the census of population is held at ten-year intervals it will usually be desirable to relate the results of a research project to more up-to-date population figures wherever these are appropriate.

Population estimates classified by sex and age

Official annual population estimates by age and sex are produced routinely for each of the countries within the UK (England, Wales, Scotland and Northern Ireland) which are aggregated to produce figures for Great Britain and the UK. Estimates are also produced for each local government authority area and each health authority area within these countries and also for aggregates such as standard regions and metropolitan counties. The annual population estimates are produced from a census-based starting figure which is then moved forward by adding births, subtracting deaths and adjusting for inward and outward migration.

The estimates are of the population usually resident in each area. 'Usual residence' is not always a straightforward concept and the relevant publications should be consulted to see how groups such as students, members of the armed forces and people living in institutions are dealt with in the population estimates.

Estimates relate to 30 June each year and are produced for each sex and are classified by age. 'Age' means age last birthday, not year of birth. Though calculated by single years of age estimates are often more readily available in grouped ages (such as 0–4, 5–9, 10–14). If mid-year estimates are used to calculate rates, it is often satisfactory to use events during the corresponding calendar year as the numerator of such rates. However, care may be needed when dealing with events which fluctuate markedly over short periods of time (such as cases in some epidemics).

Annual population estimates classified by marital status as well as by sex and age are produced at the national level. They are not available sub-nationally because of the absence of data on the usual residence of those marrying or divorcing and the lack of migration data classified by marital status. 'Marital status' refers to legal marital status. Thus those separated are included among the married.

Annual population estimates become available about a year after the date to which they relate. In theory, short-term projections could be used instead but in practice their use is likely to involve a greater risk of error. Population projections at the national level are produced, usually in alternate years, by the government actuary's department, in consultation with OPCS and the registrars general of Scotland and Northern Ireland. The projections are published by OPCS and the registrars general of Scotland and Northern Ireland. At longer intervals sub-national population projections for the larger local government authority areas and for health authority areas are produced and published by OPCS for England, by the Welsh Office for Wales and by the registrars general of Scotland and Northern Ireland for those countries. Some local authorities produce their own population projections which should be enquired about locally.

Further information, including the availability of unpublished results, may be obtained from:

Sub-national Projections and Demography Unit
OPCS
St Catherine's House
10 Kingsway
London WC2B 6JP
Tel: 0171-396 2180

or, in respect of national population projections, from:

Government Actuary's Department
22 Kingsway
London WC2B 6LE
Tel: 0171-242 6828

In the making of intercensal population estimates, OPCS and the General Register Offices for Scotland and Northern Ireland make use of data on births, marriages and deaths derived from the civil registration system. A range of regular and occasional statistical publications is produced from this material. For England and Wales, such statistics first appear in a monitor and later in an annual volume. Series of monitors and/or annual volumes have been established in the following topic areas:

- births (FM1 series)
- marriages and divorces (FM2 series)
- deaths:
  — general (DH1 series) giving key summary figures
  — deaths by cause (DH2 series)
  — infant and perinatal mortality (DH3 series)
  — deaths from accidents etc (DH4 series)
  — deaths by area (DH5 series)
  — deaths in childhood (DH6 series).

There are also the registrar general's decennial supplements which provide analyses of mortality by occupation and area, and life tables. For example, the *Decennial Supplement on Occupation Mortality* brings together census and registration data to examine the relationships between occupations and causes of death. A full description of such publications is beyond the scope of this chapter and for detailed information you should enquire at your local reference or college/university library.

Publications

The population offices publish information on population estimates and projections and on the components of population change, such as births, deaths and migration. For England and Wales the series of publications are as follows.

OPCS monitors

- PP1 series; population estimates at the national level and for local government and health areas.
- PP2 series; population projections at the national level.
- PP3 series; population projections for local government and health areas in England.

Monitors are available from

OPCS Information Branch (Publications)
St Catherine's House
10 Kingsway
London WC2B 6JP
Tel: 0171-396 2243/2208

Reference volumes in the population estimates and projections field

- VS/PP1 series; key population and vital statistics.
- PP2 series; national population projections.
- PP3 series; population projections for health areas and larger local government area in England.

Reference volumes are available from HMSO book shops or accredited agents or from:

HMSO Publications Centre
PO Box 276
London SW8 5DT
Tel: 0171-873 9090

Much information, sometimes in more detail, can also be made available on computer printout or on floppy disk. Further enquiries should be made to OPCS.

Similar population statistics for Scotland are available from:

General Register Office (Scotland)
Ladywell House
Ladywell Road
Edinburgh EH12 7TF
Tel: 0131-314 4301

Their main publications include:

- *Population Estimates, Scotland (Mid-1994)* (HMSO).
- *Annual Report of the Registrar General for Scotland*, published by GRO(S).
- *Population Projections, Scotland* published by GRO(S).

Enquiries for similar information about Northern Ireland should be made to:

General Register Office (Northern Ireland)
Oxford House
49–55 Chichester Street
Belfast BT1 4HL
Tel: 01232 252031

It should be noted that summary statistics of population estimates and projections for the UK and its constituent countries are published in the Central Statistical Office *Annual Abstract of Statistics*.

Other intercensal population statistics

The census of population also provides a benchmark for series of estimates of sub-groups of the total population such as the working population, the school population, the population of pensioners and so on. These populations are outside the scope of this chapter and if further information is needed enquiries should be made to the relevant government department.

International population statistics

This chapter is concerned mainly with population statistics of the UK. If it is necessary to relate results to population statistics of other countries, the most convenient source of information may well be the volumes of population statistics produced by the Population Division of the United Nations or by EuroStat (for countries in the EU). The main reference for the United Nations is the *United Nations Demographic Yearbook*, of which the latest edition is for 1992 with the 1993 edition due soon. For countries in the EU one can refer to the EuroStat publication *Demographic Statistics – Population and Social Conditions* (published by the Office for Official Publications of the EU) of which the latest edition is for 1994.

Conclusion

This brief survey focuses on the routinely produced and published standard information. If you need information on population statistics which is non-standard, it is always worth making enquiries of the producers of standard statistics who may sometimes have more detailed or non-standard data readily available or who may be able to give useful advice and guidance.

Many public reference libraries and university/college libraries will have available at least a selection of the publications referred to above. Such libraries will usually be able to borrow, through interlibrary lending arrangements, copies of publications they do not hold themselves. Researchers should enquire at their local library, allowing time for material that has to be obtained from elsewhere. Copies of official publications which are still in print may be purchased, prices being available from HMSO of the office publishing the relevant item.

20
Instrumentation for experiments

Andrew Penney

Introduction

Instrumentation is the name given to devices which can be used to deliver quantitative information about the world we live in. An example in everyday use would be the wristwatch, which measures the passing of time. In research there is a need to measure information pertaining to experiments, production runs, materials testing and so on. Different experimental circumstances and other factors such as budgets will dictate differing measurement techniques. The basics however remain the same. Instrumentation is all about the capture of data, and its subsequent analysis, display and storage. Often, the need for repetitive measurement and the handling of large amounts of data means that computers and more recently personal computers are being extensively used in instrumentation.

Examples of instruments are as vast as the scope of scientific research itself. Across many disciplines there is the need to measure common physical properties such as temperature, pressure and flow rate. Electrical characteristics such as voltage, current and resistance are also commonly measured – either because these parameters are of interest in their own right (in circuit analysis for example) or because transducers and sensors provide information about the parameter they are measuring through electrical signals. A thermocouple is used to measure temperature and is widely used in research and throughout industry. In this case, variation in temperature is mirrored by a small voltage which needs to be carefully measured using high gain amplification. Because the thermocouple has a non-linear response to temperature, the measured signal must be linearised before a meaningful reading is obtained. This, then, is a typical instrumentation application.

Signal types

Before considering the types of instruments that are used in the research. It is useful to consider the types of signals which are typically found in instrumentation applications. See Fig. 20.1.

Analogue input–output

Analogue signals vary in amplitude with time. An example is an audio signal. These types of signal can be further sub-divided depending on their frequency content and how they are analysed. Direct current (DC) signals vary only very slowly with time (such as a temperature), time domain signals (such as a sawtooth wave form) and frequency domain (such as a radio carrier).

Digital input–output

Digital signals vary between one of two stable states, zero or one, on or off, true or false.

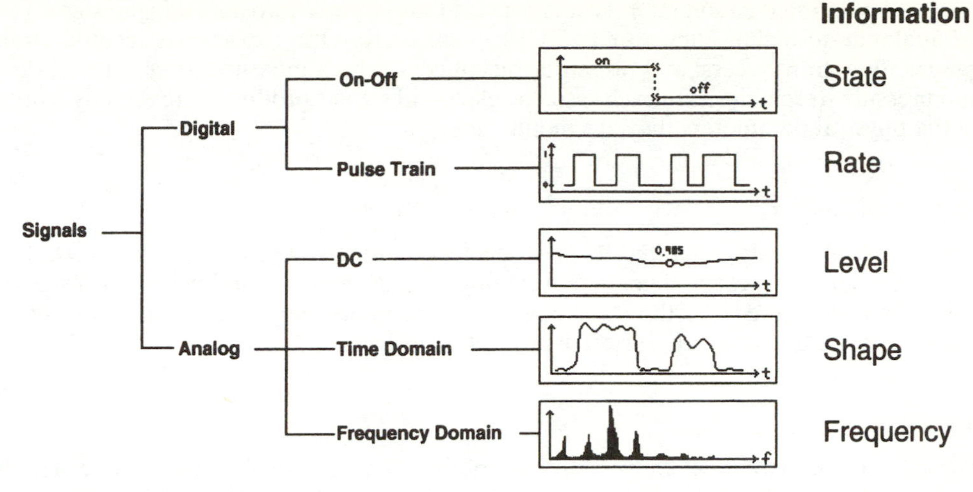

Figure 20.1 Classification of signals. Classification is based on the type of useful information the signal conveys.

Timing input–output

Where the frequency or count of a digital pulse train is of interest, digital counter technology is used.

Safety

Much instrumentation work will involve electrical connections, including the wiring of transducers and sensors. When configuring or operating such apparatus, appropriate precautions should be taken to avoid the risk of electric shock.

Instrumentation basics

There are many differing form factors for instrumentation used in the laboratory and some of the major categories of instrument will be detailed later. First however, we will detail some common aspects of measuring systems. Depending on the type of instrument system these elements may be contained in one or more pieces of equipment.

Common elements

Common elements of measuring systems are:

- transducers
- signal conditioners
- amplifiers
- isolators
- filters
- exciters
- linearisers.

Transducers

Transducers sense physical phenomena and provide electrical signals that the instruments can accept. For example, thermocouples, resistive temperature devices (RTDs), thermistors, and integrated circuit (IC) sensors convert temperature into an analogue signal that an analogue-to-digital converter (ADC) can measure. Other examples include strain gauges, flow transducers, and pressure transducers, which measure force, rate of flow, and pressure, respectively. In each case, the electrical signals produced are directly related to the physical parameters they are monitoring.

Signal conditioners

The electrical signals generated by the transducers must be converted into a form that the instrument can accept. Signal-conditioning accessories can amplify low-level signals, and then isolate and filter them for more accurate measurements. Signal conditioning can also excite and linearise certain types of transducers.

Amplifiers

The most common type of conversion is amplification. Low-level thermocouple signals, for example, should be amplified to increase resolution and optimise noise performance. For the highest possible accuracy, the signal should be amplified so that the maximum voltage swing equals the maximum input range of the ADC.

Isolators

Another common application for signal conditioning is to isolate the transducer signals from the computer for safety purposes. The system being monitored may contain high-voltage transients that could damage the computer. An additional reason for needing isolation is to make sure that the readings are not affected by differences in ground potentials or common-mode voltages. When the instrument input and the signal being acquired are each referenced to ground, problems occur if there is a potential difference in the two grounds. This difference can lead to what is known as a ground loop, which may cause inaccurate representation of the acquired signal, or if too large, may damage the measurement system. The use of isolated signal-conditioning modules will eliminate the ground loop and ensure that the signals are accurately acquired.

Filters

The purpose of a filter is to remove unwanted signals from the signal that you are trying to measure. A noise filter is used on DC-class signals like temperature to attenuate higher frequency signals that can reduce the accuracy of your measurement. Alternating current (AC)-class signals like vibration often require a different type of filter known as an anti-aliasing filter. Like the noise filter, the anti-aliasing filter is also a low-pass filter; however, it also has a very steep cut-off rate, so that it almost completely removes all frequencies of the signal that are higher than the input bandwidth of the board. If the signals were not removed, they would erroneously appear as signals within the input bandwidth.

Exciters

Signal conditioning also generates excitation for some transducers. Strain gauges, thermistors and RTDs, for example, require external voltage or current excitation signals. Signal-conditioning modules for these transducers usually provide these signals. RTD measurements are usually made with a current source that converts the variation in resistance to a measurable voltage. Strain gauges, which are very low-resistance devices, typically are used in a Wheatstone bridge configuration with a voltage excitation source.

Linearisers

Another common signal-conditioning function is linearisation. Many transducers, such as thermocouples, have a non-linear response to changes in the phenomena being measured. Responses can be linearised in either hardware or software.

Basic considerations of analogue inputs

The analogue input specifications can give you information on both the capabilities and the accuracy of the data aquisition (DAQ) product. Basic specifications, which are available on most DAQ products, tell you:

- number of channels
- sampling rate
- resolution
- input range.

Number of channels

The number of analogue channel inputs will be specified for both single-ended and differential inputs on boards that have both types of inputs. Single-ended inputs are all referenced to a common ground point. These inputs are typically used when the input signals are high level (greater than one volt), the leads from the signal source to the analogue input hardware are short (less than 15 ft) and all input signals share a common ground reference. If the signals do not meet these criteria, you should use differential inputs. With differential inputs, each input has its own ground reference. Noise errors are reduced because noise can be rejected at the input to the instrumentation amplifier.

Sampling rate

The sampling rate determines how often conversions can occur. A faster sampling rate acquires more points in a given time and can therefore often form a better representation of the original signal. For example, audio signals converted to electrical signals by a microphone commonly have frequency components up to 20 kHz. To properly digitise this signal for analysis, the Nyquist sampling theorem tells us that we must sample at more than twice the rate of the maximum frequency component we want to detect. So, a board with a sampling rate greater than 40 kS/s (thousand samples per second) is needed to properly acquire this signal (see Fig. 20.2).

Multiplexing

Multiplexing is a common technique for measuring several signals with a single ADC. The ADC samples one channel, switches to the next channel, samples it, switches to the next channel, and so on. Because the same ADC is sampling many channels instead of one, the effective rate of each individual channel is inversely proportional to the number of channels sampled.

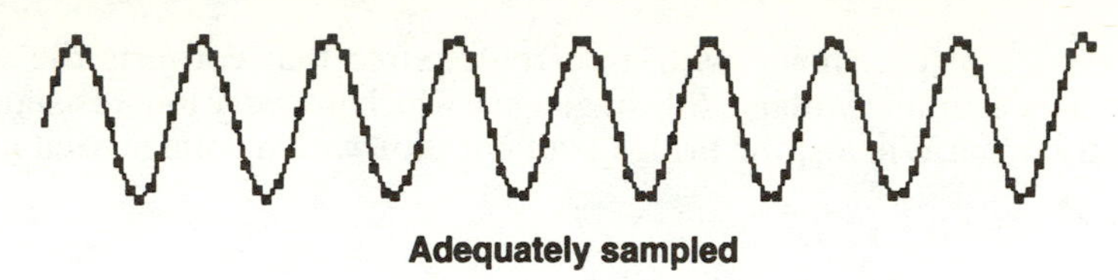

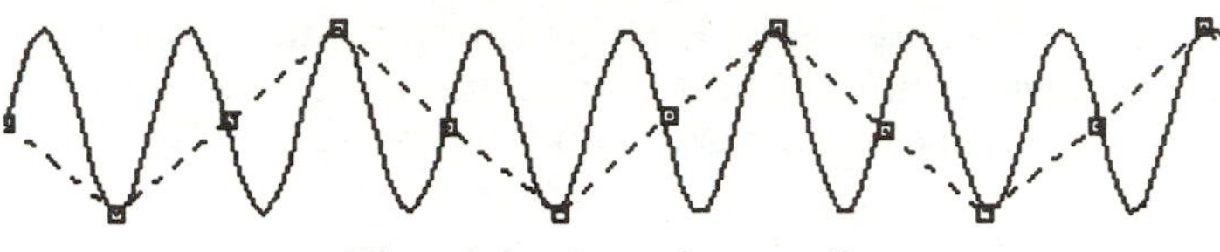

Figure 20.2 AD sampling rate. Nyquist criteria – sample rate > 2 × highest frequency.

Example:

$$(100 \text{ kS/s})/(10 \text{ channels}) = 10 \text{ kS/s per channel}$$

You can often use external analogue multiplexors to increase the numbers of channels an instrument can measure. The sampling rate is reduced proportionately by this additional external multiplexing.

Resolution

Resolution is the number of bits that the ADC uses to represent the analogue signal. The higher the resolution, the higher the number of divisions the range is broken into, and therefore, the smaller the detectable voltage change. Figure 20.3 shows a sine wave and its corresponding digital image as obtained by a three-bit ADC. A three-bit converter (which is actually seldom used but a convenient example) divides the analogue range into 23 or eight divisions. Each division is represented by a binary code between 000 and 111. Clearly, the digital representation is not a good representation of the original analogue signal because information has been lost in the conversion. By increasing the resolution to 16 bits, however, the number of codes from the ADC increases from eight to 65,536, and you can therefore obtain an extremely accurate digital representation of the analogue signal if the rest of the analogue input circuitry is designed properly.

Range

A signal range is described by the minimum and maximum voltage levels that the ADC can quantify. The multifunction DAQ boards offer selectable ranges so that the board can be configured to handle a variety of different voltage levels. With this flexibility, you can match the signal range to that of the ADC to take best advantage of the resolution available to accurately measure the signal. The range, resolution, and gain available on a DAQ board determine the smallest detectable change in voltage. This change in voltage represents one least significant bit (LSB) of the digital value, and is often called the code width. The ideal code width is found by dividing the voltage range by the gain times two raised to the order of bits in the resolution. With a voltage range of zero to ten volts, and a gain of 100, the ideal code width is:

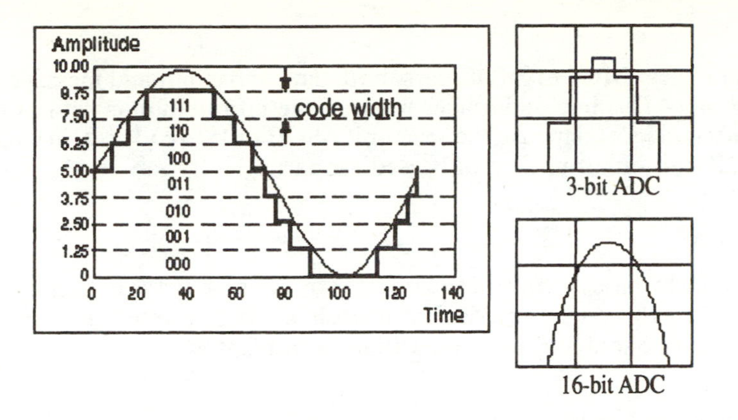

Code Width = smallest detectable change in voltage

$$= \frac{\text{input range of ADC}}{\text{gain} * 2^n} \qquad n = \text{\# of ADC bits}$$

Figure 20.3 A/D resolution.

$$10 \text{ V}/1.5\mu\text{V} = 100 \times 2^{16}$$

Therefore, the theoretical resolution of one bit in the digitised value is 1.5 μV.

Types of instrument systems in research

Plug-in data acquisition (DAQ)

Because of the popularity of the personal computer or PC, data acquisition (DAQ) cards are increasingly popular. By plugging DAQ boards into standard PCs, researchers have a cost-effective and modular instrument system. The impressive software tools available for such systems (LabVIEW for example) also adds to the attractiveness of this solution. Some of the more popular expansion busses available for plug-in DAQ include ISA, EISA, PCI, NuBus and PCMCIA. This last bus (PCMCIA) has been designed for notebook computers and offers an excellent way to add instrumentation capability to portable computers.

General-purpose interface bus

The general-purpose interface bus (GPIB) can be used to connect stand-alone instruments to a computer. It was later formalised as IEEE Standard 488-1975, and has evolved to ANSI/IEEE Standard 488.2-1987. The GPIB interface system consists of 16 signal lines and eight ground-return or shield-drain lines. The 16 signal lines, discussed below, are grouped into:

- data lines (eight)
- handshake lines (three)
- interface management lines (five).

Data lines

The eight data lines, DIO1 to DIO8, carry both data and command messages. The state of the attention (ATN) line determines whether the information is data or commands. All commands and most data use the seven-bit ASCII or ISO code set, in which case the eighth bit, DIO8, is either unused or is used for parity.

Handshake lines

Three lines asynchronously control the transfer of message bytes between devices. The process is called a three-wire interlocked handshake. It guarantees that message bytes on the data lines are sent and received without transmission error.

Interface management lines

Five lines manage the flow of information across the interface.

1. ATN (attention)
   The controller drives ATN true when it uses the data lines to send commands, and drives ATN false when a talker can send data messages.
2. IFC (interface clear)
   The system controller drives the IFC line to initialise the bus and become CIC.
3. REN (remote enable)
   The system controller drives the REN line, which is used to place devices in remote or local program mode.
4. SRQ (service request)
   Any device can drive the SRQ line to asynchronously request service from the controller.
5. EOI (end or identify)
   The EOI line has two purposes: the talker uses the EOI line to mark the end of a message string; the controller uses the EOI line to tell devices to identify their response in a parallel poll.

Talkers, listeners, and controllers

GPIB devices can be talkers, listeners, and/or controllers. A talker sends data messages to one or more listeners, which receive the data. The controller manages the flow of information on the GPIB by sending commands to all devices. A digital voltmeter, for example, is a talker and is also a listener. The controller usually addresses (or enables) a talker and a listener before the talker can send its message to the listener. After the message is transmitted, the controller may address other talkers and listeners. The controller function is usually handled by a computer, most often a PC.

VXI (VME extensions for instrumentation)

The demand for an industry-standard instrument-on-a-card architecture has been driven by the need for physical size reduction of rack-and-stack instrumentation systems, tighter timing and synchronisation between multiple instruments, and faster transfer rates than the one Mbytes/s rate of the eight-bit GPIB. The modular form factor, high bandwidth, and commercial success of the VME (Versabus Module rev. E) bus made it particularly attractive as an instrumentation platform. The tremendous popularity of the GPIB also made it attractive as a model for device communication and instrument control protocols.

The VXI bus specification adds the standards necessary to combine the VME bus with the GPIB to create a new, modular instrumentation platform that can meet the needs of future instrumentation applications.

VXI connection

Many VXI users migrate from GPIB-based systems. As a result, a GPIB-VXI interface is a popular way to control VXI instruments from a GPIB controller. An increasingly popular way to control VXI, however, is to use a custom VXI computer that plugs directly into the VXI. This embedded approach is technically attractive because the computer communicates directly with the VXI bus and is tightly coupled to the instruments.

A third alternative for VXI control is a high speed serial link to a standard computer. In the last decade, specialised instrument controllers have rapidly declined. General-purpose PCs and workstations, with their vast array of software and accessories, have assumed this role. The mainframe extension for instrumentation (MXI) bus was designed as a way to connect industry standard PCs and workstations to VXI-based systems, and also as a way to link multiple chassis configurations together.

Physically, a VXI bus system consists of a mainframe chassis that has the physical mounting and back plane connections for plug-in modules. The VXI bus uses the industry-standard IEEE-1014 VME bus as a base architecture to build upon. Because of the limited amount of front panel area on a VXI instrument, software is very important in providing an operating interface and means of control.

CAMAC

Like VXI, CAMAC is a system of instrumentation on a card which offers a modular approach to instrumentation. It is found in research applications and is popular in high energy physics applications.

Serial instruments

Many laboratory based devices for example the spectrophotometer can be connected to a computer via a serial (RS-232) link. This enables the computer to control the instrument and to capture a series of results.

Software

Software plays a vital role in developing automated data acquisition and instrument control systems. The software used in these systems spans a broad range of functionality, from device drivers for controlling specific hardware interfaces, to application software packages for developing complete systems.

You can use application software to develop each of the three primary elements of your system: data acquisition; data analysis; and data presentation. Your software needs are affected by many factors, including application requirements, computer hardware and operating system, and your instrumentation hardware. Other system requirements might dictate fast execution speed, high channel count, or real-time signal processing.

Instrument drivers

At the core of all instrumentation systems are the instrument drivers. Instrument drivers drastically reduce software development costs because developers do not have to spend time programming their instruments. The drivers can also be reused in a variety of systems and configurations.

Disclaimer

While the information conveyed here is believed by the author to be accurate, no responsibility is held for its application in practice. It should be especially noted that instrumentation used in clinical applications, directly connected to a human or animal subject is not covered here. While much of the information contained herein is pertinent to such applications, specific restrictions and safeguards apply. It is not within the scope of this work to cover these issues.

Further reading

Computer-based Measurement and Instrumentation (January 1996, reference 350034B-01). *Instrapedia – CD-ROM Encyclopedia of Information* (January 1996, reference 500256C-01). Technical seminar notes obtainable from National Instruments, 21 Kingfisher Court, Hambridge Road, Newbury, Berkshire RG14 5SJ.

21 Interviewing

Mark Hughes

Introduction

Interviews play an important part in the lives of most people. While there are similarities between research interviews and other interviews (job interviews for example) there are also differences. As a researcher there is a need to be rigorous and methodical whilst doing research interviews (subsequently referred to as interviews). The aim of this chapter is to introduce you to interviewing and the choices that you will need to make if using interviewing in your research.

The chapter starts with a discussion of the appropriateness of interviewing as a research method. The different types of interview and interview questions are introduced and this is followed by a discussion of interviewing as a process. Analysing interview data is discussed and the chapter ends with a summary in the form of a series of key questions and answers.

It is not easy to define an interview because of the variety of types of interview. The following is a useful starting point.

> *The purpose of interviewing is to find out what is in and on someone else's mind. We interview people to find out from them those things we cannot directly observe.*
>
> Patton (1980, p. 196)

The central theme throughout the chapter will be that there is no one best way to conduct an interview, but that you choose those methods that are most appropriate to your particular research. In considering the appropriateness of this method of data gathering it is worthwhile considering the strengths and weaknesses of the method.

Rationale for interviewing

The appropriateness of interviewing on a particular project may be considered in terms of the strengths and weaknesses of this form of data collection (Marshall and Rossman, 1989).

Strengths of interviewing

The strengths of interviewing are:

- face-to-face encounter with informants
- obtains large amounts of expansive and contextual data quickly
- facilitates cooperation from research subject
- facilitates access for immediate follow-up data collection for clarification and omissions
- useful for discovering complex interconnections in social relationships
- data are collected in natural setting
- good for obtaining data on non-verbal behaviour and communication

- facilitates analysis, validity checks, and triangulation
- facilitates discovery of nuances in culture
- provides for flexibility in the formulation of hypotheses
- provides background context for more focus on activities, behaviours, and events
- great utility for uncovering the subjective side, the *native's perspective* of organisational processes.

Weaknesses of interviewing

The weaknesses of interviewing are:

- data are open to misinterpretation due to cultural differences
- depends on the cooperation of a small group of key informants
- difficult to replicate
- procedures are not always explicit or depend on researcher's opportunity or characteristics
- data often subject to observer effects; obtrusive and reactive
- can cause danger or discomfort for researcher
- dependent on the honesty of those providing the data
- dependent on the ability of the researchers to be resourceful, systematic, and honest; to control bias.

The strengths demonstrate why the interview is a favoured method among social scientists for gathering data. In particular, the emphasis upon cooperation with research subjects and an emphasis upon the *natives' perspective* are important elements for the social scientist.

This position can appear contradictory to the natural scientist's concerns with objectivty and hard facts. However, within the weaknesses of interviewing some of the problems of working with accounts of people's subjective experiences become apparent. The above polarised depiction does not fully acknowledge the diversity of disciplines which have employed the interview method: accountancy; sociology; anthropology; and medicine.

Types of interview

In deciding upon the type of interview, the phrase 'continuum of formality' (Grebenik and Moser, 1962) captures the options open to interviewers. At one end of the continuum is the informal conversational interview and at the other end the closed quantitative interview. Patton (1980, p. 206) defines these four types of interview and their strengths and weaknesses.

Informal conversational interview

Questions emerge from the immediate context and are asked in the natural course of things: there is no predetermination of question topics and wording.

(+) Increases the salience and relevance of questions.
(–) Different information collected from different people with different questions.

Interview guide approach

Topics and issues to be covered are specified in advance, in outline form. Interviewer decides sequence and wording of questions in the course of the interview.

(+) The outline increases the completeness of the data and makes data collection systematic for each respondent.
(–) Important and salient topics may be inadvertently omitted.

Standardised open-ended interview

The exact wording and sequence of questions are determined in advance. All respondents are asked the same basic questions in the same order.

(+) Respondents answer the same questions, thus increasing comparability of responses. Data are complete for each person on the topics addressed in the interview.
(–) Little flexibility in relating the interview to particular individuals and circumstances.

Closed quantitative interviews

Questions and response categories are determined in advance. Responses are fixed: respondent chooses from among these fixed responses.

(+) Data analysis is simple. Responses can be directly compared and easily aggregated.
(–) Respondents must fit their experiences and feelings into the researcher's categories. May be perceived as impersonal, irrelevant and mechanistic.

These different interview types offer a range of options for the researcher. At the outset of research informal conversational interviews can be effective for refining and tightening the focus of your research. As the research progresses more standardised forms of interview can be used.

It is possible within an interview to use more than one type of interviewing. For instance a personal preference is not to use too structured an approach at the beginning of an interview. The type of interview must be appropriate for your research. The structured nature of the closed quantitative interview may be inappropriate for interviewing a senior person within an organisation, but could be appropriate if you wanted to compare the views of 50 employees within an organisation.

Three permutations of the above are worth considering when selecting the most appropriate method for your particular research.

Telephone interviewing

Telephone interviewing keeps costs down if respondents are geographically spread out. However, it may prove difficult to develop a rapport over the telephone.

Group interviewing

Group interviewing might be used when the respondents are part of a group or when the collaboration of the respondents in the research is an objective (for a discussion of group interviewing see Mullings, 1985).

Elite interviewing

An elite interview is a specialised treatment of interviewing that focuses on a particular type of respondent. Elites are considered to be the influential, the prominent, and the well-informed people in an organisation or community (Marshall and Rossman, 1989, p. 94).

Interview questions

While the content of the interviews will depend on the particular research and the interview type, the following ideas may aid the creative process. Patton (1980) suggests that there are basically six types of question that can be asked of people.

1. Experience/behaviour questions: what a person does or has done.
2. Opinion/value questions: to understand the cognitive and interpretive processes of people.
3. Feeling questions: to understand the emotional responses of people to their experiences and thoughts.
4. Knowledge questions: to discover factual information the respondent has.
5. Sensory questions: questions about what is seen, heard, touched, tasted and smelled.
6. Background/demographic questions to identify characteristics of the person being interviewed.

Each of these questions can be asked in the present, past, and future tense. Do not place too much emphasis upon questions about the future as this requires an element of crystal-ball gazing.

While the interview type and type of questions are important, you should understand the interview as a process rather than a single event.

Interview process

Ackroyd and Hughes (1981) place interviewing within context and provide a fascinating account of the interview process. They challenge the position that methods are more or less tools. The tools are important in the conduct of research but have little meaning if divorced from the context of the research. This leads to the following suggestion for good interviewing.

> *That is the prescription for good interviewing should be read as a set of propositions on how the interviewer and the respondent should interact to achieve the aims of scientifically collecting verbal data.*
>
> Ackroyd and Hughes (1981, p. 93)

We can think of the interview as a process. While the process may vary with the interview type and the setting, the successful interview involves a series of linked activities rather than a single event.

Stage one: preparations

Access to respondents is likely to influence interviews (Bell, 1987; Buchanan *et al*, 1988). Respondents are likely to start forming impressions about you when first introduced to your research. In this sense, the interview process begins before any questions have been asked.

There are frequent warnings in the methods literature about the demands interviewing makes upon time (Wragg, 1980; Bell, 1987), these pragmatic considerations should not be overlooked, particularly in terms of preparation time required.

For an effective interview you should:

- gather background information on host organisations
- check interview guides;
- check tape recorders
- have maps of locations
- plan to arrive on time.

Stage two: introductions

Variables such as, age, race, gender and social class have been identified as having a bearing upon interviews. This is expressed by Kane (1990, p. 68) as:

> *... the closer the interviewer is to the respondent in class, sex, age and interests, the greater chance the interviewer has of being successful.*

This is an ideal position, rather than a prerequisite for interviewing.

Body language plays a significant role in interviews. The respondent will be observing your body language and you will be observing the body language of the respondent. At this stage, friendly smiles are important to allay any fears of a forthcoming inquisition (for an introduction to body language see Pease, 1993).

At the outset of the interview, there is a need to establish the purpose of the interview. Whilst, the research will have been introduced at the time of setting up the interview, this non-threatening element can make for a neutral starting point.

The research interview may be a new experience for the respondent. Whenever possible it is worth agreeing the format of the interview with the respondent. Respondents usually agree to your format, but you should ask.

Stage three: the uneven conversation

The conventional social rule of 'you speak, then I speak' is suspended. The interviewer's role should be to listen.

> *The most important thing a researcher should remember to do in an interview is to listen. Interviews are primarily a way to gather information, not a conversational exchange of views.*
>
> Howard and Peters (1990, p. 29)

Sometimes a respondent gives a misleading impression, which is not necessarily intentional. Whyte (1960) has identified three explanations for this: ulterior motives; a desire to please the interviewer; and idiosyncrasy.

Respondents occasionally give answers to questions which you might challenge in the course of a conventional conversation. However, part of the uneven conversation is about listening to the respondent's view. There is a need for neutrality which is effectively captured by Patton.

> *Rapport is a stance* vis-à-vis *the person being interviewed. Neutrality is a stance* vis-à-vis *the content of what that person says.*
>
> Patton (1980, p. 231)

Such a view of the interview process allows the necessary rapport to develop without the interviewer introducing personal bias.

Stage four: the ending

To end an interview would initially appear straightforward. However, if sufficient rapport has developed during the interview, an abrupt exit becomes impossible and undesirable.

Questions can send out subtle signals. A useful question at the end of the interview is to ask: *are there any questions you would like to ask?* As well as signalling the conclusion of the interview, such a question sometimes leads to further information being provided.

Stage five: after the interview

Great emphasis is placed upon field notes (Patton, 1980, p. 251), particularly those notes made immediately after the interview. Contact summary sheets and data accounting sheets (Miles and Huberman, 1994) provide a vehicle for a summary of the main points of interviews and to check that research questions have been addressed.

Interviews can be intense experiences, requiring complete attention and frequent thinking on your feet. The fact that respondents disclose personal details is a considerable responsibility, but can leave you feeling exhausted. There can be a feeling of uneasiness after interviews which it is difficult to rationalise. This stems from the respondents opening up and apparently sharing their inner thoughts. After opening up these areas to gain the information the researcher disappears. A letter of thanks can demonstrate appreciation for the time the respondent provided for the questions. Also, it is in the research tradition of helping to keep the door open for future research (Bell, 1987, p. 79).

Recording interview data

Before you conduct an interview you should decide how your interviews will be recorded. Three possible options are: note-taking, tape recording or a combination of the two. Do keep it in mind that the respondent may not allow you to tape record the interview.

The following personal example highlights the significance of this choice. I had obtained access to seven senior managers at the head office of a large organisation. At the outset of one of the interviews, I asked if I could tape record the interview. The manager agreed and we conducted the interview with the tape recorder running. We concluded the interview and went for lunch. Over lunch the manager confided that he would have answered questions differently if the tape recorder had not been present. Instead of portraying a successful organisation, he would have described the low morale and the problems within the organisation.

The following questions should help to determine which form of recording is appropriate for your research.

- How sensitive are the issues you wish to address? If the roles were reversed would you feel comfortable being tape recorded?
- Do you have the ability to record the interview in note form?
- Does the type of interview that you are adopting require you to make notes to act as probe about certain topics as they arise?
- Do you have the resources to have the tape recordings transcribed? This can prove very costly in terms of your own time or paying somebody to transcribe the interviews.

If you decide to use a tape recorder the following supplementary notes may be useful.

Buchanan *et al* (1988, p. 61) found refusals to allow tape recording to be rare, describing tape recorders as now 'accepted technology'. While respondents agreed to be tape recorded there were times when the tape recorder was switched off and they spoke off-the-record (Howard and Sharp, 1989).

While there can be a strong desire to use tape recorders from the start of the interview, the recorders can make respondents suspicious. Consequently, it is often necessary to avoid using the tape recorder for at least the first five minutes. During this time the tape recorder remains out of sight.

In terms of transcription, Patton (1980, p. 249) offers a series of helpful tips on how to keep transcribers sane. Alternatively, Buchanan *et al* (1988, p. 61) provide a strong rationale for typing your own transcripts.

Analysing interview data

While there may be a desire to partition analysis from the conduct of interviews, Miles and Huberman (1994) among others, encourage interweaving analysis and data collection. They suggest that there are five main stages to qualitative data analysis, and that the researcher has a range of options to conduct each stage.

1. collect the data
2. data reduction
3. data display
4. drawing conclusions
5. verification of findings.

By interweaving data collection and data analysis it is possible to test the effectiveness of your interviewing and make amendments where necessary.

Silverman (1993), in his book subtitled *Methods for Analysing Talk, Text and Interaction,* is critical of much work done in this area, but offers a range of methods that can be used.

Summary

Here are some key questions to consider when you use interviews as a method of data collection.

What is the rationale for interviewing?

The strengths of interviewing as a method of data collection include the following:

- face-to-face encounter with informants
- obtains large amounts of expansive and contextual data quickly
- useful for discovering complex interconnections in social relationships
- data are collected in natural setting
- good for obtaining data on non-verbal behaviour and communication
- great utility for uncovering the subjective side, the native's perspective of organisational processes.

What types of interview exist?

Patton (1980) identifies four main types:

- informal conversational interview
- interview guide approach
- standardised open-ended interview
- closed quantitative interviews.

Variations include telephone interviewing, group interviewing and elite interviewing.

What type of questions can be asked?

- experience/behaviour questions
- opinion/value questions
- feeling questions
- knowledge questions
- sensory questions
- background/demographic questions.

What are the implications of the interview as a process?

The interview process has the following broad stages:

1. preparations
2. introductions
3. the uneven conversation
4. the ending
5. after the interview.

If you consider each of these stages your interview should be effective in gathering the data that you require.

How can I record the interview data?

- tape recorder
- notes
- tape recorder and notes.

What options exist for analysing interviews?

There are many options, but it important to consider analysis as five linked stages:

- collect the data
- reduce the data
- display the data
- draw conclusions
- verify findings.

Postscript

People's accounts of aspects of their lives can be fascinating, as each respondent has his or her own unique story to tell. You must be interested in people if you are to conduct effective interviews.

> *I'm personally convinced that to be a good interviewer you must like doing it. This means taking an interest in what people have to say. You must yourself believe that the thoughts and experiences of the people being interviewed are worth knowing.*
>
> Patton (1980)

The chapter has introduced the main tools in the interviewers tool-kit and introduced the interview process. However, for this particular research method you, as the interviewer, are an integral element.

References

Ackroyd S and Hughes J A (1981) *Data Collection in Context*. Longman, London.
Bell J (1987) *Doing your Research Project*. Open University Press, Milton Keynes.
Buchanan D, Boddy D and McCalman J (1988) Getting in, getting on, getting out and getting back. In Bryman A (ed.), *Doing Research in Organisations*. Routledge, London.
Grebenik E and Moser C A (1962) Society: problems and methods of study. In Welford A T, Argyle M, Glass O and Morris J N (eds), *Statistical Surveys*. Routledge & Kegan Paul, London.
Howard K and Peters J (1990) Managing management research. *Management Decision*, Vol. 28(5).
Howard K and Sharp J (1989) *The Management of a Student Research Project*. Gower, Aldershot.
Kane E (1990) *Doing your own Research*. Marion Boyars, London.

Marshall C and Rossman G B (1989) *Designing Qualitative Research*. Sage, Newbury Park, CA.
Miles M B and Huberman A M (1994) *Qualitative Data Analysis*. Sage, Beverley Hills, CA.
Mullings C (1985) *Group Interviewing*. Centre for Research on User Studies, University of Sheffield.
Patton M Q (1980) *Qualitative Evaluation Methods*. Sage, Beverley Hills, CA.
Patton M Q (1990) *Qualitative Evaluation and Research Methods*. Sage, Newbury Park, CA.
Pease A (1993) *Body Language*. Sheldon Press, London.
Silverman D (1993) *Interpreting Qualitative Data: Methods for Analysing Talk, Text and Interaction*. Sage, London.
Whyte W F (1960) Interviewing in field research. In Burgess R G (ed.) (1982) *Field Research: A Source Book and Field Manual*. George Allen & Unwin, London.
Wragg E C (1980) Conducting and analysing interviews. *Rediguide 11*. Nottingham University.

Part Five

DATA: ANALYSIS

22
Elementary statistics

David Hand

Introduction

Statistics is an unusual word because it has two meanings. It refers both to (typically) numerical data describing some phenomenon and to the science of collecting and analysing those data. The first of these meanings is, of course, subject specific: what the numbers are, what they mean, and the implications of them, will depend on what they are describing. In contrast, the other meaning, sometimes expanded to **statistical science**, describes methods which can be applied in any domain. It is this second meaning which is the concern of this and the next chapter.

Statistics is a vast subject. It is far too large for any single person to be able to be an expert in all its various sub-domains. It follows that what we can hope to do in these few pages is try to orient you; give you an idea of the motivation behind statistical methods; show you some very basic tools; and give you pointers to further reading which will provide the details of what you need to know. Before we get down to it, however, a word or two about the nature of modern statistics is appropriate.

Statistics has long suffered from a bad press in two regards. One is the view that it, like politicians, can bend the facts to suit any purpose. And the second is that it is a tedious and boring subject. The first of these criticisms arises because it is easy to mislead people who are statistically naive. By misrepresenting the facts – presenting only some of the numbers rather than their entirety, focusing on only particular aspects of the results, and generally distorting the data to be analysed – of course one can mislead. But here it is the improper use of statistical methods which is the appropriate target of the criticisms rather than the statistical methods themselves. It is unfortunate that the mud flying around from such misuse has sometimes adhered to the tools being used rather than to those using them improperly.

The second criticism may once have had some truth. In the days when even a relatively simple analysis involved endless hours of mechanical numerical manipulations by hand, how could the subject be regarded as exciting? But things are not like that nowadays. Now we have the computer to takeover the tedium. The computer enables us to concentrate on the higher level tasks, the seeking for patterns and structures in data, the comparison of our theories with the data, without subjecting ourselves to the tedium of endless arithmetic. This has two consequences. The first is that the reputation of statistics as boring is now completely wide of the mark.

A geologist recently commented to me that he envied me. 'Statisticians', he said, 'have the best part of scientific research. They don't have to put up with the boredom and repetition of collecting the data. They come in at the most exciting stage – when one is looking at the data one has collected.' He was forgetting the role of statisticians in deciding how to collect data (in experimental design and sampling) and he was also failing to recognise that many statisticians prefer to get some hands-on experience as it can give a better idea of the aims and difficulties of the study. Nevertheless, what he said had some truth in it. And when you have reached the stage of analysing your data, of looking to see if the information it contains matches the ideas you have had, you will also find that statistics is entering at the most exciting stage.

The second consequence of the development of the computer is that, at least at this introductory level, we do not have to dwell on the algebra and arithmetic of statistical methods. For all but the most basic of operations on very small data sets, you will use a calculator or computer. Hence, in what follows, we have attempted to focus on statistical concepts and the properties of statistical methods rather than on the mechanics which electronics will do for you.

This is the first of three chapters on statistics. It introduces the basic ideas. The other two chapters cover, respectively, more advanced statistical methods and computer software.

Scales

Data come in several forms and it is useful to distinguish between them. First we can distinguish between **numerical** and **non-numerical** measurements. Examples of the former are a person's age or weight and the size of a family. Examples of the latter are the position of a mark on a scale indicating one's extent of agreement with some statement and scores of mild, moderate, or severe on a pain scale. Numbers can, of course, be used to score the latter examples, but they do not have quite as strong empirical force as the numbers used in the former cases and sometimes one must be careful about how one analyses such data.

A second distinction we can make is between **continuous** and **categorical** data. Continuous data are measurements which can (at least, in principle) take any value within a certain range. Age, weight, and the agreement score above are examples of this. Categorical variables, in contrast, can take only one of (normally) a few values: size of family and the pain scale illustrate this. Sometimes it is not clear which class a variable is in (age, for example, might be recorded just to the nearest year) but this does not usually cause any problems.

It is often useful to divide categorical variables further, into **nominal** and **ordinal** variables. The former have no natural ordering – like the various religions, for example – while the latter do (as in the pain scale example).

Basic measures

If one looks at a raw table of numbers it can be difficult to see what is going on. Consider Table 22.1, for example. This (Frets, 1921) shows head lengths (in mm) of 25 men. Even in such a small table it is not easy to see what head size is typical and what is extremely large or small. It is not easy to see if large extremes are rarer than small extremes. It is not easy to see if there are striking exceptions to the general size of head. This being the case with only 25 values, the difficulty of coping with larger data sets will be obvious.

Table 22.1 Male head lengths (mm)

191	174	189	181	208
195	190	197	175	186
181	188	188	192	183
183	163	192	174	197
176	195	179	176	190

First let us find a way to describe 'typical' head length. A number of statistics are in common use. (And this introduces us to a third use of the word *statistics* – as the plural of 'statistic'. A statistic is some summary value calculated from or derived from a set of data.) We shall consider three here: the **mean**, the **median**, and the **mode**. These are different kinds of **averages**.

Given a sample of n numbers adding to a total T, the *mean* of the sample is that number (often denoted) $\bar{x}$ such that n copies of it also add up to T. Obviously $\bar{x}$ is smaller than the larger numbers in the sample and it is larger than the smaller ones. It is an average or *representative* value. Numerically it can be found simply by calculating the total in the sample and dividing by the sample size. For Table 22.1, we find that the mean is:

$$(191 + 174 + ... + 190)/25 = 185.72$$

The *median* of a sample is the mid-point of the sample: half the sample values are smaller than it and half are larger. In fact, we have to specify things a little more precisely than this. We order the sample in terms of size and if there is an odd number of values in the sample, then the median is the middle one. In the above example the reordered sample is 163, 174, 174, 175, ... 197, 208 and the middle one has value 188. The median is thus 188. If, however, there is an even number of values in the sample then there isn't a 'middle one'. In this case we define the median as being half way between the two middle ones. For example, in the ordered sample – 10, 14, 15, 17, 18, 18 – the median is half way between 15 and 17. That is 16.

Finally, the *mode* is the most commonly occurring value in the sample. In our example above, there is in fact no single value which is the most common. Several values occur twice, but no value occurs more than twice. Such a data set it said to be **multimodal** – it possesses several modes. In circumstances like this the sample mode is of limited use as a summary of the data. (More sophisticated methods based on identifying the values of the several modes can be useful, but they are beyond the scope of this chapter.) For continuous data, in which the sample values are all different, each value will occur once – hence, again, there will be no unique mode.

Sometimes averages such as those above are termed **measures of location**. This arises from the idea of viewing the sample values as plotted points on a number line. The mean, median, and (if it exists) the mode, are also points on the line, showing, in some sense, the 'average' position of the sample of points. Such a representation cannot be used with nominal data and is of doubtful value with non-numerical data.

The median tells us that sample value which has half the sample below it. Sometimes it is also useful to know what value has a quarter, 25 percent of the sample below it. This value is called the **lower quartile**. A similar definition, with 75 percent of the sample below it, gives the **upper quartile**.

So much for measures of location. These tell us the rough size of the sample of values we are dealing with, but they do not tell us how widely distributed those values are. Perhaps the data are very closely clustered about the mean; or perhaps the mean is a very poor summary of the values because the constituents vary greatly. We need to supplement the mean (or median or mode) with another value giving an idea of the **spread** or **dispersion** of the sample.

One simple measure is the **range**. This is simply the difference between the largest and smallest values in the sample. For the data in Table 22.1 the range is 208 – 163 = 45.

As a measure of spread the range is easy to interpret and understand. But it has the disadvantage that it is not very *stable*. Another sample from the same population is quite likely to have a substantially different range because, in calculating the range, only the two most extreme values are considered. An alternative measure of dispersion which overcomes this variability and is less variable from sample to sample is the **interquartile range**, defined as the distance between the upper and lower quartiles.

Another measure of spread which overcomes this variability, and which is the most popular measure of dispersion, is the **standard deviation**, often abbreviated to **sd**. This is most straightforwardly defined as the square root of the **variance.** So first we have to define variance. The variance of the sample is the average size of the squared difference between the sample values and their mean.

Often the divisor $(n-1)$ is used in place of n when calculating the mean of the squared differences because the resulting statistic has attractive properties. We need not go into these here – and, in any case, for even moderately large n the difference will be negligible. In our example above, and using the $(n-1)$ divisor, the sample variance is 95.29.

It follows that the sample standard deviation is the square root of this, namely 9.76. Taking square roots means that the sample sd is measured in the same units as the raw data. (The variance will be in square units – kg^2, inches squared, or whatever, because of the squaring operation in its definition.)

So far we have discussed measures of location and measures of dispersion. One other simple summary statistic, important because it often indicates when common statistical methods may be reliably used, is **skewness**. The skewness of a distribution of values is a measure of **asymmetry**. For example, does it have many small values and few very large values (*positively* or *right skewed*) or does it have lots of large values and few small values (*negatively* or *left skewed*). Income and wealth distributions, with a very few people earning or owning a lot, tend to be positively skewed. Negatively skewed distributions are less common.

You will notice that we have slipped the word **distribution** in here. This is a technical term in statistics, which fortunately has much the meaning one might expect it to. One can speak of the distribution of a sample, and also of the distribution of a **population** – the set of values which might have been chosen for the sample. We remarked above that the computer and calculator allowed us to side-step the arithmetic details in a brief exposition such as this. For this reason we shall not give a formula for skewness. If you are keen on that sort of thing then you can find calculation details in the books in the bibliography.

One of the attractive features of modern statistical technology is the ease with which plots and diagrams can be produced. Again the computer has taken all the drudgery out of the exercise. An example is given in Fig. 22.1. The data summarised here (Cox and Lewis, 1966) are time intervals between 800 successive pulses along a nerve fibre, measured in seconds. The diagram shows a **histogram**, produced by grouping the observations according to their size and plotting vertical bars whose heights indicate how many fall into each group.

Simple distributions

We have already introduced the term **distribution**. Some distributions are particularly important in statistics, perhaps because they are very common or because they have attractive mathematical properties (quite what this means we will see in a moment!).

Bernoulli distribution

The *Bernoulli* distribution arises when there are two possible outcomes, with probabilities p and $1-p$, for example. A classic example is the toss of a coin – and if the coin is 'fair' both p and $1-p$ will equal 1/2. Other examples are an application to college – which can be successful or unsuccessful – and whether it will rain or not on your birthday.

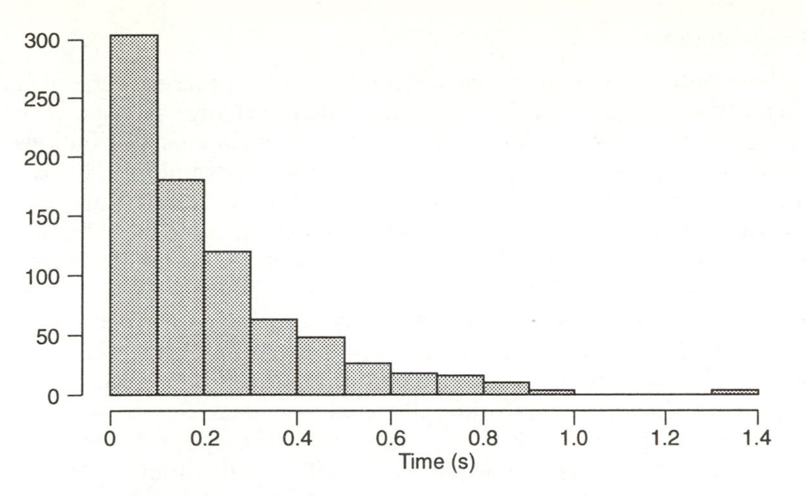

Figure 22.1 Histogram of 799 interpulse waiting times.

Binomial distribution

An extension of this is the **binomial distribution**. This arises as the sum of a number, for example, n, Bernoulli outcomes. Instead of being interested in whether a single toss of a coin will come up heads or tails, we might be interested in the proportion of 100 tosses which come up heads – what is the probability that 50 will produce heads? Or 49? And so on. Or we might be interested in how many of the ten applicants to college from a particular school will be successful. Or we might want to know on how many of your next ten birthdays it is likely to rain.

For situations like this to be modelled by the binomial distribution certain conditions have to be satisfied. In particular, we require that the individual events have *the same probability of occurring* and are *independent*. If the probability that the coin came up heads differed from throw to throw (because, for example, someone tampered with it by attaching chewing gum to one side), then the binomial distribution would not be an adequate model for the physical situation. **Independence** is a rather more complicated – and very important – concept. Two events are said to be independent if the occurrence of one does not affect the probability that the other will occur. This is likely to be satisfied for tosses of a coin. But suppose we had asked about the number of the next ten days on which it rained. Rain on consecutive days is not independent: if it was dry today it is more likely to be dry tomorrow than if it was wet today. The binomial distribution is unlikely to be a good model for this situation.

Poisson distribution

The binomial distribution is one possible model for **counts**. Another one is the **Poisson** distribution. Suppose, again, that n independent events are considered, with each of them having two possible outcomes (A and B, for example), with the same probability that A will occur each time. Then, if n is large and the probability of A is small, the total number of times A occurs can be well approximated by a *Poisson* distribution. Such a distribution might be a good model for the number of misprints in a book: there is a small chance that any particular letter will be misprinted, but there are a lot of them. Of course, we will need to consider whether or not the conditions of independence and equal probability are sufficiently closely satisfied.

Normal distribution

So far we have only considered *discrete* distributions – the outcome can only take one of a discrete set of values (e.g. 0, 1, 2, ...). In other situations any value can occur (perhaps from a certain range) – and, as above, these are called **continuous** distributions. The most important example of these is the **normal** or **Gaussian** distribution. This is a unimodal distribution (one peak) which tails off symmetrically to high values and to low values. So, scores which are r units above the mean have exactly the same probability of occurring as scores which are r units below the mean. The most likely values are those around the mean, and very large or small values are very unlikely.

The normal distribution is often a sufficiently accurate approximation to empirical distributions which occur in real life, and this is one reason for its importance. Another is that, when large samples are involved, the normal distribution is often a good approximation to the distributions of statistics calculated from data. For example, suppose that a sample of 1000 people were asked their ages and an average (the mean) calculated. If this was repeated – 100 times – we would obtain 100 slightly different mean values, and we could study the distribution of these 100 values. We would find that the majority were clustered together, with few large values and few small values. Moreover, similar numbers of samples would have very small values and very large values – the distribution of the 100 means would be roughly symmetric. To summarise, the distribution of the 100 means, each based on a sample of 1000 people, would be roughly normal.

This sort of thing often occurs, especially when the means (or other statistics) are based on large samples (like the 1000 in our example). This striking property of the normal distribution makes it of fundamental importance in statistical theory: if exact distributions cannot be worked out, maybe a reasonable approximating normal distribution can be used.

Estimating parameters

The statistics calculated from samples can often be regarded as **estimates** of **parameters** of the populations from which the samples were drawn. For example, consider the population of the UK. This entire population will have an age distribution and it will have an average value (a parameter – *a defining characteristic* – of the population). We could discover this average value by asking everyone their age and calculating the average (though we would have to move quickly – people are getting older all the time!). Alternatively, we could take a sample of far fewer people and simply find its average. This sample average is a statistic. It turns out that the variation in such a sample average is inversely related to the square root of the sample size. So, by taking a large enough sample, we can make the variation in our sample average as small as we like. Put another way, we can make the sample average as close as we like to the population average (or, at least, we can specify an interval – a **confidence interval** – such that the probability that the interval contains the true value can be as high as we like).

Testing hypotheses

These basic ideas can be used to test theories in the following way. Suppose we want to compare two treatments for some disease – treatment A and treatment B, for example. We give A to one group of people and B to another group (the two groups being selected by **random allocation**, so as to ensure that there is no bias in terms of the likely recovery rates). And we compare the recovery rates in the two groups. If the two treatments were equally effective (the **null hypothesis**) we could work out the probability of obtaining any particular difference between the proportions recovering under A and under B.

In particular, we can work this out for differences as large as or greater than the difference we actually obtained. This tells us how likely we are to obtain a difference as large or larger than that actually observed if the groups are really equally effective (if the null hypothesis is true). If this probability is very small it casts doubt on our initial assumption that the two treatments are equally effective. If it is small enough, it may lead us to reject that initial assumption (that is, to reject the null hypothesis). What is 'small enough' here will depend on the experimental situation and the investigators carrying out the work. Common values chosen are one percent, five percent, and ten percent.

In the above, we used sample statistics to make an inference about a population value – about the difference between the proportion who would recover if everyone received A and the proportion who would recover if everyone received B. Such a technique is called a **hypothesis test** because we are testing a hypothesised value of a population parameter. This basic idea can be extended to a vast number of situations. A fairly straightforward extension of the above situation allows one to compare the means of two groups, using what is called a ***t*-test.**

An entirely different situation is as follows. Suppose we have two categorical variables – the three-valued pain scale introduced above, and the sex of the patients in the study. Data arising from such a situation can be arranged in a three by two **cross-classification** called a **contingency table**. We might wish to know whether resistance to pain differs between the sexes – whether the distribution of female patients across the three pain categories is the same as the distribution for male patients. (This question can be expressed in several alternative but equivalent ways. Is the ratio of the number of males to females the same in each of the three pain categories? Are the two categorical variables, pain and sex, independent?) A hypothesis test of this question begins by assuming that the distributions in the populations are the same and, based on this assumption, derives the probability that one would obtain a sample difference between the observed distributions as great or greater than that actually obtained. Again, if this probability is sufficiently small, one will feel that the initial assumption (the null hypothesis of independence) is untenable. This forms the basis of a **chi-squared** test for independence.

Sometimes particular tests make fairly stringent assumptions about the distributions involved in the situation being investigated. For example, the *t*-test mentioned above assumes that the populations from which the samples are drawn follow normal distributions. Such tests are called **parametric** tests. Other tests – **non-parametric** or **distribution-free** tests – relax these assumptions, and so are more generally applicable. Having said that, however, if the assumptions of the parametric test are justifiable, they are generally more **powerful**. This means that they are more sensitive to departures from the null hypothesis.

Conclusion

Given the limited length of this chapter, it is obvious that we could not go into great depth here. This means that we have only been able to scratch the surface of the ideas we have outlined and present a very few central ideas. The books listed in the bibliography go into more detail and if you are involved in analysing data in your work you are encouraged to obtain one.

To conclude, however, there is a further general point which should be made. This is that modern statistics, although a tremendously exciting discipline, is also a complex one. Research is very seldom a question of looking at the data, deciding what technique to apply, running the computer to produce a single numerical value, and writing things up. Typically all sorts of complications arising as the complexities of real life intervene.

Because of this you are advised to seek statistical advice both before collecting your data and before analysing it. Such a precaution could prevent a great waste of time and a good deal of mental anguish.

Bibliography

Cox D R and Lewis P A W (1966) *The Statistical Analysis of Series of Events*. Chapman & Hall, London.

Daly F, Hand D J, Jones M C, Lunn A D and McConway K J (1995) *Elements of Statistics*. Addison-Wesley, Wokingham.

Frets G P (1921) Heredity of head form in man. *Genetica*, Vol. 3, pp. 193–384.

23
Further statistical methods

David Hand

Introduction

Chapter 22 outlined some basic statistical ideas and methods. Here we move on to describe some more advanced techniques. Given their advanced nature, the best we can do is scratch the surface. However, we hope that this will be sufficient to indicate the sort of thing that can be accomplished using modern methods. Naturally, if you do have data which need such methods, you are advised to seek professional advice. Although statistical software is now very readily available, statistical understanding is not so easy to come by. A few minutes discussion with an expert could save endless weeks of frustration.

We begin by distinguishing between techniques for **prediction**, where there is a *criterion* variable which has to be predicted from the other variable or variables, and techniques where the variables are all equivalent. The former is used for prediction as well as for understanding, while the latter is mainly used for understanding what is going on in a set of data – for deciphering the relationships between variables and objects.

Regression analysis

Regression analysis is a statistical model-building technique. It relates a single **response** or **dependent** variable to one or more **predictors** or **independent** variables. The model, a mathematical equation, can be used as a summary of the relationship between the response and the predictors and it can also be used to predict the value of the response given values of the predictors.

We shall illustrate using data on the output of wind-powered generators of electricity (Joglekar *et al*, 1989). The predictor here is wind velocity (in mph) and the response variable is direct-current output. Then our regression equation will permit us to answer questions such as:

- On average, for the generator being studied, what extra current output results from an extra 1 mph in wind velocity?
- Given a wind velocity of 3 mph, what average current output should we expect?

Such a model is constructed as follows. We begin with a set of data. In this case, we will need a sample of measurements of wind velocities and the associated current outputs. Let us denote the values of the response variable, current, by y and the values of the predictor variable, wind velocity, by x. Then our sample provides us with a collection of pairs of values, one pair for each measurement occasion. The data are shown in Table 23.1.

Now, let us conjecture that a suitable model relating output to velocity is that output is 0.25 times velocity. A glance at the data shows that this is certainly not a perfect model. The first pair has a velocity of 2.45 so that $0.25 \times 2.45 = 0.613$, which is not equal to the

Table 23.1 Wind velocity and direct current output

x	y	x	y
2.45	0.123	6.00	1.822
2.70	0.500	6.20	1.866
2.90	0.653	6.35	1.930
3.05	0.558	7.00	1.800
3.40	1.057	7.40	2.088
3.60	1.137	7.85	2.179
3.95	1.144	8.15	2.166
4.10	1.194	8.80	2.112
4.60	1.562	9.10	2.303
5.00	1.582	9.55	2.294
5.45	1.501	9.70	2.386
5.80	1.737	10.00	2.236
		10.20	2.310

given output of 0.123. Nevertheless, examination of the other values shows that multiplying velocity by 0.25 gives current outputs in the right ball park. But how can we find a better value?

What we need is some overall measure of **goodness-of-fit** between our model and the data. Each pair gives us a separate measure of goodness-of-fit: the difference between the output value predicted from the model (0.613 in the above example) and the observed output value (0.123 above). So for this pair alone, its goodness-of-fit measure, is 0.613 – 0.123 = 0.5. But how can we combine the separate such results for each pair into a single overall measure?

We could add them up, but this would have the problem that predictions which were too small (and, so, which had negative goodness-of-fit measures) would tend to cancel out with predictions which were too large (and which therefore had positive goodness-of-fit values). This problem is overcome by squaring the individual goodness-of-fit measures – so making them all positive.

So, the overall goodness-of-fit measure is the sum of squared differences between the output predicted from the model and the observed value.

We can calculate such a measure for a range of conjectured values for the constant relating output to velocity. For example, in addition to 0.25 we could try 0.2, 0.21, ... , 0.3, ... , 0.4. And then we could choose that which best fitted the data.

In fact, there is a better way. The sum of squared differences criterion is a common measure in statistics and the optimum value of the constant can be found mathematically. In practice nowadays, of course, it will normally be found by a computer (see Chapter 24). The value of the constant which defines the relationship between output and velocity is called the **regression coefficient** for the regression of output on velocity. Based on the sample of data we have, our computer will have given us an estimate of that coefficient.

We started the above example by saying that we needed a sample of pairs of measurements. Great care must be taken when this sample is selected. We can distinguish two situations.

In the first situation, the sample is a **random** sample from the population of interest. If this is the case, then hypothesis tests, as outlined in Chapter 22, can be done. So, in our example, we might take a random sample of pairs of measurements. Then we can test whether or not there is any relationship between velocity and current in general. You might ask: 'how likely is it that, if there is no general relationship between velocity and current, we would have obtained the value for the *sample* regression coefficient which was actually obtained?' (The answer is: 'unlikely'. Less than five percent would lead you to reject the hypothesis of no relationship between velocity and current.) If the sample has not been randomly selected then the basis of the hypothesis test to answer this question is invalid. If we had deliberately chosen pairs of values which had low current associated with high velocity and high current associated with low velocity then we might have found an apparent inverse relationship: more wind meaning less output, on average. This would have been nonsensical in terms of statements about the overall relationship between wind velocity and current output.

The second way to choose the sample arises when you can exercise control over the predictor variable. For example, we might want to know whether increasing the concentration x of a chemical X leads to an increase in product Y. To explore this we could do a series of experiments, measuring output Y for various *predetermined* X levels. As it happens, the statistical inferences described above are also valid with this approach.

So far we have described what is called **simple regression**. Simple regression involves a single predictor variable, wind velocity in the above example. **Multiple regression** extends this to model the simultaneous effect of several predictors on the response variable. So, to illustrate, in the above example we might have been concerned about the effect of wind velocity, air temperature, and humidity on the current output. The mathematical model would then involve a separate constant multiple for each of these predictor variables.

Put in mathematical terms, the predicted value of the response is expressed as a weighted sum of the values of the predictors, where the weights are the regression coefficients.

We could have done three separate simple regressions, one for each of the three predictor variables. It is important to understand the distinction between this and the multiple regression involving all three simultaneously. The regression coefficients from the former tell us how the response will change when there is unit change in one of the predictors. The latter tells us how the response will change when there is unit change in one of the predictors *keeping the other predictors constant*. The multiple regression tells us the unique effect of each predictor over and above that of the others.

At first exposure this idea, like so many others in statistics, is not easy to grasp. But it is an important one. Which of the two types of analysis to do, whether to report several separate univariate analyses or a single multivariate analysis, will depend on the research objectives.

Logistic regression

In the preceding section, the model formulated to predict the value of a response variable had the form of a simple weighted sum of the predictor variables. This is probably the most ubiquitous statistical model that there is: it has a long history and has been used in just about every area of human endeavour. However, despite its power, it is not a universal answer. It has been extended in a number of ways and this section and the next consider two of these extensions. Here we look at what happens when the response can take values only in a certain range, and in the following section we look at what happens when the response is restricted to just two possible values.

Suppose we were exploring the relationship between the dose of a drug and the probability of being cured. One way we could set about this would be to make up several different doses of the drug and administer each of these doses to a group of patients. This is the same sort of method of drawing the sample as the second method described in the previous section on regression analysis.

People vary, so in each group we would expect some would recover and others would not. If the chance of cure increases with increasing dose, then we might expect more people to get better in those groups which were receiving the larger dose. This is beginning to look rather similar to the set-up described in the last section: we could, perhaps, model 'chance of recovery' as the response variable in a regression with 'dose' as the predictor.

This is a perfectly reasonable thing to do and is often done. However, it does have a drawback. This is that probability can not be greater than one or less than zero. (A probability of one means certainty, and you can't get more probable than that! Similarly, a probability of zero means impossible, and you can't get less probable than that.) If we model the probability of recovery as some constant times the dose, then:

1. for sufficiently large doses the predicted probability of recovery would be greater than one
2. for sufficiently small doses the predicted probability of recovery may be less than zero. (Actually, this will depend on the data. Whereas there is, in principle, no limit to the maximum dose which can be given, there is a limit – zero – to the smallest.)

To cater for this, so that our model will only give reasonable predictions, we modify it slightly. Instead of using the weighted sum (in our case, just the constant times the dose) to predict probability of recovery, we **transform** the model so that its predictions always lie between zero and one. The most common type of transformation used is the **logistic** transformation. If the raw, untransformed model predicts a value above one, the transformed version predicts a value below one – but the nearer to one the greater the untransformed prediction. A similar thing applies to low predictions.

Because of the central role of the logistic transformation in such models, this technique is called **logistic regression**.

Discriminant analysis

Prognosis means determining the likely future outcome, and is important for people who have suffered a head injury. It will depend on many factors, including age, response to stimulation, and change in neurological function over the first 24 hours after the injury. We would like to predict the future outcome, for example recovery or not, on the basis of some of these variables. Again we shall build a model to do this, and again we will base our model on a sample of people. In particular, we will have a sample of people who have known predictor variable values (such as age) and whom we have followed-up so we know their outcome.

One way to approach the problem is as follows. First consider the 'recovered' group. Such people will have a range of ages, a range of responses to stimulation, and so on. In a word, they will have a *distribution* across the possible predictor variables. The same applies to the other, non-recovered, group. If these distributions differ it means that for some combinations of age and response to stimulation, one of the two 'recovery' categories is more likely than the other. By modelling the distributions we can identify the combinations which are most likely to correspond to membership of each of the two 'recovery' groups.

There are many ways in which the distributions might be modelled, but the most common assumes a particular class of forms for the distributions. This method leads to what is known as **linear discriminant analysis**. It is called this because it leads to a predictive model which has the form of a weighted sum (a *linear combination*) of the predictor variables. Large values of this weighted sum are associated with combinations of the predictor variables which characterise one of the recovery groups and small values are associated with combinations which characterise the other recovery group. That is, the type of model used in linear discriminant analysis, for permitting one to allocate an object to one of two classes, has the same basic structure as the models discussed in the preceding paragraphs.

Analysis of variance

In the above we described several **predictive** statistical techniques and pointed out that they all had the same underlying form – a weighted sum of the predictor variables. Another very common type of technique, which also has this underlying structure, though this is often concealed in elementary descriptions, is **analysis of variance**. This is aimed at describing the differences between groups of objects. (So it is closely related to the discriminant analysis of the previous section. In fact, there are various different kinds of links between all of the techniques described in this chapter. That is one of the exciting things about statistics: its methods are not isolated tools; they form a complex and interlinked system of ideas and methods.

Suppose that each of the predictor variables is categorical – that is, they can each take one of only a few possible values. For example, age might have been partitioned into young, middle, old; sex will be male or female; and some other measure might be graded as bad, impaired, or good. We shall suppose that the response variable is numerical.

The cross-classification of the predictor variables forms a set of **cells**, for example, young males with impairment; old females with impairment. And now we can ask questions such as: do the males differ from the females? Does the response decrease as one moves from young to old? Does the effect of age differ between the two sexes?

To answer such questions, a linear model (a weighted sum) is again constructed and hypothesis tests are done. However, since designs involving categorical variables in this way are so common, special ways of describing the results have been created, and such analyses are summarised by means of an **analysis of variance table** which shows the influence of (in this example) sex, age, and the other predictors on the response, as well as how the effect of each of these predictors differs according to the levels of the others.

Other methods

So far all of the techniques we have discussed are predictive in the sense that they seek to determine the likely value of one variable given an object with known values of the other variables. Not all questions are like this, however. Another whole class of models is concerned with describing the relationships between variables and objects when no variable can be separated out as a response. **Principal components analysis** is one such. This technique allows us to determine which combinations of variables explain the most differences between the objects in the sample.

In a study of the patterns of consumption of psychoactive drugs (Huba *et al*, 1981), data were collected on 1634 students showing the extent to which they had used cigarettes, beer, wine, spirits, cocaine, tranquillisers, drug store medications used to get high, heroin, marijuana, glue and other inhalants, hallucinogenics, and amphetamines. Study of this data-set using principal components analysis showed that the greatest range of differences between the students could be explained in terms of the overall extent to which they used the substances. After this, most of the remaining differences between the students could be explained in terms of whether or not they used illegal substances. In effect, what principal components analysis has done is reduce the very complex array of data to a simple and comprehensible description of the main features which distinguish between the students.

Time series

All of the methods we have outlined above involve multiple measurements on each object. Typically there will only be a few (or a few tens) of such measurements and there will be several or many objects. In a sense, such situations lie in the middle of a continuum at one end of which lie *univariate* problems, with single measurements on each object (methods for analysing data of that kind are described in Chapter 22). At the other end of this continuum lie **time series**.

Time series are characterised (at least, in their simplest form) by having just a single object but on which many measurements have been taken – the responses at each of a set of consecutive times.

Time series are ubiquitous forms of data. Examples are: stock closing prices at the end of each trading day; temperature at a particular location measured at midday each day; daily rainfall; and an individual's body weight measured at 8.00 am each day.

Such data-sets are important for several reasons. **Forecasting** is an obvious one. It would be immensely useful if we could predict tomorrow's, next week's, or next year's FT index! Understanding is equally important: is the economy showing an underlying upward trend or are the short-term figures deceptive?

There are several approaches to modelling time series. Some focus on modelling them in terms of underlying components such as trend, seasonality, and superimposed random terms. Others focus on the probability that a particular value will occur in the next period given that the current (and, perhaps, preceding) periods have the observed values. More complex models include the effects of other variables on the score at each time.

Other techniques

Statistics is a vast domain, with methodological research going on all the time; new methods for new problems and improved methods for old problems are being developed. Recently, stimulated in part by the possibilities presented by the growth in computer power, new classes of flexible multivariate techniques have attracted a great deal of interest. These include **neural networks**, **projection pursuit regression**, **radial basis function models**, and **multivariate adaptive regression splines**. To some extent these methods avoid the need to think carefully about what kind of model might be appropriate for the problem in question. But this is a two edged sword: some insight into what is going on can prevent problems and also lead to better classes of models.

Further reading

Huba G J *et al* (1981) *Journal of Personality and Social Psychology*, Vol. 40, pp. 180–193.

Joglekar G, Schuenemeyer J H, and LaRiccia V (1989) Lack-of-fit testing when replicates are not available. *American Statistician*, Vol. 43, pp. 135–143.

Krzanowski W(1988) *Principles of Multivariate Analysis*. Clarendon Press, Oxford.

Lovie P and Lovie A D (1986) *New Developments in Statistics for Psychology and the Social Sciences, Volume 1.* British Psychological Society, London.

Lovie P and Lovie A D (1991) *New Developments in Statistics for Psychology and the Social Sciences, Volume 2.* British Psychological Society, London.

24
Data analysis by computer

C E Lunneborg and David Hand

Introduction

Statistics, as a scientific discipline, is evolving rapidly. In large part this rate of evolution is a consequence of the development of the computer. Just 20 years ago data analysis took place either by means of a rudimentary (by today's standards) calculator or on a centrally-located computer. Nowadays, most analysis takes place inside a personal computer located on the user's desk (or, via a network, on some other personal computer or workstation). Punched tape and cards have given way to direct interaction with the computer. Reams of printed output have been abandoned in favour of direct read-out from a screen.

But, from a data analytic viewpoint, these are all superficial changes. The fundamental changes relate to what data analyses can be done, the way in which data analysis is done, and the speed with which data analysis can be done. All of these are, again, results of the development of powerful computer hardware.

For example, traditional methods, which might have taken months of painstaking hand calculation, can now be tackled effectively instantaneously. One consequence of this is that one does not need to be so sure one is doing the right thing before undertaking it. This can be both good and bad: one can fit several, or even many, different models to a set of data, which is good; but one can also overfit the data (find a model which fits the data so well that it does not generalise very well), which is bad. There is also a very real danger here: that modern sophisticated statistical methods can be used without a proper understanding of the (often deep) theory underlying them. This has obvious implications for the validity of any conclusions one might draw. It seems that the development of accessible software has led not to the redundancy of statisticians (as was once feared might happen) but to an even greater need for them.

Computer power has also provided impetus for the invention of entirely new statistical methods, methods which would have been completely impracticable or even inconceivable before such machines were available. Examples of such methods are:

1. Resampling methods, such as jackknife, bootstrap, and cross-validation methods. Such tools repeatedly analyse resamples drawn from the original sample, and so get an idea of the variability which results from sampling.
2. Non-linear models. These are often analytically intractable, and need rapid optimisation methods to estimate the parameters of the models.
3. Stochastic optimisation methods permit global optima to be found, even in the presence of many local optima. Simulation allows one to explore the properties of estimators or models which cannot be solved using analytic methods.
4. Recursive partitioning (tree) algorithms are becoming popular for classification and prediction problems. These are very different from traditional methods based on distribution theory.

5. Non-parametric smoothing and curve estimation methods, such as spline and kernel approaches, are becoming increasingly important.
6. New kinds of statistical model, such as graphical models, based on a combination of graph theory and distribution theory, and neural networks, effectively a flexible non-linear model, are being developed.

These represent novel kinds of statistical tools, but even deeper and more fundamental changes are occurring. Contrary to what is perhaps a widespread misunderstanding, statistics, just like any other modern science, is in a constant state of growth and development. What might properly be termed new *schools* of statistics are developing.

For example, computer power permits practical application of Bayesian methods. These are based on an interpretation of 'probability' different from that adopted in most commercial statistical software. (A 'subjective' rather than a 'frequentist' approach is adopted: the probability that an event will occur is one's degree of belief that it will occur, rather than an objective property of the event.) Until recently, widespread practical application of these ideas was prevented by the difficulty of carrying out the high dimensional integrations involved. Recent theoretical developments coupled with advances in computer power, have made such methods practicable and commercial software is beginning to appear.

The Gifi school of non-linear multivariate analysis, based on ideas of optimal scaling, also represents a new school. The aim here is to find transformations of the raw data which optimise some criterion, and then to display the result in a way which can be grasped and interpreted by the human user. This is quite different from the conventional approach to statistics, with its emphasis on inference, and moves away from probability to focus on algorithms.

Graphical methods are becoming increasingly important. Whereas, not so long ago, producing a graph was a slow and painful process, now accurate and revealing displays can be produced with ease. This has led to a new philosophy of informal data analysis, moving away from formal inference to informal sifting and examination of data and making use of the immense power of the human eye to detect patterns.

Interactive data analysis in general is becoming central to the way much data analysis is done. We need no longer search through the mountains of line-printer output seeking the one number we want. Now those numbers will appear on the screen on command, along with graphs and pictures. And this leads us to make a distinction between two types of statistical software: statistical **packages** versus statistical **languages**.

Statistical packages consist of suites of programs, typically collected together so that they can be used from a common command language. They will be aimed more at the potential researcher from another discipline than at a professional statistician. And they may involve constructing a sequence of commands according to some template. As we shall illustrate in the following section, such systems are well suited to structured forms of analysis such as multivariate analysis of variance, where the command lines can match the structure of the problem and the data. In contrast, as we shall illustrate later, in the section on statistical languages, these are more like programming languages. They will have simple commands for basic statistical operations, and will require the user to put these together to undertake more sophisticated analyses. Since they have been built from smaller building blocks, they are more flexible than packages. They also often have an emphasis on graphics.

The distinction between the two software types is not always clear. Some packages have their own internal programming language or permit an interface to FORTRAN commands. Some languages have attracted a considerable user base, and these have built up libraries of macros for performing sophisticated and structured analyses. Moreover, as time goes on the two types seem to be growing closer together. This may be a natural evolution from two different starting places: packages having commenced life as statistical analysis systems in the days of punched cards, and statistical languages having developed from programming languages. It is also clearly stimulated by the advent of software systems such as Windows.

In the next two sections we look at packages and languages in turn, illustrating with some currently available systems.

Statistical packages

Although, as noted above, packages are moving towards interactive modes of data analysis, they have their origins in batch mode analysis, in which a complete program was written and submitted. This is in fact a convenient way of undertaking structured analyses, as we shall illustrate. We begin by looking at an example from BMDP.

BMDP is a suite of programs covering a wide range of analyses, including simple data description, analysis of variance and covariance, multiway frequency tables, log-linear models, factor analysis, stepwise discriminant analysis and regression analysis, Box–Jenkins times series analysis, life tables and survival functions, cluster analysis, non-linear regression, unbalanced repeated measures models with structured covariance matrices, correspondence analysis, and many others. Each program is called and run separately and has its own syntax, though these have been put into a uniform format.

BMDP2V, for example, performs analysis of variance and covariance with repeated measures. A BMDP2V program for analysing a repeated measures study is shown in Fig. 24.1. As with other packages of this kind, BMDP groups together commands into paragraphs comprised of sentences. The /INPUT paragraph specifies the file containing the data and describes the number of variables and the format of the data. The /VARIABLE command names the variables. The /GROUP command describes the codes used to describe the groups in the data, and gives the groups names. The /DESIGN paragraph represents the core of the analysis, specifying the dependent variables (there are five of them, being repeated measures of the response at five different times). Note that the structured nature of these commands makes it straightforward to specify that

```
/INPUT FILE = '\bmdp\eg5.dat'.
   VARIABLES = 6.
   FORMAT = FREE.

/VARIABLE NAMES = GROUP, TIME0, TIME1, TIME2, TIME4, TIME6.

/GROUP CODES(GROUP) = 1, 2.
   NAMES(GROUP) = PLACEB, LECITH.

/DESIGN GROUPING = GROUP.
   DEPENDENT = TIME0 TO TIME6.
   LEVEL = 5.   NAME = TIME.
   ORTHOGONAL.
   POINT(1) = 0, 1, 2, 4, 6.

/PRINT LINESIZE = 80.

/END
```

Figure 24.1 Example of a BMDP program.

we want to take orthogonal contrasts (mean, linear component of change over time, quadratic component of change over time) of the five measures and that the five measures are not equally spaced over time.

As will probably be clear from this small program segment, care must be exercised to ensure that the program is doing what you want it to. Slight changes to the commands can lead to large changes in the requested analysis, and hence to large changes in the output.

SPSS, like BMDP, is a suite of programs covering a wide range of analyses. They include: frequency distributions, univariate and multivariate analysis of variance, loglinear models, correlation analysis, regression analysis, factor analysis, cluster analysis, reliability analysis, and so on. A particular program is invoked by specifying its name, and, as with BMDP and other systems of this kind, optional sub-commands direct precisely what analysis will be performed. Figure 24.2 illustrates the commands to perform a basic discriminant analysis. The GROUPS sub-command specifies which groups are to be discriminated between and the VARIABLES sub-command says which variables are to be used.

The options which can be requested include control of how the variables should be entered in a stepwise analysis. So, you can choose all simultaneously, that variable which minimises the overall Wilks's lambda, that which maximises the Mahalanobis distance between the two closest groups, that which maximises the smallest F ratio between pairs of groups, that which minimises the sum of unexplained variation between groups, or that variable which gives the largest increase in Rao's V.

As will be clear, to understand these choices and their implications requires a more than passing familiarity with the technique. What seems to happen in practice is that users find (perhaps in the manual accompanying the software) an example similar to the analysis they want to do and modify that example appropriately.

Packages usually come in different versions according to the environment in which they will be run. Thus, for example, there are mainframe, PC, and Windows versions of SPSS, with very slightly different features. The last of these, naturally, makes use of menus. Not only does this mean you can avoid digitally typing-in command names, but it also means that, at least at a basic level, you can avoid learning the SPSS command language.

If you regard statistical packages as being aimed at the researcher who wants to analyse data, rather than at someone who wants to develop novel statistical methodology, it will come as no surprise to discover that they often have comprehensive data entry facilities. This can be particularly useful (and avoid errors) in applications such as clinical trials, where large quantities of highly formatted data need to be entered accurately. Models of the *Case Record Forms* can be set up electronically and data keyed in to match the layout of the forms.

Although different systems tend to be used in different domains (such as **SAS** in medical applications and SPSS in social science applications) the larger packages are so comprehensive that they can be adopted equally well in any field.

Statistical languages

The professional statistician has need for doing non-standard analyses and for developing new techniques. To meet this need several statistical computing platforms have been developed, each rooted in a statistical programming language. These languages, in turn,

```
DISCRIMINANT GROUPS = DRUG(1,2)/
   VARIABLES = RESPONSE1 RESPONSE2 RESPONSE3/
```

Figure 24.2 Using SPSS for a discriminant analysis.

were developed from general purpose programming languages such as FORTRAN, Pascal, and C. In any of these general programming languages, a matrix multiplication might be expressed in a form something like this

```
for i = 1, n {
  for j = 1, m {
    z[i,j] = 0
      for k = i, p {
        z[i,j] = z[i,j] + x[i,p] * y[p,j]
      }
  }
}
```

it clearly makes life easier to have a programming language that recognises that X and Y are matrices and that, provided they are of the correct dimensions, allows the programmer to construct a third matrix, the result of post-multiplying X by Y in this very natural fashion,

$$Z := X * Y,$$

assigning the result of the multiplication the name Z.

In addition to providing such shortcuts modern statistical programming languages operate in computing environments which provide three other important features:

1. *Interactivity*. The requested computations are done instantaneously and are available for inspection and further computation. The statistician-programmer interacts with the data, making choices as the analysis proceeds.
2. *Graphical capability*. Because the interaction between statistician and data now occurs through a reasonably high-definition computer screen, those data can be described graphically as well as numerically in the course of analysis.
3. *Local expansibility*. Users may write programs of their own, made up of commands known to the computing language. The resulting program, known as a macrocommand, user function, or external procedure, can be executed repeatedly exactly as if it were one of the program's more basic commands.

It is this last feature which gives statistical computing languages their greatest appeal. New statistical algorithms may be assembled quite quickly from existing language commands and then, after careful checking, can be made available to other users of that language.

We provide brief descriptions of four widely-used statistical computing platforms and a fuller illustration of the capabilities of a fifth. The four differ to some extent in how they make use of external procedures as well as in the range of statistical tasks that may be accomplished by built-in functions. The statistical programming languages appear to be moving closer to consumer-market statistical packages as more and more standard statistical analyses are incorporated, either as built-in commands or as standard distribution external procedures.

Perhaps the most widely-used statistical computing platform among academic statisticians is **Splus**, a commercial version of the language (**S** or **New S**) developed by and for scientists at the Bell Laboratories in the US. It is available for Windows as well as Unix computers. Splus features highly tailorable graphics, useful for presentations as well as for data exploration. Certain common built-in statistical tasks have been made object-oriented, able to choose the correct computational algorithm based on the nature of the data. External procedures are contributed to a third-party computer base from which they may be downloaded, rather than being distributed with the program itself. Three textbooks have been published recently, aimed at a growing student use of Splus (Everitt, 1994; Venables and Ripley, 1994; Spector, 1994).

GAUSS as a statistical programming language for either Unix or DOS platforms is firmly grounded in a set of highly optimised linear algebra and related matrix handling routines. As the number of built-in commands is small, relative to other statistical computing languages, GAUSS may be easier to learn. The basic capabilities are extended by collections of add-on external routines, each addressing a particular area of statistical computation such as linear regression, curve fitting, and time series models. Some of these applications, particularly in econometrics, are distributed on behalf of third party developers. As with Splus, there is an Internet discussion group for GAUSS users.

SPIDA (Statistical Package for Interactive Data Analysis) is also available for DOS and Unix computers. As distributed it includes a sizeable collection of both basic (such as singular-value decomposition, data selection, and plot) and applied (such as Poisson regression, mean contrasts, and principal components) commands. The program is supported by an extensive manual, providing an excellent introduction to SPIDA capabilities, including the programming of SPIDA macrocommands. At present, user-developed macrocommands are neither collected nor distributed though those judged most useful tend to get incorporated as built-ins to later versions of the package. A textbook with accompanying student version of SPIDA has been published (Lunn and McNeil, 1991).

Splus, GAUSS and SPIDA all operate on one or more elements of a collection of objects (e.g., vectors, matrices and character strings) which make up an active workspace. Our fourth statistical programming language, **Stata**, like a spreadsheet program concentrates its efforts on a particular data array, which may be quite large. Stata is equipped with a large number of data analysis capabilities, the bulk of which are distributed as external procedures in the form of so-called ADO files. Stata, in DOS and Unix form, is supported by a well-developed (but, extra cost) newsletter and diskette system for distributing newly developed external routines. Several textbooks have been written to facilitate the use of Stata, both in its full and student versions (such as Hamilton 1990, 1993).

We single out **SC** (Statistical Calculator) for particular attention as it has remained closest in form to a statistical programming language. SC's emphasis is on providing a large number of procedures (and functions) both built-in and external, each of which carries out a specific task. Many tools are available, then, to the user who wants to develop and then carry out his or her particular data analysis. Version 1.2 distributed in November 1994 has several hundred built-in and external procedures.

Figure 24.3 gives the text of a purpose-built external function, called mn_adp2( ). The function was written specifically to explore the sampling distribution of trimmed means computed for resamples from the following set of ten observations:

$$0, 6, 7, 8, 9, 11, 13, 15, 19, 40$$

The function takes a single argument, the name of a ten-element data vector, and returns a trimmed mean. We shall trace the flow through this function. At line 3 the local( ) command identifies objects which are local to the function; these objects will be erased from the workspace when the function returns its response. Within the function, the input data vector is the first (and only) argument and is known as 1. Thus, at line 4 tmp becomes a working version of the data vector. The elements of this working vector then are sorted smallest to largest (line 5) and the mean of the ten elements assigned to xx (line 7). The loop of instructions from line 9 through line 14 is executed as many times as there are elements of size 40 in the vector. Each time it is executed an additional smallest and largest element in the vector are trimmed before a new mean is computed to replace that in xx. The limit( ) and delimit( ) commands at lines 11 and 13 serve to

```
func mn_adp2( ) {
 # computing an adaptive trimmed mean for the Sprent example
 local(tmp,xx,i,bot,top)
 tmp=$1$
 sort(tmp)
 i=10
 xx=mean(tmp)
 while((tmp(i) == 40) && (i > 6)) {
  top= i - 1
  bot= 11 - top
  limit(tmp,bot,top)
  xx= mean(tmp)
  delimit(tmp)
  i= i - 1
 }
 return xx
}
```

Figure 24.3 An exemplar SC external function: mn_adp2( ).

trim the requisite number of small and large values and then to restore the entire vector. The additional condition in the while( ) command, that the index i not be reduced below 7, insures that at least the two middle-most observations, tmp(5) and tmp(6), contribute to a trimmed mean. When the while loop is exited, the current value of xx is returned. Thus, this SC command

TM= mn_adp2(V)

would place the appropriately trimmed mean of the vector V in TM.

The function, mn_adp2( ), cannot only be invoked on its own in this way but used as an argument for a higher order function or procedure. Figure 24.4 describes a built-in SC procedure, boot( ), which takes a function such as mn_adp2( ) as one of its arguments.

If the vector V is given the above set of ten values,

V:= 0,6,7,8,9,11,13,15,19,40

and B is declared as a vector with 1000 elements,

vector(B,1000)

then executing the boot( ) command in this way,

boot(B,mn_adp2,V)

```
p boot(B,f,V) - bootstrap SC function f on data vector V
   B <- vector of bootstrap resample statistics
    f -> SC function used to compute statistic
     V -> input data vector

! boot( ) carries out replacement sampling from the elements of V. For each such sampling a statistic
is computed by the function f and stored as a sequential element in the vector B. Sampling continues
until B is filled.
```

Figure 24.4 A description of the SC procedure boot( ).

will fill B with adaptive means computed for 1000 bootstrap resamples from the vector V. A bootstrap resample is a sample of the same size drawn with replacement from a basic sample.

That is, within boot( ) is SC language similar to this:

```
vector(tmp,sizeof($3$))
i= 0
repeat(sizeof($1$)){
 i++
 rsample($3$,tmp)
 $1$[i]= !$2$!(tmp)
}
```

The vector tmp is created to be the same size as V, boot( )'s third argument. Then, a repeat loop is executed 1000 times, the size of the vector B, boot( )'s first argument. Each time this loop is executed:

1. a replacement sample from V is placed in tmp
2. the function mn_adp2( ), whose name is given as the second boot( ) argument, is called to compute an adaptive mean for the sample presently in 'tmp', and
3. the resultant statistic is stored as the next sequential element in the vector B (i++ advances 'i' by one each execution of the repeat loop).

Subsequently, the collection of 1000 trimmed means in the vector B could be examined in a variety of ways. For example, mtsk(B) would report the skewness and kurtosis of the distribution and hist(B) would yield a histogram.

While extremely flexible, statistical programming languages like SC can be difficult to master. They are ideal, however, where new forms of analysis are to be tried. For the user who is going to perform only standard analyses one of the standard statistical packages may be a better choice.

Conclusion

The availability of high-speed computing, inexpensive enough to be available on virtually every researcher's desk, has changed data analysis. Not only can standard analyses be carried out more rapidly, but we can also investigate our data, logically, numerically, and graphically, in ways unimaginable just decades ago. And these changes will continue, at an accelerating pace, as statisticians and computer scientists work together to take even more advantage of modern computing facilities.

We have referred to only a sample of general-purpose data analysis programs – either packages of standard routines or more flexible statistical programming languages. In addition there is a growing number of spreadsheet packages for the data analyst. These include extensions of spreadsheet programs ideal for those researchers whose data are naturally represented in that form, as well as programs for the design of experiments and for the analysis of time series.

The statistical computing landscape is constantly changing and readers may want to monitor the software review columns of *Applied Statistics* or *Statistics and Computing* or the statistical computing section of *The American Statistician*. Finally, regular computing journals such as *Byte* or *Personal Computing* periodically publish valuable comparative reviews of the more mainstream statistical packages.

References

Further details about the software products described in this chapter may be obtained from:

BMDP Statistical Software, Inc
1440 Sepulveda Boulevard
Suite 316
Los Angeles, CA 90025
USA
Tel: 213 479 7799
Fax: 213 312 0161

SPSS Inc.
444 N. Michigan Avenue
Chicago, IL 60611
USA
Tel: 312 329 3500

SAS Institute, Inc
SAS Circle
Box 8000
Cary, NC 27512-8000
USA

SAS Software Ltd
Wittington House
Henley Road
Medmenham
Marlow
Buckinghamshire SL7 2EB

GAUSS
Aptech Systems, Inc
23804 Kent-Kangley Road
Maple Valley, WA 98038
USA
Tel: 206 432 7855
FAX: 206 432 7832
email: info@aptech.com

SC Mole Software
23 Cable Road
Whitehead
County Antrim BT38 9PZ
Northern Ireland
Tel/fax: 01960 378988

SPIDA
Statistics Laboratory
MacQuarie University
Sydney NSW
Australia

Stata
Computing Resources Centre
Los Angeles, CA
USA
Tel: 213 470 4341
Fax: 213 470 0235

SPlus
Statistical Sciences
Suite 500
1400 Westlake N.
Seattle, WA
USA

Splus
Statsci Europe
52 Sandfield Road
Headington
Oxford OX3 7RJ

Examples of books providing introductions to some of the programs described above are:

Everitt B (1994) *Handbook of Statistical Analyses using S-Plus*. Chapman & Hall, London.
Foster J J (1992) *Starting SPSS/PC+*. Sigma Press, Winslow.
Hamilton L (1990) *Modern Data Analysis: A First Course in Applied Statistics.* Brooks/Cole, Belmont, CA.
Hamilton L (1993) *Statistics with Stata-3*. Brooks/Cole, Belmont, CA.
Lunn A D and McNeil D R (1991) *Computer-intensive Data Analysis.* John Wiley, Chichester.
Spector P (1994) *An Introduction to S and S-Plus*. Duxbury, Belmont, CA.
Venables W and Ripley B (1994) *Modern Data Analysis with S-Plus*. Springer-Verlag, London.

Part Six

SPECIAL TOOLS

25
Mathematical models and simulation

Andrew Metcalfe

Introduction

We all use simple mathematical models in our everyday lives. For example, how much wallpaper would we need for our lounge? We probably model it as a box shape, work out the wall areas, and add them up. If there is a window bay, we may modify the model a bit.

The mathematical models needed for research programmes are more complicated but the sequence of: stating the problem; formulating a mathematical model; obtaining the solution; and interpreting the solution in the practical context, is common to all applications. We will also remember the accuracy of our solution, and then refine our model to improve subsequent predictions. Rooms do not change shape, but we often do need to model change and very many mathematical models include dynamics. It is also often necessary to make explicit allowance for random events. The calculus is about modelling dynamics, and probability theory deals with chance.

It may be some time since you took any mathematics courses, and *Engineering Mathematics Exposed* (Attenborough, 1994) would be a good starting point. My own introductory statistics book, *Statistics in Engineering* (Metcalfe, 1994), emphasises the modelling aspects of the subject. The *Guide to Mathematical Modelling* (Edwards and Hamson, 1989) would be a useful supplement to these. Burghes and Wood (1980) have written a book on mathematical modelling in the life sciences. A collection of essays, edited by Bondi (1991), gives an indication of the wide range of topics that can be usefully described in mathematical terms.

If you are an engineer, mathematician or physicist you will already have had considerable experience of mathematical modelling, and have the necessary concepts to teach yourself new techniques relatively easily. Even so, finding a relevant course, at either postgraduate or undergraduate level, should help you learn new methods more quickly. Kreyszig's book, which is now in its seventh edition (Kreyszig, 1993), is a comprehensive general reference work. Jeffrey's books, including *Mathematics for Engineers and Scientists* (4th edition) (Jeffrey, 1989), are written in a clear and accessible manner. *Classical Mechanics and Control* (Burghes and Downs, 1975) gives useful insight into mathematical modelling. The books by Clements (1989), Haberman (1977) and Murthy *et al* (1990) are examples of texts on mathematical modelling rather than specialist areas of mathematics.

A good mathematical model will be as simple as possible, while including the essential features for our application. If we want to calculate an escape velocity for a space rocket we can model it as a point mass, that is, its physical extent is ignored in the simplest calculation. If the planet has an atmosphere we should allow for air resistance. As a first attempt we could model the rocket as a cylinder, and we might then improve on this by using a cone to model the front end. A more accurate answer should be between the two extremes, but closer to that given by the improved model. Recovering our astronauts requires more detailed modelling. For instance, we need to model the heat distribution in the surface layers of the command module during re-entry. The success of a model is usually judged in terms of the accuracy of predictions made using it, but we also try to capture something, at least, of the way we imagine the world to be.

I have chosen the following examples of mathematical models to give an indication of the ranges of both techniques and areas of application. Although we perceive time and space to be continuous, and variables that vary over space or time (field variables) as discrete or continuous, we do not have to model variables in the same way. For example, the Markov chain model for dam storage treats both volume of water and time as discrete. That is: volume of water is measured in multiples of, typically, twentieths of the total capacity of the dam; and time jumps from the end of one dry season to the end of the next one.

You may find the summary of some of the main techniques of mathematics modelling shown in Tables 25.1 and 25.2, by category, helpful. It is somewhat subjective, and by no means complete. For instance, I have classified the finite difference method as a discrete space/time model, but you may prefer to think of it as an approximate solution procedure for an underlying continuous space/time model. Many differential equations can only be solved numerically, in discrete time steps. Techniques, some of which extend over two categories, are shown in bold, followed by possible research situations in round brackets, an introductory reference is given in square brackets. Examples follow in this chapter and in Chapters 26 and 27.

Chaotic dynamics

Models whose solutions include chaotic dynamics are a link between deterministic and stochastic models which include random variation. It may also be possible to imagine

Table 25.1 Deterministic models

		Space/time	
		Discrete	**Continuous**
Field variable	**Discrete**	**discrete dynamics** (population dynamics – particularly of insects, bacteria and viruses. Epidemiological applications) [Stewart, 1990] **dynamic programming** (deterministic production and inventory models)	**catastrophe theory** will model a sudden change (marketing) **Fourier series** will approximate a step change [Jeffrey, 1989]
	Continuous	**finite difference method** (groundwater pollution modelling) **dynamic programming** (control of systems) [Jacobs, 1993]	**differential equations** (vibrations of rotating machinery) **partial differential equations** (fluid dynamics) **finite elements** (stress analysis, heat conduction) **catastrophe theory** (sociology, chemistry, e.g. phase transition) [Poston and Stewart, 1978]

Table 25.2 Stochastic (random) models

		Space/time	
		Discrete	**Continuous**
Field variable	**Discrete**	**Markov chain** (Markov Chain Monte Carlo (MCMC) methods in image processing. Dam storage.)	**point processes** (modelling rainfall)
	Continuous	**time series models** (streamflow) **Image processing** (medical instruments) **Kriging** (geostatistics)	**field models** (MTB model) **spectral and wavelet analysis** (earthquake records) **Ito calculus** (stochastic differential equations) [Bhattacharya and Waymire, 1990]

chaos models generating the random variation. Examples include: turbulent motion; the motion of molecules in a gas, and roulette wheels. Ian Stewart's book (Stewart, 1990) is an enjoyable and authoritative introduction to chaotic dynamics and discusses the links with fractals. The book by Thompson and Stewart (1986) is an accessible mathematical account. Tong (1990) gives a time series, that is discrete time, account. Possible research applications vary from quantum theory to astronomy.

Whatever your speciality, it is worthwhile checking other fields. For example, marine technology and aeronautical engineering have much in common. The mathematical techniques associated with engineering control are also used in econometrics. Many research topics, in disciplines as different as civil engineering and medicine, involve the identification and analysis of non-linear systems, and there are many other such examples.

Conclusion

Mathematical models are the basis for any quantitative theory in engineering, science or social science, and I hope the following chapters will give you some indication of their scope.

The choice of model depends on the context:

- **deterministic models** have been used to model non-aqueous pollutants in groundwater (Faust *et al*, 1989)
- **stochastic simulations** have been used to model the migration of radioactive waste from burial vaults (Mackay *et al*, 1996, for example)
- **fractals** have been used to study the interaction of oil and water, in particular the phenomenon known as viscous fingering, in oil wells (Stewart, 1990).

The choice of model also depends on:

- time, which we perceive as continuous, may be more conveniently modelled as discrete
- population size, which is a discrete whole number, may be more conveniently represented as continuous.

Such approximations are likely to be negligible when compared with others made when physical processes are modelled.

References

Attenborough M (1994) *Engineering Mathematics Exposed*. McGraw-Hill, Maidenhead.

Bhattacharya R N and Waymire E C (1990) *Stochastic Processes with Applications*. John Wiley, New York.

Bondi C (1991) *New Applications of Mathematics*. Penguin, London.

Burghes D N and Downs A M (1975) *Classical Mechanics and Control*. Ellis Horwood, Chichester.

Burghes D N and Wood A D (1980) *Mathematical Models in the Social, Management and Life Sciences*. Ellis Horwood, Chichester.

Clements R R (1989) *Mathematical Modelling*. Cambridge University Press, Cambridge.

Edwards D and Hamson M (1989) *Guide to Mathematical Modelling*. Macmillan, London.

Faust C R, Guswa J H and Mercer J W (1989) Simulation of three-dimensional flow of immiscible fluids within and below the unsaturated zone. *Water Resources Research*, Vol. 25(12), pp. 2449–2464.

Haberman R (1977) *Mathematical Models: Mechanical Vibrations, Population Dynamics and Traffic Flow*. Prentice-Hall, Englewood Cliffs, NJ.

Jacobs O L R (1993) *Introduction to Control Theory* (2nd edn). Oxford University Press, Oxford.

Jeffrey A (1989) *Mathematics for Engineers and Scientists* (4th edn). Chapman & Hall, London.

Kreyszig E (1993) *Advanced Engineering Mathematics* (7th edn). John Wiley, New York.

Mackay R, Cooper T A, O'Connell P E and Metcalfe A V (1996) Contaminant migration in heterogeneous porous media: a case study. *Journal of Hydrology*, March.

Metcalfe A V (1994) *Statistics in Engineering – A Practical Approach*. Chapman & Hall, London.

Murthy D N P, Page N W and Rodin E Y (1990) *Mathematical Modelling*. Permagon Press, Oxford.

Poston T and Stewart I (1978) *Catastrophe Theory and its Applications*. Pitman, London.

Stewart I (1990) *Does God Play Dice? The New Mathematics of Chaos*. Penguin, London.

Thompson J M T and Stewart H B (1986) *Nonlinear Dynamics and Chaos*. John Wiley, Chichester.

Tong H (1990) *Non-linear Time Series: A Dynamical System Approach*. Oxford University Press, Oxford.

26
Deterministic models
Andrew Metcalfe

Vibration control of an out-of-balance rotor

Background

Although the simplest method of reducing unwanted vibration is to add dampers to the system, Victorian engineers also designed clever mechanical devices, known as vibration absorbers, which were tuned to counteract forces causing the disturbance (Den Hartog, 1956).

An interesting recent application to a wobbly footbridge is described by Jones *et al* (1981). Both dampers and vibration absorbers, which are made up from small auxiliary masses and springs, are passive devices, in so much as they do not require any auxiliary sensors, actuators, or power supplies. This is still a significant advantage, despite microprocessors and advances in sensor and actuator designs. However, there are specialist applications where the advantages of active systems outweigh the drawbacks. The possibility of applying control forces with contactless electromagnets is crucial for some of these. Another advantage is their potential to supply energy when required as well as dissipate it, whereas a passive system can only dissipate and temporarily store energy. They can also produce a force at a point in such a way that the force is dependent on signals that are received from sensors which may be far removed from that point. In contrast, passive systems produce a local force which is only related to local variables. A further advantage is that active systems can be adapted to different operating conditions, as these conditions occur, without any outside intervention.

Applications include: car suspensions; reduction of vibration of circular saws, which can result in considerable financial savings (Ellis and Mote, 1979); high speed centrifuges on magnetic bearings; and the reduction of vibrations of rotating shafts. The modelling of rotating shafts illustrates many of the general issues which arise in mathematical modelling.

Model

A diagram of the rotor, excluding the details of the bearings, is shown in Fig. 26.1. The rotor is a continuous body, but Burrows and Sahinkaya (1970) and Metcalfe and Burdess (1986, 1990) were content to concentrate on the displacements of nine points on the rotor measured at nine stations. The first step is to imagine the rotor to be made up of nine parts, with the mass of each part concentrated at its centre of gravity, and these point masses being connected by massless beam elements. The theory which describes this situation has been thoroughly worked out (see Meirovitch, 1986 for example), and the end result is a set of linear differential equations, with constant coefficients, which describe the free vibration of the undamped system. Let (x_i, y_i) represent the displacement of the ith point mass (m_i), resolved in the x and y directions. If we now define an 18×1 displacement vector q by

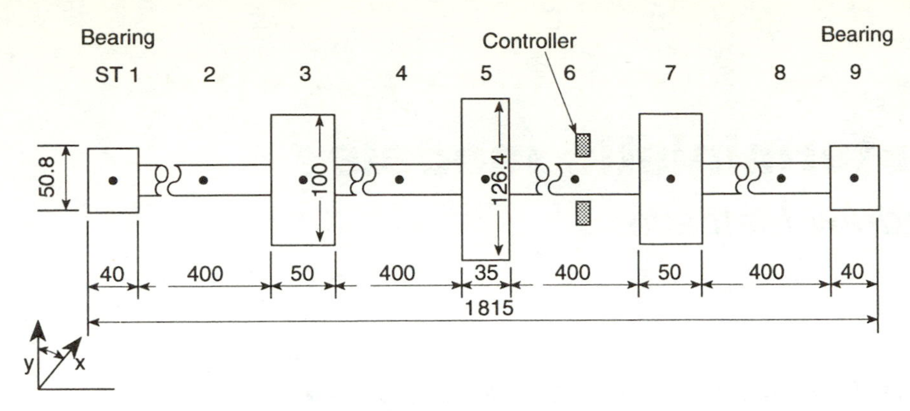

Figure 26.1 Rotor bearing model.

$$q^T = (x_1, \dots, x_9, y_1, \dots, y_9)$$

an 18×18 mass matrix M by

$$M = \begin{bmatrix} \mathrm{diag}(m_1, \dots, m_9) & 0 \\ 0 & \mathrm{diag}(m_1, \dots, m_9) \end{bmatrix}$$

and an 18×18 structural stiffness matrix,

$$K_s = \begin{bmatrix} \mathcal{K}_s & 0 \\ 0 & \mathcal{K}_s \end{bmatrix}$$

where the only non-zero elements of the 9×9 banded matrix $\mathcal{K}_s$ are of the form $k_{i,i-1}$ or $k_{i,i}$, $k_{i,i+1}$, then the equations of motion are

$$M\ddot{q} + K_s q = 0$$

At this stage we have modelled the rotor by an 18-degree-of-freedom system (nine displacements in the x direction and nine in the y direction). We will return later to the question of whether 18 degrees of freedom will suffice. The general theory of such systems is covered in any text on vibrations or linear differential equations (for example Jeffrey, 1990; Thomson, 1993; Weaver *et al*, 1990). There will be nine characteristic frequencies (ω_i) (natural frequencies) and nine corresponding mode shapes in the plane containing the x-direction and the axis of the shaft, and a similar set of nine frequencies and mode shapes for the y-direction. The mode shapes corresponding to the three lowest frequencies are shown in Fig. 26.2. A *précis* of the theory follows.

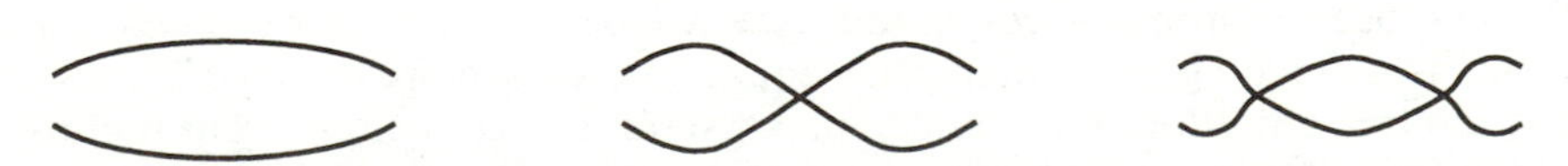

Figure 26.2 Typical first three mode shapes.

Rearrangement of the equations of motion gives:

$$\ddot{q} + (M^{-1} K_s)q = 0 \tag{26.1}$$

Now assume a simple harmonic motion so that

$$\ddot{q} = -\lambda q$$

where $\lambda = \omega^2$. Then the equation of motion becomes

$$[(M^{-1}K_s) - \lambda I]q = 0 \tag{26.2}$$

For this equation to have any interesting (non-zero) solution we require,

$$\det(M^{-1}K_s - \lambda I) = 0 \tag{26.3}$$

This is known as the characteristic equation of the system, and the solutions λ_i are called **eigenvalues.** The natural frequencies are determined by

$$\lambda_i = \omega_i^2$$

If we substitute λ_i into equation (26.2) the corresponding eigenvector q_i describes the mode shape. If the rotor is flexed in some way and then released, the resulting motion will be modelled as some linear combination of the mode shapes vibrating at their natural frequencies. According to the model, this vibration will continue forever because there is no damping. In practice, there will always be damping, caused by factors such as internal friction and air resistance, and the vibration will die down. However, such light damping would not significantly affect the natural frequencies or mode shapes.

The objective of our project was to reduce the vibration of a rotor caused by its not being perfectly balanced. This mass unbalance produces a disturbing force when it rotates, which will be harmonic at the frequency of rotation (Ω). If an undamped linear system is forced at a frequency which equals one of its natural frequencies, the amplitude of vibration will increase to infinity. This is known as resonance. Despite the light natural damping in structures, resonance is a real phenomenon which causes noise and wear, and is frequently dangerous. Operating speeds of rotors must be well away from, and preferably below, their natural (critical) frequencies. We were modelling a rotor in journal bearings, and the oil film provided some damping (C_o) and increased the stiffness (K_o). Holmes (1960) presented an elegant model for a journal bearing, and showed that these effects depend on the speed of rotation. So, our model became

$$M\ddot{q} + C_o\dot{q} + (K_s + K_o)q = Hf \tag{26.4}$$

where $f^T = (\cos \Omega t, \sin \Omega t)$, H is an 18×2 matrix determined by the assumed mass unbalance distribution, and the matrices C_o and K_o depend on Ω. The next step was to model the transducer and actuator. Both were assumed to be positioned at station six. No transducer dynamics were modelled, and we just assumed that x_6 and y_6 were measured continuously and instantaneously. The actuator output (u) was modelled as a first-order system, and feedback control was implemented. So, for the x-direction,

$$\dot{u}_x = -au_x + \theta(x_6 - r_x) \tag{26.5}$$

where r_x was the output of a second-order analogue circuit which provided the control signal,

$$\ddot{r}_x + \alpha\dot{r}_x + \beta r_x = x_6 \tag{26.6}$$

The rationale for choosing a, α, β and θ is described in Metcalfe and Burdess (1986, 1990).

State space description

Any linear system can be described by

$$\dot{\boldsymbol{x}} = A\boldsymbol{x} + B\boldsymbol{u}$$
$$\boldsymbol{y} = C\boldsymbol{x} + D\boldsymbol{u}$$

where $\boldsymbol{x}$ is a vector representing the states of the system, $\boldsymbol{y}$ is a vector representing the observations, $\boldsymbol{u}$ is a vector representing input signals or disturbances, and the matrices A, B, C and D, which can be time varying, are of the appropriate dimensions. In our application, one obvious choice for the state vector, was

$$\boldsymbol{x}^T = (x_1, \dot{x}_1, \dots, x_9, \dot{x}_9, y_1, \dot{y}_1, \dots, y_9, \dot{y}_9, r_x, \dot{r}_x, u_x, r_y, \dot{r}_y, u_y)$$

There is a detailed theory of such systems, and of their control, most of which is implemented in the computer package MATLAB, and its associated toolboxes. Useful texts include Barnett and Cameron (1985) and Borrie (1992).

Simulation

In the engineering literature, a numerical solution to a mathematical equation which purports to model a specific piece of equipment is often referred to as a (computer) simulation study. This distinguishes it from a scale model, for example. Results of simulations at frequencies from 10 up to 240 radian/s in 10 radian/s intervals, with the parameter a set at 50,000, and θ set at 10^6 are shown as a solid line in Fig. 26.3. The 'sum of squared displacements' is the sum of squares of the x and y displacements at the nine stations sampled over a two-second interval, which is long enough for the transient response to become negligible. The vertical scale is, 20 times logarithm base ten of, the ratio of the controlled to the uncontrolled response (dB). Thus negative values reflect an attenuation and any values above zero would correspond to a detrimental effect. At a speed of 80 radian/s the vibrations are reduced by a factor of nearly 100. The controller also stabilised the rotor up to at least 400 radian/s. The dotted line shows the results when a is reduced to 1000, and θ is reduced to 5×10^5.

Discussion

It was not necessary, for this application, to model explicitly the rotation of the shaft. The harmonic forces generated by the shaft's rotation could be represented by the circular functions (sine and cosine). However, one drawback of a relatively simple model for the complex dynamics of an out of balance rotor is that it does not allow for the possibility of whirling (defined as the rotation of the plane containing the bent shaft and the line of centres of the bearings). Thomson (1993) explains how this can be modelled for a single disc.

Another simplification was to think of the rotor as made up of nine point masses. This restricted the analysis to nine mode shapes and nine natural frequencies. We are assuming that the ignored modes are of no importance. Although high order modes can often be neglected safely, they can have unexpected and potentially disastrous effects.

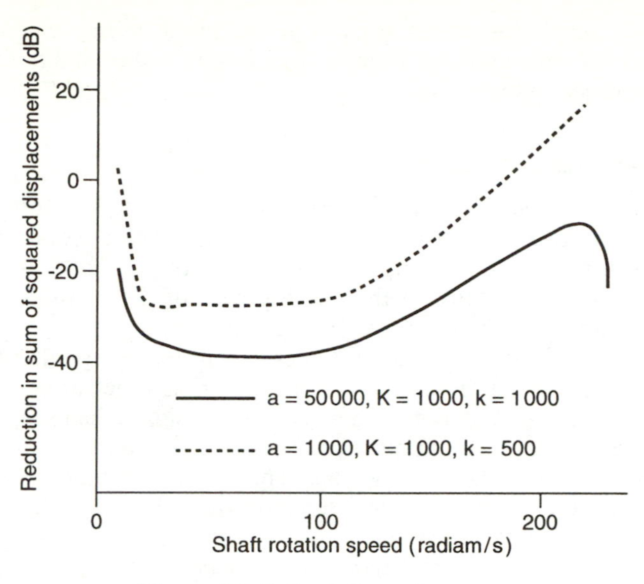

Figure 26.3 Controller performance.

Individual skin panels in an aeroplane fuselage can resonate due to acoustic excitations from the jet engines. This phenomenon is known as drumming and can lead to fatigue failure. The mathematical model for the fuselage does not include high frequency modes of individual panels. In practise, the interaction between the gobal bending of the fuselage and natural frequencies of the panels is usually small enough for the latter to be analysed separately.

There are other mechanisms by which high order modes, ignored in the analysis, can play havoc.

Feedback control systems designed for a finite mode approximation to a physical system, can be unstable in practice (Bàlàs, 1982). This happens when the controller affects lightly damped modes which were ignored in the model, referred to as control spillover. The movements of these modes are detected by the sensors, observation spillover, and this may cause the controlled system to become unstable. Another potential pitfall is that the essential features of a structure may not be adequately described by a linear model. In 1940, four months after the Tacoma Narrows Bridge was opened, it collapsed in a mild gale. It was a slender bridge and, although some benign longitudinal oscillations had been allowed for in the design, the torsional vibrations that destroyed it were quite unexpected. These are no longer held to be an example of forced resonance. Billah and Scanlan (1991), and Doole and Hogan (1994) present a convincing non-linear model which accounts for the mechanism of the collapse. I suggest either the book by Thompson and Stewart (1986) or that of Drazin (1992) if you wish to follow up the mathematics.

Unfortunately, it is easier to point out the hazards of assuming linear models and ignoring high-order modes than it is to give general advice about when not to do so. It is unwise to rely on computer simulations. Laboratory experiments with test rigs, and scale models, are often the next step. The recent publicity about flawed computer chips, and the possibility of bugs in complex computer software, should also be borne in mind.

The rotor example demonstrates the main concept of the **finite element** model, which is to divide a continuum into a finite number of bits which have simpler geometry than the original. Our bits were hypothetical point masses, joined by massless beams. In general each bit (finite element) has a number of nodes that determine the behaviour of the element. Since displacements or stresses at any point in an element depend upon those at the nodes, we can model the structure by a finite number of differential equations describing the motions of the nodes.

The finite element method

The finite element method is a computational method which is used routinely for the analysis of stress, vibration, heat conduction, fluid flow, electrostatics and acoustics problems. Practical applications of the method rely on the power of digital computers, which have provided the impetus for the development of the method over the past 40 years, since early papers by, for example, Turner *et al* (1956).

Original work in elasticity, early in the nineteenth century, led to differential equations which described the stress–strain relationship for materials which are assumed to behave in a 'linear' manner. These equations can be solved for simple bar shapes. The essence of the finite element method is that quite complex and realistic structures can be made up from such simple 'elements', see Fig. 26.4. The use of continuous elements was the next step, and applications have included those shown in Figs 26.5–26.9. The following example is taken from Appleby (1994a, b). His booklet is an ideal introduction to the method, and the associated software FINEL. This is easy to use, and is capable of solving fairly large problems of static stress analysis, and field problems such as heat conduction and potential flow for two-dimensional and axisymmetric three-dimensional situations. It can also be used to solve two- and three-dimensional frame structures. The book by Woodford *et al* (1992) is a very clear introduction to the more sophisticated PAFEC finite element software, and Zienkiewicz (1989) is a standard reference work.

A plane stress problem

A simple triangular bracket, attached to a wall, is loaded at its tip. The bracket is made

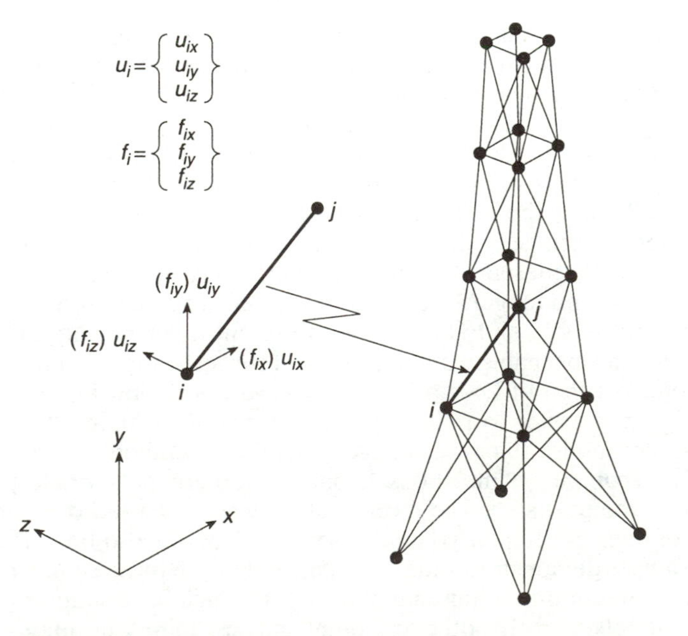

Figure 26.4 A simple bar type structure and its typical 'element'.

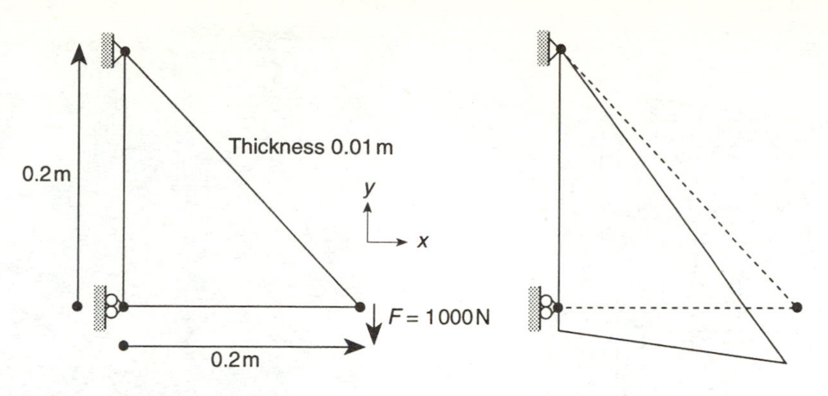

Figure 26.5 One-element model of a bracket, Appleby, 1994).

from thin-sheet mild steel, and the load is in-plane, so the equations of plane stress apply. The simplest model of this structure is a single element which deforms in a linear way, that is the displacements are linear functions of x and y, and the deformed plate remains a straight-sided triangle. With this assumption about the way the structure behaves, we know at the outset that the deformed shape must be something like that shown in Fig. 26.5.

Note that we have not so far done any calculation, but simply imposed on the problem a chosen form of solution. Normally, we would choose a form that we expect to be capable of representing the true solution to reasonable accuracy.

In this case, the bracket should really bend much more near the tip than we have allowed, and it is not surprising if our model is much stiffer than it should be. We can improve things by using four elements, each of which behaves linearly, but which together permit more complex behaviour (see Fig. 26.6).

In practice, we often use more than one mesh, of progressively increasing refinement, to help us assess the likely accuracy of the solution. If the increased resolution makes little difference, it suggests that we may have reached convergence to a practical solution. It is important to realise that convergence may be very slow, so that we are much further from the limiting values than it appears, and convergence to a seriously misleading solution is also possible! We should then use a more accurate form of model, rather than increase the resolution of a simple one. We must have independent verification, if only from experience or an approximate estimate, before we can rely on any results.

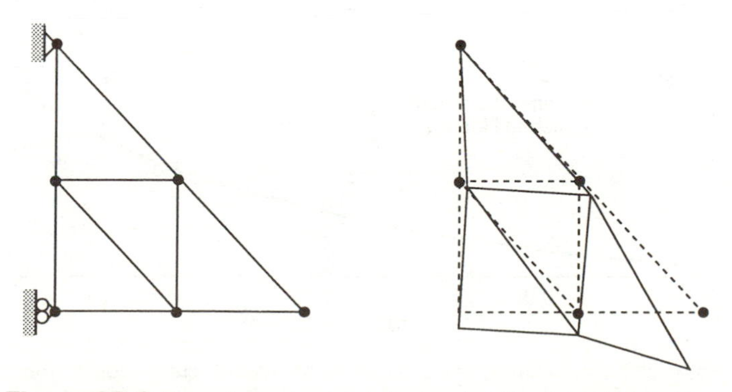

Figure 26.6 Four-element model of a bracket (Appleby, 1994a, b).

Figure 26.7 (a) Model of gas platform – beam representation (courtesy of M Kirkwood, British Gas Research Station, Loughborough, UK). (b) Model of gas platform – tube and plate representation (courtesy of M Kirkwood, British Gas Research Station, Loughborough, UK).

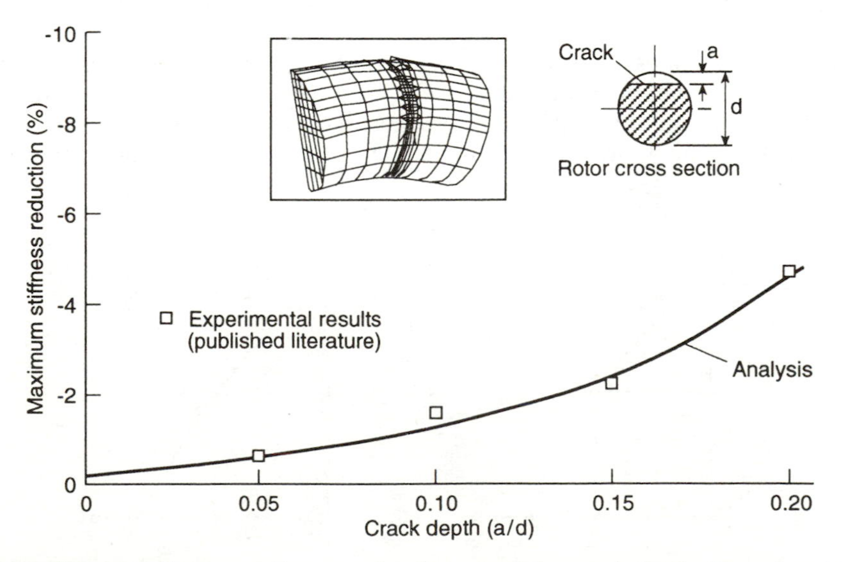

Figure 26.8 Rotor stiffness variation as a function of relative crack depth. Reproduced by kind permission of Imam *et al* (1989).

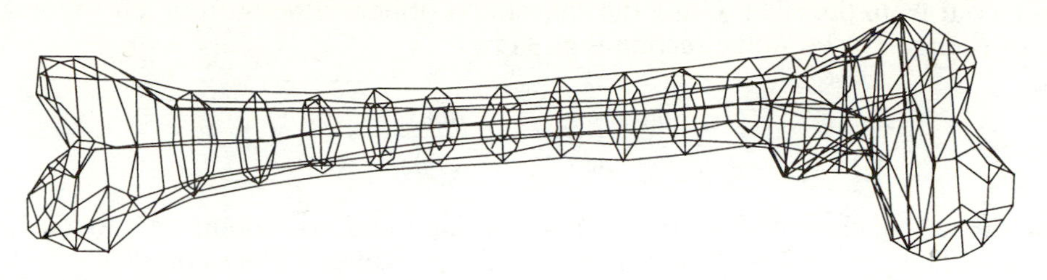

Figure 26.9 A problem of biomechanics. Plot of linear element form only – curvature of elements omitted. Note degenerate element shapes.

The finite difference method

The finite element method applies the exact equations for the idealised elements to a model of the system, made up from these elements. The **finite difference** method approximates the differential equations describing the original system, by replacing derivatives with ratios of small, rather than infinitesimally small, changes in the variables. The two methods are conceptually different, and although many problems can be solved with either method, the solutions will not, in general, be identical. Goltardi and Venutelli (1993) compare the two methods for solving the Richards equation, which describes flow in unsaturated soil. The essential idea behind the finite difference method can be seen in the following example.

One-dimensional heat conduction

Imagine a cylinder, perfectly insulated along its length so that heat can only flow along it (Fig. 26.10). Let $u(x, t)$ represent the temperature at a point, a distance x from the left-hand end, at a time t. We know from experiments that the amount of heat flowing per unit time and per unit cross-section area (Q) is proportional to the difference in temperature across the section of width (h). That is, assuming positive flow from left to right

$$Q = -K \frac{u(x + h, t) - u(x)}{h}$$

for small h, and in the limit

$$Q = -K \frac{\partial u}{\partial x}$$

where K is the coefficient of thermal conductivity. The rate of change of heat in the section is equal to the rate of heat flowing in from the left minus the amount of heat

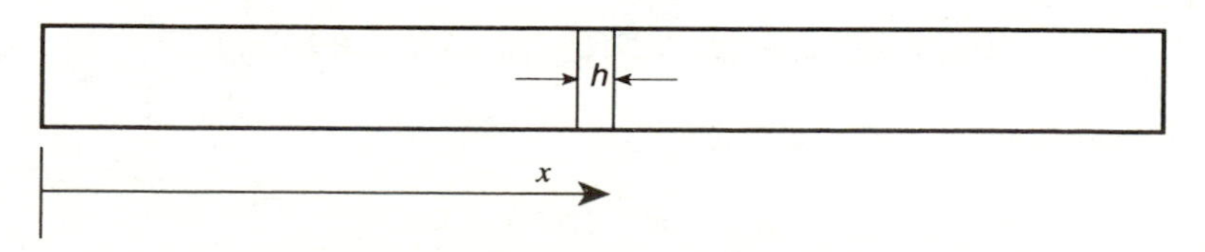

Figure 26.10 Heat flow along a cylinder.

flowing out from the right. Using the convention of heat flowing from left to right, the rate of change of heat in the section is given by

$$K\frac{\partial u(x+h,\,t)}{\partial x}-K\frac{\partial u(x,\,t)}{\partial x}$$

Now, the rate of change of temperature in the slab ($\partial u/\partial t$) is proportional to the rate of change of heat and inversely proportional to the thickness of the slab. So

$$K\left(\frac{\partial u(x+h,\,t)}{\partial x}-\frac{\partial u(x,\,t)}{\partial x}\right)\bigg/h=c\frac{\partial u(x,\,t)}{\partial t}$$

where c is the heat capacity of the material, and in the limit

$$\frac{\partial^2 u}{\partial x^2}=k\frac{\partial u}{\partial t}$$

This equation is known as the heat-flow equation in one spatial direction. A practical problem will include initial conditions,

$$u(x,0)=f_0(x)$$

and boundary conditions

$$u(x_1,t)=g_1(t)$$
$$u(x_2,t)=g_2(t)$$

where x_1 and x_2 are the ends of the cylinder.

The three-dimensional version of this equation was first derived over 200 years ago. Jean-Baptiste Fourier produced an analytic solution, for which he developed Fourier series, and published his main work, *Théorie Analytique de la Chaleur*, in 1822. But, despite the practical successes of the model, Stewart (1990) says, slightly provocatively, 'according to at least one expert, Clifford Truesdell, whatever good the classical heat equation has done for mathematics, it did nothing but harm to the physics of heat'. The simplest difference analogue to equation (26.7) is obtained by replacing $\partial^2u/\partial x^2$ by its second-difference evaluated at t_n and replacing $\partial u/\partial t$ by its forward difference analogue (Peaceman, 1977). Let U_{in} be the temperature $u(i\Delta x, n\Delta t)$.

$$\frac{U_{i+1,n}-2U_{i,n}+U_{i-1,n}}{\Delta x^2}=\frac{U_{i+1,n}-U_{i,n}}{\Delta t},\ 1\le i\le I-1$$

which we can solve explicitly for $U_{i,n+1}$:

$$U_{i,n+1}=U_{i,n}+(\Delta t/\Delta x^2)(U_{i+1,n}-2U_{i,n}+U_{i-1,n})$$

with initial condition

$$U_{i,0}=f_0(x_i)$$

and boundary conditions

$$U_{0,n} = g_1(t_n)$$
$$U_{I,n} = g_2(t_n)$$

Alternative formulations could use backward differences, for example,

$$(U_{i,n} - U_{i,n-1})/\Delta t$$

or centred differences, for example,

$$(U_{i,n+1} - U_{i,n-1})/(2\Delta t).$$

The pros and cons of different schemes are discussed in books on the subject, such as those by Davis (1986) or Smith (1985).

Groundwater pollution modelling

Modelling the impact of pollutants, such as oil spills or seepage from dumps, on groundwater is a topical area of applied research. The models are expressed in terms of partial differential equations, and these are solved for specific scenarios. The whole process can be thought of as simulating the effect of pollution incidents.

The basis of the mathematical description of multiphase fluid flow in porous media is the conservation for mass and momentum for each fluid phase.

Faust *et al* (1989) present a two-phase flow model based on a three-dimensional, finite difference formulation. They use it to investigate the flow of immiscible, denser than water, non-aqueous fluids, from two chemical waste landfills near Niagara Falls, into the groundwater. One of the research issues was the effectiveness of clay as a geological barrier. The mathematical model was a simplification of the three-phase fluid flow equations used for petroleum reservoir simulation (Peaceman, 1977). Despite the 'simplification', the equations still look quite daunting. Subscripts n and w refer to the non-aqueous phase and water phase, within the ground, respectively. The symbols used are: k for intrinsic permeability (units of length squared (L^2); k_r for dimensionless relative permeability; ρ for density (ML^{-3}); μ for dynamic viscosity ($ML^{-1}T^{-1}$); P for fluid pressure ($ML^{-1}T^{-2}$); g for gravitational acceleration (LT^{-2}); D for depth (L); q' for mass source or mass sink $ML^{-3}T^{-1}$); ϕ for dimensionless porosity; and S for dimensionless volumetric saturation. Also ∇ is the differential operator,

$$\frac{\partial}{\partial x}\mathbf{i} + \frac{\partial}{\partial y}\mathbf{j} + \frac{\partial}{\partial z}\mathbf{k}$$

where **i**, **j**, and **k** are unit vectors in the x, y and z directions, and t is time. The depth D need not be measured parallel to any of the (x, y, z) co-ordinate axes but is frequently identified with z. The equations in the article are given as:

$$\nabla \cdot \left[\frac{k\rho_w k_{rw}}{\mu_w}(\nabla P_n - \rho_w g \nabla D)\right] - \nabla \cdot \left[\frac{k\rho_w k_{rw}}{\mu_w}\nabla(P_n - P_w)\right] + q'_w = \frac{\partial(\phi\rho_w S_w)}{\partial t}$$

$$\nabla \cdot \left[\frac{k\rho_n k_{rn}}{\mu_n}(\nabla P_n - \rho_n g \nabla D)\right] + q'_n = \frac{\partial[\phi\rho_n(1 - S_w - S_a)]}{\partial t}$$

where the sum of the volumetric saturations equals one. That is

$$S_n + S_w + S_a = 1$$

To make progress, Faust *et al* assume that phase densities and viscosities are constant, which is reasonable in this context, and make use of five known additional relationships. These are of the following form, but the details of the relations are omitted here.

k_{rw} depends on S_w
k_{rn} depends on S_w and P_n
$(P_n - P_w)$ depends on S_w
S_a depends on P_n
ϕ depends on P_n

The equations can then be solved for P_n and S_w as functions of time and space. Some of the results are shown in Fig. 26.11.

New mathematical developments

Finite element and finite difference methods are ubiquitous in applied research, because so many mathematical models are expressed in terms of partial differential equations

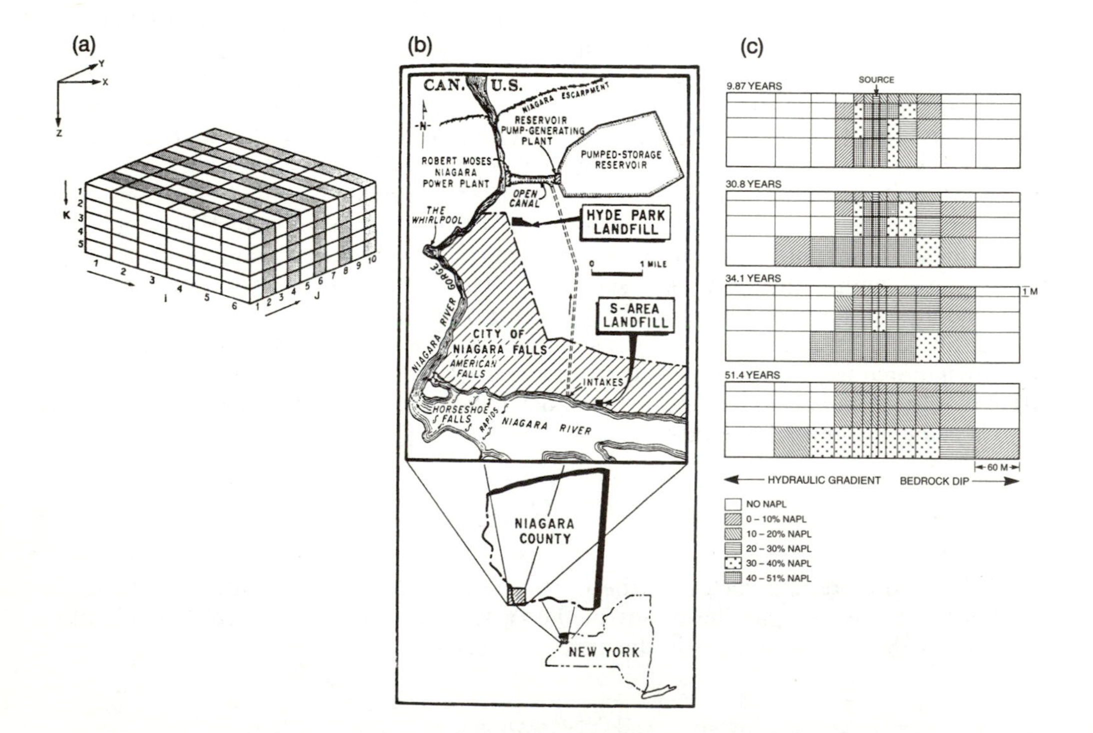

Figure 26.11 (a) Schematic representation showing typical three-dimensional, finite difference grid with alternate slices shaded for SSOR (odd–even) method. Reproduced by kind permission of Faust *et al* (1989). (b) Locations of the Hyde Park and S-Area Landfills, Niagara Falls, New York. Reproduced by kind permission of Faust *et al* (1989). (c) NAPL saturation distribution with time in vertical section through the centre of the source area and aligned with the direction of bedrock dip. Reproduced by kind permission of Faust *et al* (1989). (Faust, C R, Guswa J H and Mercer J W, *Water Resources Research*, Vol. 25, pp. 2449–2464, 1989. Copyright by the American Geophysical Union.)

which have to be solved numerically. Nevertheless, there have been exciting mathematical developments in other areas. Some outstanding examples are: catastrophe theory, fractals, chaos, and wavelets. Some of the underlying ideas may be traced back to the last century, but modern computer power has enabled dramatic progress.

René Thom's famous treatise on catastrophe theory (a part of the singularity theory), *Stabilité Structurelle et Morphogénèse*, published in 1972, was the culmination of work, by him and others, over the preceding ten years. He suggested using the topological theory of dynamical systems, originated by Henri Poincaré, to model discontinuous changes in natural phenomena, with special emphasis on biological systems (Poston and Stewart, 1978). Poston and Stewart give an accessible account of the theory, some fascinating examples, and citations of many more. The following example is loosely based on one given by Burghes and Wood (1980).

Selling T-shirts

Suppose you have started a small business selling T-shirts, with portraits of famous mathematicians and associated formulae printed on them. You wish to model student intention to buy (I). You assume intention to buy depends on personal enthusiasm E and social pressures against purchase S (Fig. 26.12). The personal enthusiasm could be improved by advertisements, which emphasise the high quality of the T-shirts and their educational value. The social pressures against such garments are left to your imagination. When these are high, the postulated 'threshold effect' (Fig. 26.13) is that an increase in enthusiasm will have little effect on intention to buy, until a certain threshold is reached when social pressures are overcome and intention to buy jumps to a new high level. The 'delay effect' is the converse when enthusiasm decreases (Fig. 26.14). The whole scenario is represented in three dimensions in Fig. 26.15. The projection of the fold onto the E–S plane is a cusp C. If the social factors against purchase are low the intention to buy increases continuously with enthusiasm (Fig. 26.12 and path 1 in Fig. 26.15). If social factors are high, as enthusiasm increases we enter the cusp region and only jump to the top leaf when we reach the right-hand side (path 2). However, as enthusiasm decreases from a high value we enter the cusp region from the right, and jump to the lower leaf when we reach the left-hand side (path 3).

A mathematical model for such a cusp catastrophe is of the following form. If we set up the coordinate system shown in Fig. 26.16, the surface M is given by

$$z^3 = xz + y$$

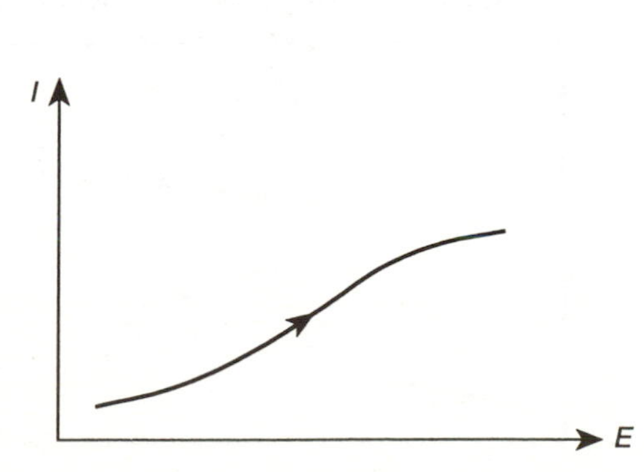

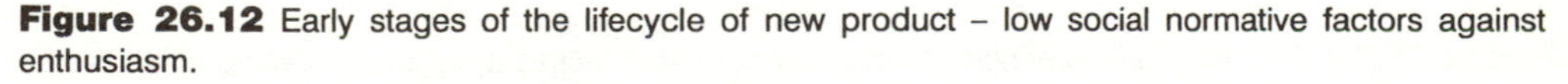

Figure 26.12 Early stages of the lifecycle of new product – low social normative factors against enthusiasm.

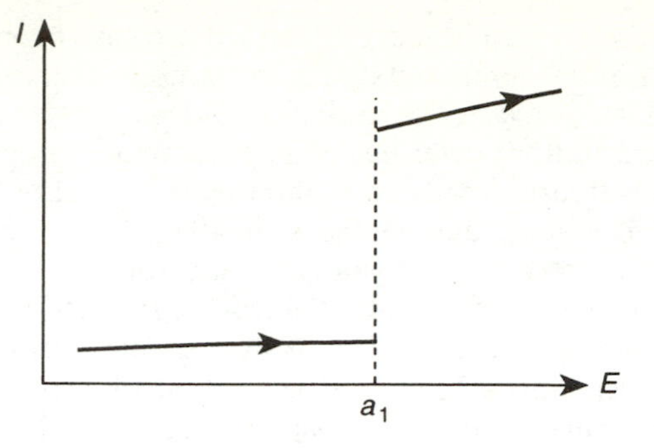

Figure 26.13 The threshold effect – high social normative factors against; increasing enthusiasm.

The projection of the fold on the x–z plane is a parabola of the form

$$z^2 = \theta x$$

for some choice of positive θ. If you now eliminate z from the two equations you obtain the equation of the projection on the x–y plane

$$(1 - \theta)^2 x^3 = \theta^3 y^2$$

The concept of structural stability, or insensitivity to small perturbations, plays an important part in catastrophe theory. Chaos is the name given to the pseudo-random behaviour of some deterministic systems, and their extreme sensitivity to small perturbations. A pseudo-random number generator is a very simple example. Stewart (1990) gives a general introduction, and Thompson and Stewart (1986) provide a mathematical treatment.

Fractal geometry, particularly associated with Benoît Mandlebrot (1982, for example), is closely allied with chaos theory. Fractals exhibit a similar structure over a wide range of scales and provide realistic models for the outlines of coastlines, the shapes of trees and snowflakes, the surfaces of viruses and many other natural phenomena. The idea is that, however much you enlarge portions of a coastline, you will still see 'bays' and 'headlands', even on individual rocks. There are many popular books on the subject, and an introductory mathematical account is given by Falconer (1990).

Dynamic programming

The ideas behind **dynamic programming** were first formalised by Bellman (1957). The method can be used for a wide variety of problems, including scheduling of work in

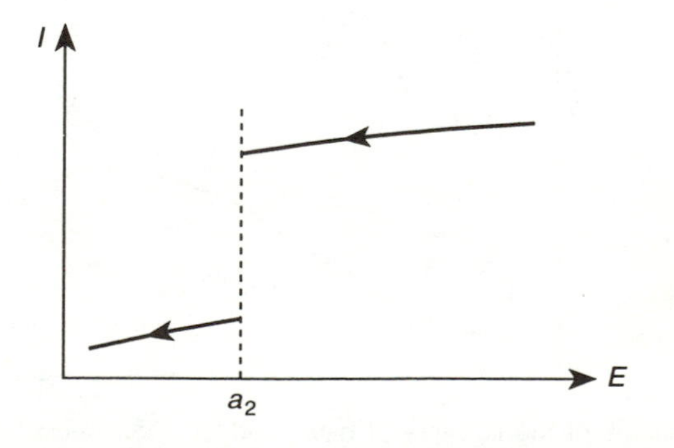

Figure 26.14 The delay phenomenon – high social normative factors against; decreasing enthusiasm.

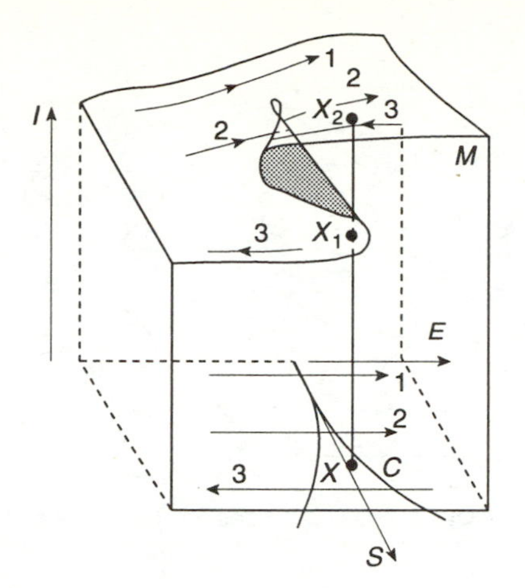

Figure 26.15 Intention to purchase depends on enthusiasm and social factors against.

factories (Smith, 1991) and optimal control (Jacobs, 1993). The technique is demonstrated in the following example (after Smith). We wish to travel from A to J in Fig. 26.17 at a minimum total cost. The cost of individual legs are shown in Fig. 26.17. The trick is to work backwards from J. The 'optimum' policy for H is the only one and costs 23 units. We shall write this as:

$$f_1^*(H)$$

where the subscript, one in this case, is the number of steps to go, the argument, H in this case, is where we are at now, and the asterisk denotes 'optimality'. In this notation

$$f_1^*(I) = 29.$$

The essential step is to notice that, for instance,

$$\begin{aligned} f_2^*(E) &= \min\{16 + f_1^*(H), 18 + f_1^*(I)\} \\ &= \min\{16 + 23, 18 + 29\} \\ &= 39 \text{ (go to } H). \end{aligned}$$

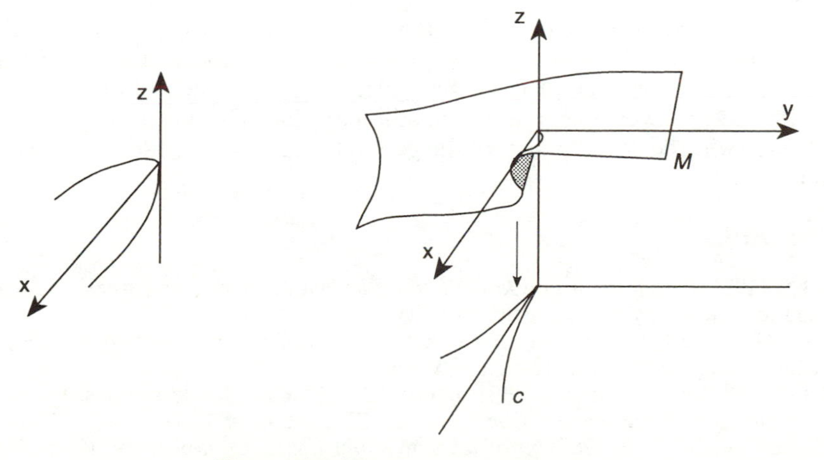

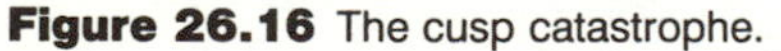

Figure 26.16 The cusp catastrophe.

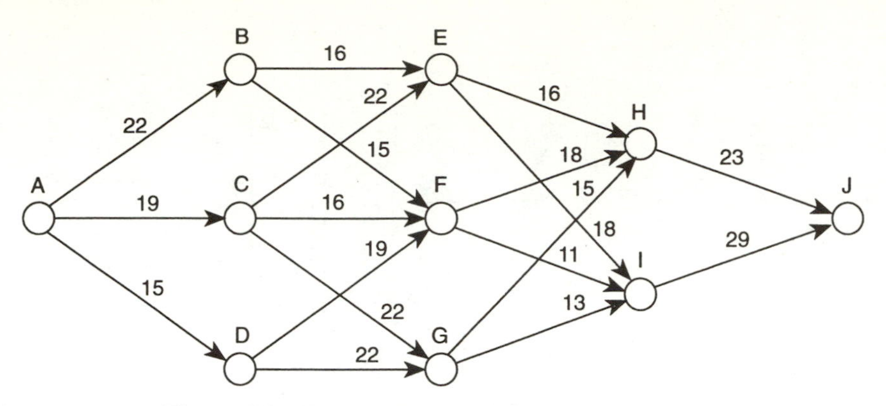

Figure 26.17 Possible routes for the business traveller.

Similar calculations give

$$
\begin{aligned}
f_2^*(F) &= 40 \text{ (go to } I) \\
f_2^*(G) &= 38 \text{ (go to } H) \\
f_3^*(B) &= 55 \text{ (either } E \text{ or } F) \\
f_3^*(C) &= 56 \text{ (go to } F) \\
f_3^*(D) &= 59 \text{ (go to } F).
\end{aligned}
$$

So finally

$$
\begin{aligned}
f_4^*(A) &= \min(22 + f_3^*(B), 19 + f_3^*(C), 15 + f_3^*(D)) \\
&= \min(22 + 55, 19 + 56, 15 + 59) \\
&= 74 \text{ (go to } D).
\end{aligned}
$$

The cheapest route is therefore $A\ D\ F\ I\ J$ with a total cost of 74 units.

The general recurrence relationship is

$$f_n^*(i) = \text{minimum } \{c(i,j) + f_{n-1}^*(j)\}$$

where $c(i,j)$ is the cost of going from i to j. It would be feasible to enumerate all the possibilities for this problem, but such a strategy would be prohibitive for more realistically sized problems. The dynamic programming algorithm is a much more efficient way of dealing with these. Nevertheless there are still many calculations to be performed, and it is easy to see why the development of the general idea was dependent on the electronic computer.

References

Appleby J C (1994a) *An Introduction to the Finite Element Method*. Department of Engineering Mathematics, University of Newcastle upon Tyne.

Appleby J C (1994b) *FINEL Software (1 MS DOS Disk) and Tutorial*. Department of Engineering Mathematics, University of Newcastle upon Tyne.

Bàlàs M J (1908) Trends in large space structure control theory: fondest hopes, wildest dreams. *IEEE Transactions on Automatical Control*, Vol. AC27, pp. 522–535.

Barnett S and Cameron R G (1985) *Introduction to Mathematical Control Theory* (2nd edn). Oxford University Press.

Bellman R E (1957) *Dynamic Programming*. Princeton University Press.

Billah K Y and Scanlan R H (1991) Resonance, Tacoma Narrows Bridge failure, and undergraduate physics textbooks. *American Journal of Physics*, Vol. 52(2), pp. 118–124.

Borrie J A (1992) *Stochastic Systems for Engineers*. Prentice-Hall, New York.

Burghes D N and Wood A D (1980) *Mathematical Models in the Social, Management and Life Sciences*. Ellis Horwood, Chichester.

Burrows C R and Sahinkaya M N (1983) Vibration control of multi-mode rotor-bearing systems. *Proceedings of The Royal Society of London*, Vol. A386, pp. 77–94.

Davis J L (1986) *Finite Difference Methods in Dynamics of Continuous Media*. Macmillan, New York.

Den Hartog J P (1956) *Mechanical Vibrations* (4th edn). McGraw-Hill, New York.

Doole S H and Hogan S J (1994) A piecewiselinear suspension bridge model: non-linear and orbit continuation. Submitted to: *Dynamics and Stability of Systems*. Preprint from http://www.fen.bris.ac.uk/engmaths/research/reports/94r4.ps.

Drazin P G (1992) *Non-linear Systems*. Cambridge University Press, Cambridge.

Ellis R W and Mote Jr C D (1979) A feedback vibration controller for circular saws. *Transactions of ASME, Journal of Dynamic Systems, Measurement and Control*, Vol. 101(1), pp. 44–48.

Falconer K (1990) *Fractal Geometry*. John Wiley, Chichester.

Faust C R, Guswa J H and Mercer J W (1989) Simulation of three-dimensional flow of immiscible fluids within and below the unsaturated zone. *Water Resources Research*, Vol. 25(12), pp. 2449–2464.

Gottardi G and Venutelli M (1993) Richards: Computer program for the numerical simulation of one-dimensional infiltration into unsaturated soil. *Computers and Geosciences*, Vol. 19(19), pp. 1239–1266.

Holmes R (1960) The vibration of a rigid shaft on short sleeve bearings. *Journal of Mechanical Engineering Science*, Vol. 2, pp. 337–341.

Imam I, Azzaro S H, Bankert R J and Scheibel J (1989) Development of an on-line rotor crack detection and monitoring systems. *ASME Journal of Vibration, Acoustics, Stress and Reliability in Design*, Vol. 111, pp. 241–250.

Jacobs O L R (1993) *Introduction to Control Theory* (2nd edn). Oxford University Press, Oxford.

Jeffrey A (1990) *Linear Algebra and Ordinary Differential Equations*. Blackwell Scientific Publications, Boston, MA.

Jones R T, Pretlove A J and Eyre R (1981) Two case studies in the use of tuned vibration absorbers on footbridges. *The Structural Engineer*, Vol. 59B, p. 27.

Mandlebrot B B (1982) *The Fractal Geometry of Nature*. Freeman, San Francisco, CA.

Meirovitch L (1986) *Elements of Vibration Analysis* (2nd edn). McGraw-Hill, New York.

Metcalfe A V and Burdess J S (1986) Active vibration control of multi-mode rotor-bearing system using an adaptive algorithm. *Transactions of ASME Journal of Vibration, Acoustics, Stress and Reliability in Design*. Vol. 108(2), pp. 230–231.

Metcalfe A V and Burdess J S (1990) Experimental evaluation of wide band active vibration controllers. *Transactions of ASME Journal of Vibration and Acoustics*, Vol. 112(4), pp. 535–541.

Peaceman D W (1977) *Fundamentals of Numerical Reservoir Simulation*. Elsevier, Amsterdam.

Poston T and Stewart I (1978) *Catastrophe Theory and its Applications*. Pitman, London.

Smith D K (1991) *Dynamic Programming – A Practical Introduction*. Ellis Horwood, New York.

Smith G D (1985) *Numerical Solution of Partial Differential Equations*. Oxford University Press.

Stewart I (1990) *Does God Play Dice? The New Mathematics of Chaos*. Penguin, London.

Thompson J M T and Stewart H B (1986) *Nonlinear Dynamics and Chaos*. John Wiley, Chichester.

Thomson W T (1993) *Theory of Vibration with Applications* (4th edn). Chapman & Hall, London.

Turner M J, Clough R W, Martin H C and Topp L T (1956) Stiffness and deflection analysis of complex structures. *Journal of Aeronautical Science*, Vol. 23, pp. 805–823.

Weaver W Jr, Timoshenko S P and Young D H (1990) *Vibration Problems in Engineering* (5th edn). John Wiley, New York.

Woodford C H, Passaris E K S and Bull J W (1992) *Engineering Analysis using PAFEC Finite Element Software*. Blackie Academic, Glasgow.

Zienkiewicz O C (1989) *Finite Element Method* (4th edn translated by Taylor R L), two volumes. McGraw-Hill.

27
Stochastic models

Andrew Metcalfe

Introduction

Stochastic models start with a deterministic model, often but not always, a simple empirical relationship, and account for deviations between the model and data by postulating **random errors**. These errors encompass:

- inherent variation in the population being modelled
- modelling error
- measurement error.

A typical research project will involve:

- thinking of reasonable models for the situation
- fitting these models to existing data and choosing one or two that seem the best, from an empirical point of view
- simulating future scenarios
- monitoring the success of predictions.

The random errors are modelled by a probability distribution. Generation of random numbers from a given probability distribution is an essential part of any (stochastic) simulation.

The book by Bhattacharya and Waymire (1990) is a good introduction to the mathematical theory of stochastic processes. The book by Bras and Rodriguez-Iturbe (1994) is also useful.

Waiting times in a doctors' surgery

Doctor White has two partners, Scarlett and Black. Patients have recently been complaining about the lengths of waiting times. White intends to model the situation, and simulate what will happen if they recruit an associate. If she simplifies the situation, it has the following features: the patient arrival pattern; the distribution of times of consultations; and the fact that her colleagues themselves seem to be ill half of the time. She now needs to collect some data, and these are summarised in Fig. 27.1.

Simulations usually rely on pseudo-random numbers. These are produced from some deterministic rule, but appear as if they are randomly distributed from some uniform distribution between zero and one. Edwards and Hamson (1989) suggest the formula

$$X_{n+1} = 97X_n + 3 \qquad (\text{mod } 1000)$$

$$R_{n+1} = X_{n+1}/1000$$

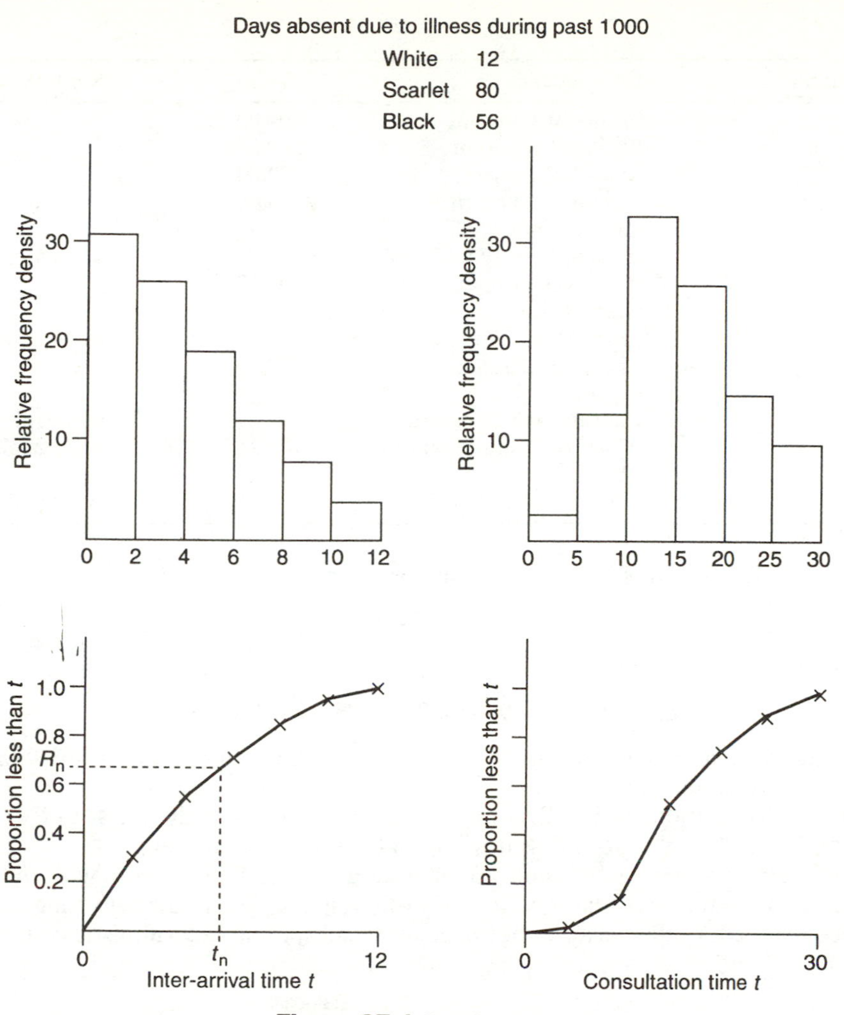

Figure 27.1 Practice data.

with an arbitrary value of X_0 start off with. Notice this must repeat after, at most, 999 numbers, but it is adequate for short simulations. For more serious work you should use more sophisticated algorithms, such as those used in reputable software or described in specialist texts. A good general reference for computer algorithms is *Numerical Recipes* by Press *et al* (1992).

We can simulate a doctor being ill by, for example, assuming Scarlett is off if R_n exceeds 0.92. A simple way to simulate consultation time is to take a time t_R such that R_n is the proporation of consultations shorter than this. Note that this will only do for short simulations, as it assumes no consultation can take longer than the longest so far recorded. It would be better to fit plausible probability distributions, and generate random deviates from these (see Cooke *et al*, 1994, for example).

To continue with the example, suppose X_0 is chosen as 538, then

$$\begin{aligned} X_1 &= 97 \times 538 + 3 \pmod{1000} \\ &= 82189 \pmod{1000} \end{aligned}$$

Table 27.1

Random number	Consequence	Time	Remark
0.189	White (W) present	09:00	
0.336	Scarlet (S) present	09:00	
0.595	Black (B) present	09:00	
0.718	Patient 1 (P1) arrives after 6 minutes, sees W	09:06	
0.649	Consult time 17 minutes		W occupied 09:23
0.956	P2 arrives 11 minutes after P1, sees S	09:17	
0.735	consult time 17 minutes		S occupied 09:44
0.298	P3 arrives 2 minutes after P2, sees B	09:19	
0.909	Consult time 25 minutes		B occupied 09:44
0.176	P4 arrives 2 minutes after P1, sees W	09:21	waits 2 minutes
0.075	Consult time 7 minutes		W occupied 09:30

and as (mod 1000) means 'take the remainder after division by 1000',

$$X_1 = 189$$

The sequence continues:

336; 595; 718; 649; 956; 735; 298; 909; 176; 075; ...

The $\{R_i\}$ sequence is obtained by dividing by 1000. A possible realisation, without an associate is shown in Table 27.1.

With the distributions in Fig. 27.1 the queue will build up. If the latest time a patient can arrive is 11:00, the expected wait will then be about half an hour.

Complicated systems can be built up on computers and used to help with the design of factories, road traffic studies and in many other applications. The following examples have been chosen, partly because they have more substantial deterministic components.

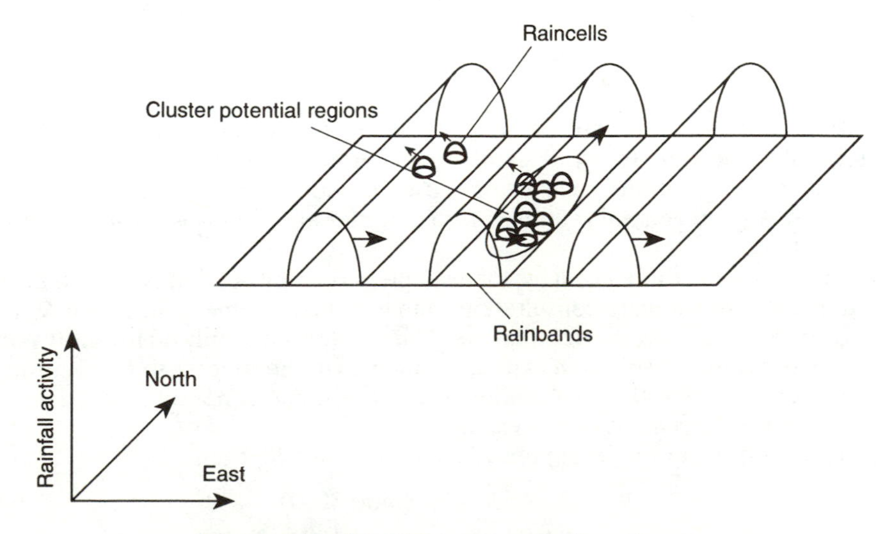

Figure 27.2(a) MTB model – definition of terms.

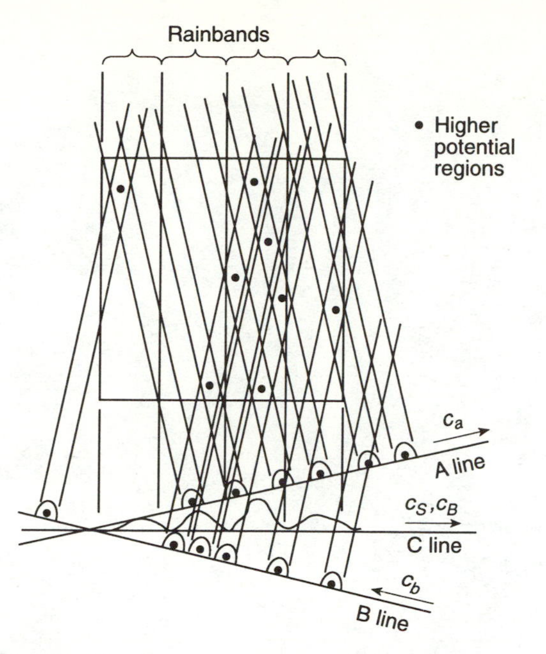

Figure 27.2(b) Generating the cluster potential regions.

Modelling rainfall

The distribution of rainfall, over time and space, is essential information for designers of water resource projects ranging from flood protection to irrigation schemes. Ideally, and provided there were no long-term climate changes, statistics could be calculated from long records over an extensive network of rain gauges. In practice, rain gauge networks are often sparse or non-existent and, even in countries with good coverage, records for

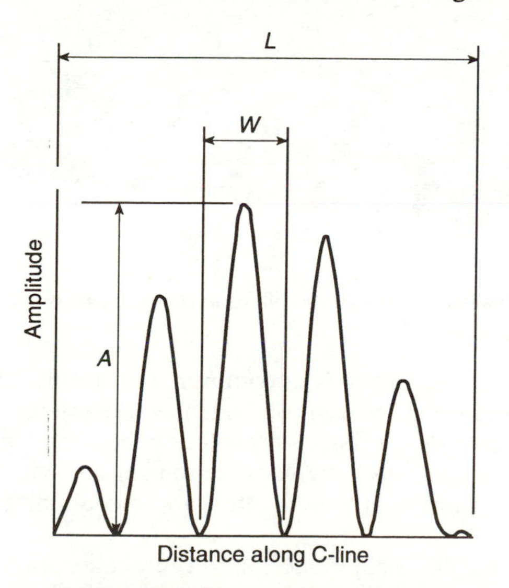

Figure 27.2(c) Modulating function.

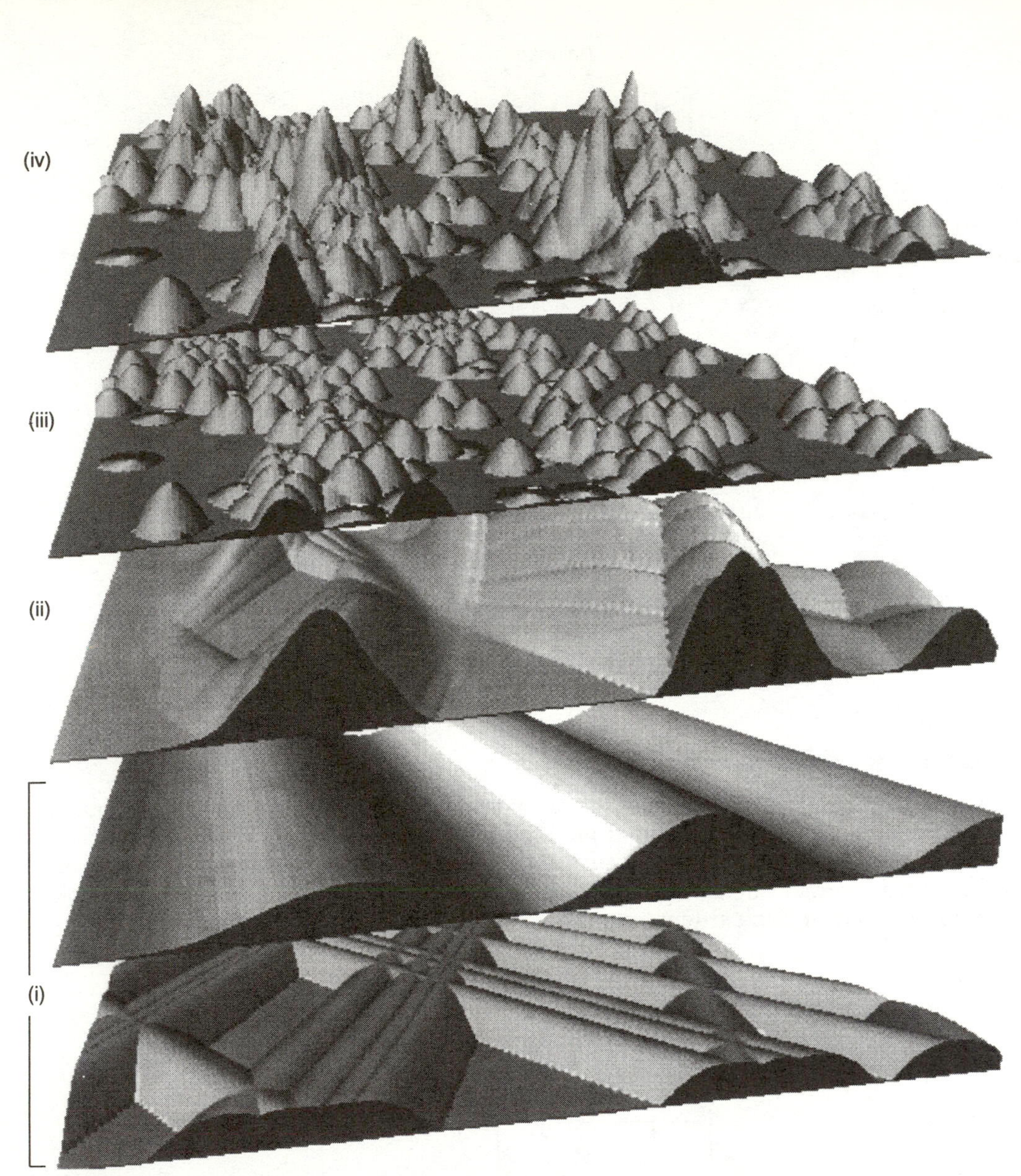

Figure 27.2(d) Realisation of MTB model (reproduced by kind permission of D Mellor).

periods exceeding 50 years are relatively uncommon. Furthermore, records usually consist of daily rainfall totals, and for some purposes, such as assessment of the hydraulic performance and pollution impact of sewers, finer resolution, down to five-minute rainfall totals, is needed. For some purposes, it may be possible to progress with rainfall at a single site. Other applications need rainfall at several sites, and more ambitious projects require a rainfall field model.

The development of rainfall field models, and their calibration from radar data, is an active research topic. The rainfall model described here will be coupled with a deterministic rainfall-run-off model of the River Brue catchment in the south west of England and will be used for flood warning, and perhaps the design of flood protection schemes.

One of the best known space–time rainfall models (WGR model) was developed by Waymire *et al* (1984). The WGR model is designed to simulate synthetic rainstorms which reproduce the statistics of real rainfall. Recent work by Dale Mellor at the University of Newcastle upon Tyne has aimed at identifying and explicitly recreating the physical aspects, such as rain bands and rain cells, within storms. This model, known as the MTB model (Mellor, 1996), has two advantages. The first is that if raincells and rainbands can be identified from radar data, real-time flood warnings should become considerably more accurate. The second is that it may lead to a better understanding of storm structures and improved weather forecasts.

The structure of the model is indicated in Fig. 27.2(a), (b). Three lines (one horizontal, the other two at prescribed angles to the horizontal) emanate from an arbitrarily chosen origin. Along each of these lines, a stochastic process is generated, and the features of this process are projected perpendicularly into the area (depicted as a square in Fig. 27.2(b)) over which a storm is to be synthesised. These projections are referred to as **bands** and relate to the original **turning bands method**, extensive modifications of which will produce realistic rainfall fields. This is the reason for the model being known as the **modified turning bands model** (MTB).

The original turning bands method, introduced by Matheron (1973) and developed further by Mantoglou and Wilson (1982), is designed to generate stationary Gaussian random fields. (In engineering the normal distribution is commonly called Gaussian, after Carl Gauss.) It uses a large number of lines with equal angles, or random angles from a uniform distribution, between them. Stochastic processes are then realised along the lines, usually using spectral methods, so that the result of projecting the generated features and summing them in the region produces a field which is approximately Gaussian, by virtue of the central limit theorem. The lines and their associated bands are collectively referred to as bands.

In the present case, on the band parallel to the bottom edge of the square (the c-band), a function (Fig. 27.2(c)) is placed which derives from the superposition of sine waves. This is used to modulate, multiplicatively, the other underlying details of the storm so that it produces, on projection into the region, a storm which decays at the edges and contains a banded structure. This function changes in time so that the storm moves in the direction of the c-band. In Fig. 27.2(b), the storm moves from west to east with speed c_S, while the rain bands move in the same direction with speed c_B.

Along each of the inclined bands, independent Poisson point processes (see Cox and Isham, 1980, for a detailed text) are realised and inverted parabolas are centred on the points, as indicated in Fig. 27.2(b). These slide along their bands with predetermined speeds c_a and c_b. These inverted parabolas are projected into the region, summed, and then multiplied by the modulating function projected from the c-band.

The field so generated is then taken to represent a time-varying spatial potential function of rain cells in the region, and is used as the rate function of a non-homogeneous Poisson process which controls the births of rain cells. That is, where the field takes on a high value there is a large probability of rain cells occurring there, and where the field is low there is a smaller chance of a rain cell occurring. In the area of interest, typically a catchment, these rain cells appear superposed on the points of the inhomogeneous three-dimensional (x, y, t) Poisson process. The rain cells themselves are described as parabolas of revolution, whose heights represent rainfall intensity at the corresponding point on the surface, with a peak that grows and decays quadratically in time. All the quadratic functions are truncated in space and time to avoid negative rainfall intensities. Once a rain cell appears it moves over the catchment with some predefined velocity, assumed the same for all rain cells, as it grows and decays in time.

A realisation of the model is given in Fig. 27.2(d) which depicts the successive stages taken in generating the rainfall field. Part (i) of 27.2(d) shows the alignments of the projected bands and modulating function, which are aggregated to give the smoothly varying field shown in part (ii). This is the inhomogeneous Poisson point process used to generate rain cells; an instantaneous snapshot of their spatial locations is shown in part (iii). The paraboloids centred on these locations are then summed to give the instantaneous rainfall field shown in part (iv).

Synthetic radar images produced with the model can be seen in Mellor's papers (1996), where they can be compared with some real images. The resulting synthetic rainfall field is seen to be a kaleidoscope of phenomena, simulating rain cells, cluster potential regions and rain bands, and the overall profile of a storm is modelled realistically. Also the features just described move about and develop in realistic ways. In particular, the rain bands enter the storm profile at the trailing edge, grow and decay as they pass through the storm, and eventually peter out at the leading edge, as observed by Hobbs and Locatelli (1978).

Short-term flood risk

The risk of flooding in the short term has an important bearing on some civil engineering decisions. For example, contractors working on a dam face, from barges or with floating cranes, will benefit from accurate estimates of flood risk. Water engineers responsible for reservoir operation, who need to balance the requirements of flood control, provision of domestic and industrial water supply, public amenity, and effluent dilution, will also benefit from up-to-date estimates of the risk of occurrence of high flows. In such cases, the risk of flooding will be influenced by prevailing catchment conditions and weather forecasts, as well as the average seasonable variation. Insurance companies might also have an interest in estimating short-term flood risks.

The most useful indicators of flood risk will be the time of year, a measure of how wet the catchment is (if it is saturated, rain will run off directly into the river rather than soak into the ground), and rain forecasts (although these are not reliable beyond one or two days in the UK, and general tendencies are incorporated in the time of year). The following model was fitted to data from the River Dearne in South Yorkshire, and then the dependence of flood risk on catchment wetness, during the summer months, was estimated by simulation.

Autoregressive models

Tsang (1991) used base flow in a river, which she approximated by daily minimum flows, as a measure of catchment wetness. An autoregressive model, so-called because past values of base flow are used to predict today's value, with rainfall as an additional explanatory variable was proposed. Let w_t represent the minimum flow in the river at Barnsley on day t, r_t represent the rainfall over the catchment on day t, and E_t represent a sequence of 'errors' which includes everything that the very simple model leaves out. The errors are to have an average value of zero, and to be independent of each other and the other terms in the model. The model was of the form

$$w_{t+1} = w_t - \Delta w + a_0 r_t + a_1 r_{t-1} + a_2 r_{t-2} + E_t \qquad (27.1)$$

The constant decrement Δw and the unknown coefficients, a_0, a_1, a_2, can be estimated from past data by using a multiple regression routine. Chatfield (1989) is a popular introduction to time series, with considerable coverage of autoregressive and similar models, which are generalised as integrated autoregressive moving average models (ARIMA). The errors are described by a normal probability distribution. Any negative values of w_t generated in the simulation are set at zero.

The treatment here is restricted to the assessment of short-term flood risk in the summer season, which is the most relevant for contractors working in rivers. The model of equation (27.1) was fitted using the regression procedure in MINITAB. The fitted model (where w is measured in cubic metres per second and r in millimetres) was:

$$w_{t+1} = w_t - 0.0202 + 0.055r_t - 0.0419r_{t-1} - 0.0059r_{t-2} \qquad (27.2)$$

and the standard deviation of the errors was estimated as 0.529. All the estimated coefficients were significantly different from zero at, or beyond, the one percent level. The negative coefficients for r_{t-1} and r_{t-2} can be accounted for by the fact that, when they are non-zero, they also influence w_t, which is itself included in the model. Addition of further terms on the right-hand side did not significantly reduce the estimated standard deviation of the errors.

Markov chain for wet and dry days

A wet day was defined as a day with rainfall exceeding 0.2 mm (anything less is likely to evaporate). Over the 21 years studied there were 1316 transitions from a wet day to a wet day (*WW*), 691 transitions from a wet day to a dry day (*WD*), 686 transitions from a dry day to a wet day (*DW*) and 1749 transitions from a dry day to a dry day (*DD*). The estimated transition matrix (***M***) was therefore

$$\boldsymbol{M} = \begin{array}{c} \\ W \\ D \end{array}\!\!\begin{array}{c} \begin{array}{cc} W & \;\;D \end{array} \\ \begin{bmatrix} 0.656 & 0.344 \\ 0.282 & 0.718 \end{bmatrix} \end{array}$$

The transition matrix is a convenient summary of the probabilities. Also if $\boldsymbol{p}^{(t)}$contains the probabilities that day t is wet, $p_W^{(t)}$, and dry, $p_D^{(t)}$,

$$\boldsymbol{p}^{(t)} = (p_W^{(t)}, p_D^{(t)})$$

then the probabilities for day $t + 1$ are given by the equation

$$\boldsymbol{p}^{(t+1)} = \boldsymbol{p}^{(t)}\boldsymbol{M}.$$

This simple model was tested by comparing frequencies for different event durations, and the results shown in Fig. 27.3(a) suggest it is adequate.

Fitting the rainfall distribution

A Weibull distribution with density

$$f(x) = \beta\alpha^{\beta}x^{\beta-1}\exp(-(\alpha x)^{\beta}) \qquad \text{for } x \geq 0$$

was found to give a good empirical fit to the amounts of rainfall (x) on wet days. Method-of-moments estimates for α and β were 0.2955 and 0.698, respectively. The fitted probability density function is compared with the histogram of daily rainfalls on wet days in Fig. 27.3(b).

The Weibull distribution provides a good fit to the marginal distribution of rainfall, but there was some evidence that the average amount of rainfall on the single wet days was less than the average daily rainfall during periods of more than one consecutive wet day. There was also some evidence that daily rainfall totals during periods of consecutive wet days were not independent. The model could be elaborated to allow for these effects, but this was not done at the time.

Fitting the conditional distribution of peak flows

An event was defined as a peak flow with associated rainfall of at least 14 mm during the preceding two days. It was assumed that smaller amounts of rainfall could not lead to flooding. Tsang fitted a Weibull distribution with α as a function of y (the total rainfall over the two days preceding the day of the event) and w (the base flow on the day before the event). Several functional forms were tried and the final relationships, fitted by maximum likelihood, were

$$\hat{\alpha} = (0.38(y - y_0) + 4.78w)^{-1} \quad (27.3a)$$

$$\hat{\beta} = 1.72 \quad (27.3b)$$

where y_0 is the rainfall threshold (14 mm) which defines an event.

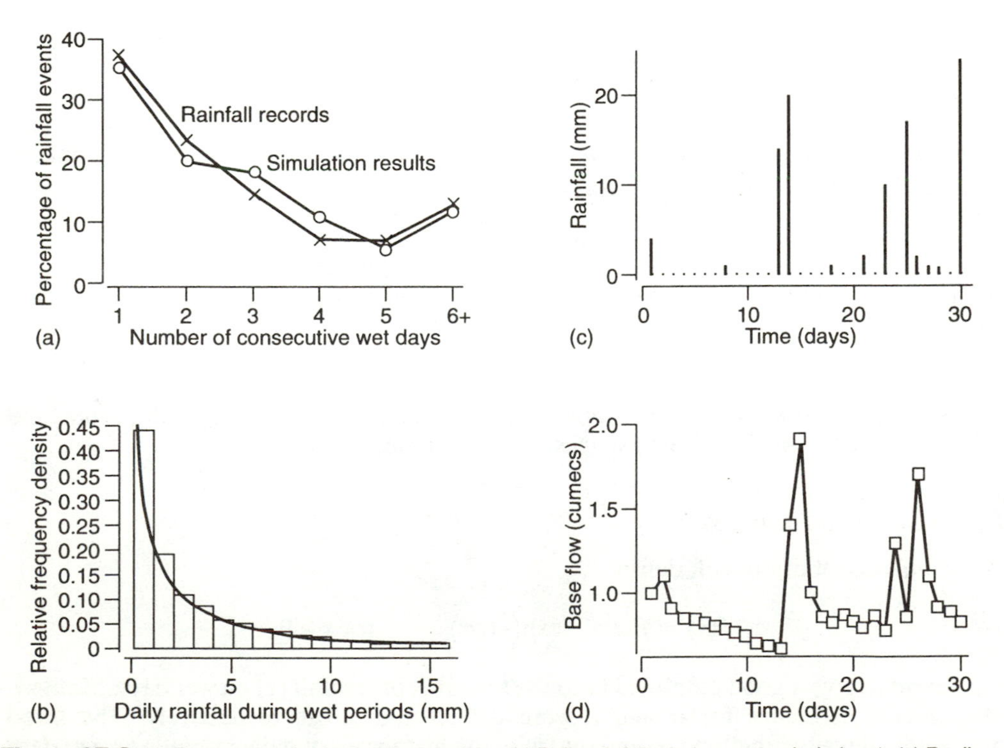

Figure 27.3 (a) Frequency of consecutive wet days. (b) Daily rainfalls during wet periods (mm). (c) Realisation of rainfall over 30 days. (d) Corresponding base flow series.

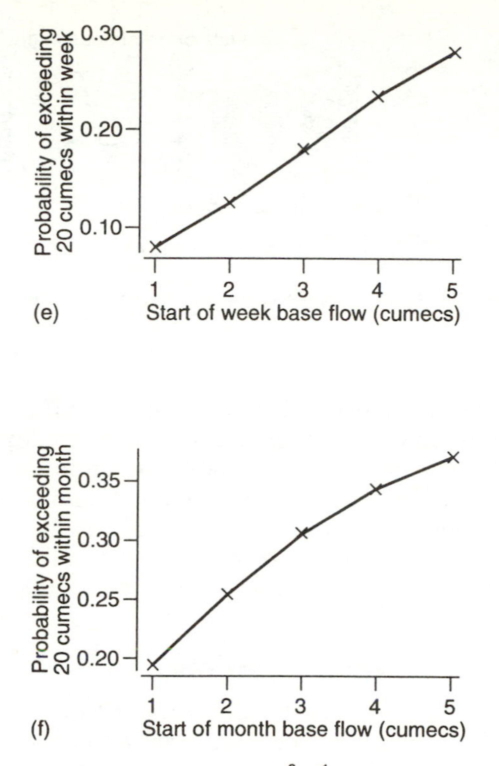

Figure 27.3 (e) Start of week base flow ($m^3 s^{-1}$). (f) Start of month base flow ($m^3 s^{-1}$).

Estimating the flood risk

The first step in estimating flood risk is to generate a sequence of wet and dry days for the required period (T), using the estimated transition probabilities. Next, a random sample of daily rainfalls is generated for the wet days. Equation (27.2) can then be used to generate a base-flow sequence. Events with two-day rainfall total exceeding the rainfall threshold of 14 mm are identified, so at this stage the number of rainfall events (m) is known. Now suppose that the probability of exceeding some critical flow (q_c) is required. For each rainfall event, the probability of not exceeding q_c can be calculated from the assumed Weibull distribution with parameters given by Equation 27.3(a, b). The risk of flooding during the period T is then estimated by:

$$\begin{aligned}&\Pr(\text{at least one exceedance of } q_c \text{ in } T \text{ days})\\&= 1 - \Pr(q_c \text{ not exceeded in first event})\\&\quad \times \ldots \times \Pr(q_c \text{ not exceeded in event } m)\end{aligned} \tag{27.4}$$

A program was written to calculate the flood risks for seven and 30 days ahead. Figure 27.3(c) shows one realisation of the rainfall inputs for 30 days ahead. The corresponding base flows, for a start-of-month base flow of 1 $m^3 s^{-1}$, are shown in Fig. 27.3(d).

Risk estimates, based on 1000 simulations, are shown in Figs 27.3(e, f). The error attributable to variability between simulations was of the order of two percent for the lowest start-of-period base flow and five percent for the highest, and this is negligible when compared with uncertainty in parameter estimates.

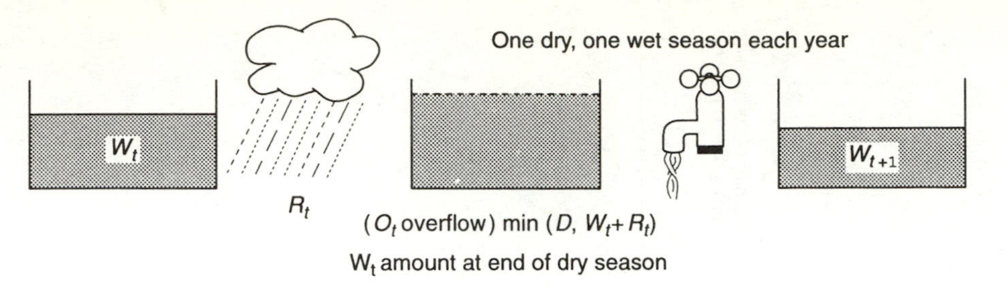

Figure 27.4 Moran's Markov chain model for storage.

Discussion

The results in Figs 27.3(e, f) show a relationship between the flood risk and start-of-period base flow, which is more pronounced for the predictions seven days ahead. The model can be explained in physical terms and is reasonably straightforward to fit. It would be possible to refine it by, for example, generating correlated random variables for rainfall sequences extending over more than one day, but this will only be worthwhile if it makes a substantial difference to risk estimates. The need for simulations to compute a flood risk is not a serious disadvantage given modern computing facilities.

Theory of dams and storage systems

Moran (1954) proposed a Markov chain model for water from dams in countries with clearly defined wet and dry seasons. A schematic is shown in Fig. 27.4. The dam has a capacity of N units of water, and the volume of water in the dam is assumed to be an integer. The transition matrix has dimensions of $N \times N$, and a choice of N of about 20 is probably adequate for most purposes. Let D be the usual demand for water during the dry season. We will assume D is fixed, but it could easily be treated as a random variable in simulations. Let W_t be the amount of water in the dam at the end of dry season t. The net inflow during the following wet season is R_t units, and inflows are assumed to be independent with an estimated probability distribution. No water is drawn from this storage dam during the wet season. At the end of the wet season the dam contains:

$$\text{minimum}(N, W_t + R_t),$$

and there may have been an overflow of

$$O_t = W_t + R_t - N.$$

During the dry season the water released is:

$$\text{minimum}(D, W_t + R_t).$$

To set up the Markov chain, working in integral multiples of a unit of water, for the probability distribution of W_t we proceed as follows:

$$R_t = 0, 1, 2, \dots \text{ with probabilities } \theta_0, \theta_1, \theta_2, \dots, \text{ respectively.}$$

$$W_t = 0, 1, 2, \dots, N - D \text{ with probabilities } p_0^{(t)}, \dots, p_{N-D}^{(t)}.$$

Suppose $N > 2D$, it follows that

$$\begin{aligned}
p_0^{(t+1)} &= p_0^{(t)}(\theta_0 + \theta_1 + \dots + \theta_D) + p_1^{(t)}(\theta_0 + \dots + \theta_{D-1}) + \dots + p_D^{(t)}\theta_0 \\
p_1^{(t+1)} &= p_0^{(t)}\theta_{D+1} + \dots \qquad\qquad\qquad\qquad + p_D^{(t)}\theta_1 + p_{D+1}^{(t)}\theta_0 \\
&\vdots \\
p_{N-D}^{(t+1)} &= p_0^{(t)}(\theta_N + \dots) + p_1^{(t)}(\theta_{N-1} + \dots) + \dots + P_{N-D}(\theta_D + \dots)
\end{aligned}$$

In matrix terms, where p is a $1 \times (N - D+1)$ array,

$$p^{(t+1)} = p^{(t)} \begin{bmatrix} (\theta_0 + \theta_1 + \dots + \theta_D) & \dots & (\theta_D + \dots) \\ (\theta_0 + \theta_1 + \dots + \theta_{D-1}) & \dots & (\theta_{D-1} + \dots) \\ \vdots & & \\ \theta_0 & & \\ 0 & & \\ \vdots & & \\ 0 & \dots & (\theta_N + \dots) \end{bmatrix}$$

It is intuitively reasonable to expect $\boldsymbol{p}^{(t)}$ to tend to some limiting distribution, for any initial condition given by $\boldsymbol{p}^{(0)}$. For example, an empty dam would be represented by

$$(1, 0, 0, \dots, 0),$$

and you could obtain the limit of $\boldsymbol{p}^{(t)}$ by iterating the last equation. But, there is a neater way which you might be able to see for yourself.

In a typical application, the θ_i and D will be estimated from past records. Simulations, using similar models for several interconnected storage dams, could be used to investigate different operational rules for water supply.

New statistical developments

The **wavelet** transform is an important development in signal analysis. Spectral analysis (see for example, Firth, 1992; Hearn and Metcalfe, 1995) gives an average frequency description of random signals, but is of limited use if statistical features of the signal change over time. Wavelets can be thought of as local Fourier analyses which track the changes. An early application was the analysis of earthquake records (Goupilland *et al*, 1984). Newland (1993) gives an introductory account and some MATLAB program listings.

Another particularly active research area is the analysis of digital images from space probes, electron microscopes, brain scanners, and so on. Noise and blur have to be removed, and there are a wide variety of statistical techniques which can be used. The images are usually considered in terms of discrete grey scale values, defined over a grid of picture elements, typically 1024×1024. The grey scale is sometimes treated as continuous. Some good introductory papers can be found in the internet directory:

http://www.stats.bris.ac.uk/pub/reports/MCMC

A collection of practical applications is now available in a book edited by Gilks (Gilks *et al*, 1995).

Kriging is a method for spatial interpolation between a few point values. It was developed in the mining industry, and this is reflected in some of the terms used, such as the 'nugget effect'. The book by Isaaks and Srivastava (1989) is a good starting point.

References

Bhattacharya R N and Waymire E C (1990) *Stochastic Processes with Applications*. John Wiley, New York.
Bras R L and Rodriguez-Iturbe I (1994) *Random Functions and Hydrology* (2nd edn). Addison-Wesley, London.
Chatfield C (1989) *The Analysis of Time Series: An Introduction* (4th edn). Chapman & Hall, London.
Cooke D, Craven A H and Clarke G M (1990) *Basic Statistical Computing* (2nd edn). Edward Arnold, London.
Cox D R and Isham V (1980) *Point Processes*. Chapman & Hall, London.
Edwards D and Hamson M (1989) *Guide to Mathematical Modelling*. Macmillan, London.
Firth J M (1992) *Discrete Transforms*. Chapman & Hall, London.
Gilks W R, Richardson and Spiegelhacter D G (1995) *Practical Markov Chain Monte Carlo*. Chapman & Hall, London.
Goupillaud P, Grossmann A and Morlet J (1984) Cycle-octave and related transforms in seismic signal analysis. *Geoexploration*, Vol. 23, pp. 85–102.
Hearn G E and Metcalfe A V (1995) *Spectral Analysis in Engineering: Concepts and Cases*. Edward Arnold, London.
Hobbs P V and Locatelli J D (1978) Rainbands, precipitation cores and generating cells in a cyclonic storm, *Journal of Atmospheric Science*, Vol. 35, pp. 230–241.
Issaaks E H and Srivastava R M (1989) *Applied Geostatistics*. Oxford University Press, New York and Oxford.
Mantoglou A and Wilson J L (1982) The turning bands method for simulation of random fields using line generation by a spectral method. *Wat. Resour. Res*, Vol. 18(5), pp. 1379–1394.
Matheron G (1973) The intrinsic random functions and their applications. *Advan. Appl. Prob*, Vol. 5, pp. 439–468.
Mellor D (1996) The Modified Turning Bands (MTB) model for space-time rainfall: I model definition and properties. *Journal of Hydrology* (in press).
Metcalfe A V (1994) *Statistics in Engineering – A Practical Approach*. Chapman & Hall, London.
Moran P A P (1954) Theory of dams and storage systems. *Australian Journal of Applied Science*, Vol. 5.
Newland D E (1993) *Random Vibrations, Spectral and Wavelet Analysis* (3rd edn). Longman, London.
Press W H, Flannery B P, Reulolsky S A and Vetterling W T (1992) *Numerical Recipes*.
Tsang W W (1991) *Estimation of Short Term Flood Risk*. MSc dissertation, University of Newcastle upon Tyne.
Waymire E, Gupta V K, and Rodriguez-Iturbe I (1984) A spectral theory of rainfall intensity at the meso-β-scale. *Water Resources Research*, Vol. 20(10), pp. 1453–1465.

Part Seven

PRESENTATION

28
Writing the thesis

Tony Greenfield

Introduction

A thesis or a dissertation is a work of scholarship which is published and made available for others to read. Even if you do not publish your work, or parts of it, as papers in academic journals, your thesis will be published by being placed in the university library, provided your examiners are satisfied with it. You may believe that the final stage of your work is the presentation of your thesis to your examiners, but that is not so. If they grant your higher degree the thesis will be available for anybody to read for all time. Through your work any reader may judge the scholarship of your department, of your university, and of the examiners. It is in the university's own interests that your work should withstand criticism. You will fail the university if in any way your work lacks quality and, if it does, the examiners will be right to refuse or defer your degree.

You must therefore ensure that in every respect your thesis or dissertation is of the highest achievable quality. There are several good texts which may help you, including other chapters in this book. Paul Levy offers some wise counsel in his entertaining chapter on presentation (Chapter 29). In this chapter I summarise a few points which I, as an examiner, see as important. These will include a suggestion, but not a rule, for overall **structure**, guidance on **style**, and advice about the **presentation of statistics**. But first I shall distinguish between a thesis and a dissertation.

A **dissertation** is a formal treatment of a subject, usually in writing. When the word is applied to undergraduate or postgraduate research it usually refers to work done, either as a review of a subject or as the application of established methods to the study of a specified problem. A dissertation is usually submitted as part of an examination for a master's degree such as an MSc, MA, or MPhil, for a postgraduate certificate and sometimes for a first degree.

A **thesis** is a dissertation resulting from original research. It is usually submitted as the only document for examination for a doctorate such as a PhD, a DPhil, or an MD. It may be submitted for a master's degree. Originality is the essential feature of a thesis and if you are submitting a thesis you must make clear which ideas were *yours*, what original work you did, and what was done by others.

In the rest of this chapter I shall refer to the thesis but most of my advice applies equally to the dissertation.

Structure

The obvious structure is: introduction; background; materials and methods; results; conclusions. This suggests that there need be no more than five chapters and indeed one academic supervisor I know insists that six should be the limit. But there is *no* limit to the number of chapters or to the number of words in your thesis unless it is imposed by your department. Brevity is better than prolixity provided you do not skimp on information.

You must interest your readers and not bore them; try to catch and keep their interest so that they will read every word of your thesis with understanding, approval and pleasure. You must convince them that you know your subject and that you have contributed new ideas and knowledge. Do not deter nor offend them with bad spelling, bad grammar, poor printing, uninformative diagrams, disorderly presentation, false or incomplete information, inadequate explanation, or weak argument.

Here in more detail, with some comments, is my guidance for structure with the headings:

- title
- summary
- keywords
- contents
- introduction
- background and choice of subject
- methods, results and analysis
- discussion
- conclusion and recommendations
- references and bibliography
- appendices
- glossary
- notation
- diary
- acknowledgements.

Title

It is not easy to devise a title for a thesis. It must be short. It must identify your work so that anybody working in your subject area will immediately recognise it. Beware of the possibility of misclassification. You may be amused to find a book about traction (for treatment of spinal injuries) in the transport section of a library but you wouldn't want that to happen to your work. Do not use abbreviations or words that may be difficult to translate into other languages without misinterpretation.

On the title page you should add:

- your full name and existing qualifications
- the name of the degree for which you are a candidate
- the names of your department, faculty and university
- the date of submission.

Summary

You must tell your readers what your work is about, why you did it, how you did it and what conclusions you reached. This is also your first opportunity to declare your originality. Can you do this briefly, in no more than 200 words, and simply with no technical terms? If you can, you will encourage your readers to read on.

Keywords

Keywords are needed for classification and for reference. Somebody sometime will be interested in your subject but will not know of your work. You want to help them to discover it through your library's on-line inquiry service. What words would their minds conjure if they wanted to find the best literature on the subject? Those must be your keywords.

Contents

In many theses that I have read the contents list has been no more than an abbreviated synopsis: a list of chapter headings. The contents list must tell me where to find the contents so it must include page numbers. This is easy with a word processor. But do not restrict this to the chapter headings. I want to be able find each of the appendices by name and by page number. It is frustrating to be told that a table of data can be found in the appendices. Even if you direct me to appendix nine, my time will be wasted and my temper frayed if I have to search.

Please, in your contents list, give the page number of every chapter, of every appendix (which should have a name), of the index, of the references, of the bibliography, of the glossary, of your diary, and of your acknowledgements. If you do that, your readers, including your examiners, will be able to find their way through your work.

Include a contents list of illustrations. For example:

Figure one **Flowchart of materials** **page 23**
and so on.

Introduction

Your introduction should be an expansion of your summary and your contents list combined. Your opening sentence should be a statement of the purpose of your research followed by the briefest possible statement of your conclusions and recommendations. Describe the path you followed to go from one to the other in terms of the contents of your thesis. Then your readers will know what to expect. Say who you are and in what department you are working; describe the resources that were available and what difficulties you met.

Review the contents of your thesis, perhaps devoting a paragraph to a summary of each subsequent chapter.

Background and choice of subject

Why did you choose the specific subject? Did you discover an interest in it because of something you had done earlier, or from your reading? Or was it suggested to you by your supervisor? What is the context of your subject: in what broader realm of knowledge does it fit? Is there a history of discovery? Who were the early workers and what did they publish? Describe the narrowing of the field, related problems that others have studied, the opportunities for further study that they have revealed and the opportunities created by scientific and technical advances. Will any human needs, other than intellectual, be met through your research?

This is where you demonstrate your ability to study the literature. Name the most important of the references you found but beware of boring your readers with a long and unnecessary account of every document that you found in your search just because it seemed slightly related. Recount any different viewpoints you may have found in the literature. State your opinion about the reliability of the evidence supporting the different viewpoint and lead into your own. You should state this in terms of questions to be answered and hypotheses to be tested.

Methods, results and analysis

In a short paper you would write a section on materials and methods followed by a section on results. In a longer study you will probably write several chapters each dealing with different aspect of the whole. It may not be easy to separate the methods from the results. Possibly each of several chapters will have a description of methods and results. In some subjects the most important part of the method will be a theoretical development. This must be rigorous and explained so clearly that any intelligent reader will understand it. There is a danger, in referring to theoretical development by another worker, that your readers will not know it. They will be frustrated if you simply give the result and a reference. That reference is unlikely to be easily at hand. They will appreciate your skill if you can explain the theory in a nutshell without re-enacting the full theoretical development.

Describe technical equipment and techniques for observation and measurement. Compare the costs, reliability and ease of alternative techniques. Did you design and administer questionnaires? Did you have constructive or investigative discussions with individuals or groups? Did you use brainstorming? Be honest about the practical work that you did and what was done by others. Describe the methods of experimental design and data analysis that you used and what computer programs you used. Examiners will not be impressed by a bland statement like '*the data were analysed by standard statistical techniques*'. In describing your experimental design you should refer to the questions and hypotheses stated in your introduction, perhaps repeating them in detail.

Results are usually best displayed in tables of data with summary statistics. If there are just a few measurements keep them in this section, but if they run into several pages put them in an appendix with clear references. Sometimes graphs will be suitable to illustrate the data but it is not always true that 'a picture is worth a thousand words'. Especially avoid routine use of bar charts and pie charts. These can be a waste of space and should be reserved for platform presentations. Simple data tables will do.

A chapter on methods and results is not the place to discuss results unless discussion of results at one stage of your research is part of your argument for developing further theory and methods for a later stage. If this is so, be sure to keep your discussion separate from the results in a section of the chapter.

Discussion

There is no need to repeat any results, but only to refer to them, interpreting and commenting on them with reference to your aims, your hypotheses, the work and opinion of others and the suitability of your theory, methods, experimental design, and analysis.

Conclusions and recommendations

Conclusions follow naturally from your discussion. Have you met your objectives? Have you found or done something new? Have you added to the store of human knowledge, even if you have only demonstrated a negative result? What did *you* contribute? Do you have any recommendations for further study? Was the course of your research smooth? Did you follow any blind alleys, any deviations from your intended study design? You should report them so that later workers don't repeat your mistakes. Similarly, if you found any negative results you should report them. It is unethical not to do so. Also be clear about what is meant by a negative result. It is rarely a demonstration that 'there is no effect'. Usually it is a statement that 'there is insufficient evidence to conclude that there is an effect', which is not the same.

References and bibliography

References and bibliography are usually lumped together but you can separate them.

In the bibliography list those books which will be useful for background reading and further study and which describe methods and theory that are widely used and accepted such as statistical methods, survey design and analysis, general mathematics, computing, communications, and history.

In the references list those books and papers to which specific reference is needed for the development of your argument within the thesis.

In both lists you should consistently follow a standard form for each reference. The most usual standard is the Harvard system.

The standard for a paper is:

authors, date, title (in quotes), journal (in italics), volume (in bold), issue, pages.

The standard for a book is:

authors, date, title (in italics), publisher (including city).

Be sure that every reference is necessary for the development of the argument and that you have truly read every one. Your examiners will not be impressed by a long list of references of which some are hardly relevant. They may even ask you about individual ones. Be sure too that every reference in your list is actually referred to in the text of your thesis and, similarly, every reference in the text is in your list. A careful examiner will check this. You will help them, and may convince them, if against every reference in your list you put, in square brackets, the pages of your thesis where the references are made.

Please don't try to impress me, if I'm your external examiner, by including references to my papers unless they are truly relevant.

Appendices

Some people collect masses of material (data forms, memoranda, correspondence, company histories, handwritten notes) and fill hundreds of pages which they describe as appendices.

Appendices are valuable only if they comprise material which is germane to your thesis and for which there is some possibility that your readers will make reference. You might describe special equipment, a detailed specification of a computer program that you used, a table of costs, a theoretical argument from other research (but any theory that you have developed should be in the main text), data collection forms, printed instructions to surveyors, maps, or a large table of raw data.

Be sure to give each appendix a name as well as an appendix number, just as you would if it were a chapter of the main text. Number the pages consecutively with the main text so that the readers can easily find any appendix when it is referenced.

Glossary

In your thesis you should highlight the first occurrence of every word and acronym that may be unfamiliar to your readers: underline it or put it in italics or bold or in a sans serif type. Define the word in your main text but then enter the word to your glossary where you should repeat the definition. If you introduce a lot of technical language your readers may quickly forget a definition and want to be reminded. Give page numbers to the glossary so that it can be found easily.

Notation

You may define notation as you like anywhere in your thesis but, as with technical words, you must keep to your definitions. The symbol π, for example, is normally used to denote the ratio of a plane circle's circumference to its diameter. It is also used occasionally to denote probability. If you use it as such, define it in the text and again in a table of notations. Whenever you use a symbol in the text that may puzzle your readers let them check its meaning in this table of notations.

Diary

How did you spend your time? Were you always at your desk or at the laboratory bench? Or were you in the library, at seminars and conferences, visiting equipment suppliers, interviewing, discussing your research with your supervisor or other researchers, or lying on the beach (and thinking)?

A diary of your work will help your examiner to understand the complexity and difficulties of your research and will help those who follow to plan their own projects.

Acknowledgements

I mention acknowledgements last, not necessarily because they should go last but because they are not part of the substantial research report. They are important and you could put them immediately after the title.

You did not work on your own. You were guided and helped by your lecturers and demonstrators, your supervisor, technicians, suppliers of equipment, surveyors and assistants, other research students with whom you discussed your work, the secretaries and administrators, and you were sustained by grant awarders and your family.

You should thank them all.

Style

At the end of this chapter is a list of some good books about scientific and technical writing. Here are a few points to ponder.

Personal pronoun, I.

There is continuing debate on whether to use the first person, singular, active 'I'. One professional institute banned its use in all reports some years ago. My daughter, as an undergraduate, assured me that she would fail if she used it in her dissertation. I was shocked but didn't interfere.

It is not simply a matter of taste. It is a question of honesty and of the credit that you deserve and need to justify the award of a higher degree. *You* are responsible for ensuring that others recognise what *you* have done, what ideas *you* have had, what theories *you* have created, what experiments *you* have run, what analyses and interpretations *you* have made, and what conclusions *you* have reached. You will not succeed in any of this if you coyly state:

- it was considered (*write*: I proposed *or* I thought)
- it was believed (*write*: I believed)
- it was concluded (*write*: I concluded)
- there is no doubt that (*write*: I am convinced)
- it is evident that (*write*: I think)

- it seems to the present writer (*write*: I think)
- the author decided (*write*: I decided).

But be careful: too much use may seem conceited and arrogant.

Spelling

Every word processor now has a spell-checker, yet many published papers and theses are littered with misspellings and misprints. One reason is that the writers are too lazy or ignorant to use the spell-checker.

But another reason is that many words, when misspelled, are other correctly spelled words. The spell-checker will not identify these. If there are any in your submission you are telling the examiners that you have not read carefully what you have written. Here is a recent example:

'Improved understanding of these matters should acid cogent presentation ... .'

The typist had misread 'aid' in poor handwriting as 'acid'.

The most common (anecdotally) of scientific misprints is the change from 'causal relationships' to 'casual relationships'. Other common ones which the spell-checker won't find are:

- fro *for* from *or* for
- lead *for* led (past tense of to lead)
- gibe *for* give
- correspondents *for* correspondence *or* corresponding ('changes in watershed conditions can result in correspondents changes in stream flow').

If you are writing for an English audience use English rather than American spelling. Words like 'color' and 'modeling' look wrong to an English eye as do words ending in -ize instead of -ise. On the other hand, if you write for an American audience use American spelling. Fortunately, word processors offer the choice of the appropriate spell-checker.

Abbreviations and points

Do *not* use points except as full stops at the ends of sentences or as decimal points.

- write BSc, PhD, *not* B.Sc., Ph.D.
- write A Brown or Albert Brown *not* A. Brown.
- write UK, UNO, WHO, ICI, *not* U.K., U.N.O., W.H.O., I.C.I.

Do ***not*** contract:

- department into dept.
- institute into inst.
- government into gov't.
- professor into prof.

Do not use:

- e.g. (write 'such as' or 'for example')
- i.e. (write 'that is')
- *et al* (except in references)
- etc (put a full stop otherwise the readers will wonder 'what are the etceteras?' and if you can't be bothered to explain them they may not want to read further).

The plurals of acronyms may be written with the initials followed by a lower case 's' without an apostrophe (ROMs).

Avoid possessives with acronyms (do not write NORWEB's) by recasting the sentence to eliminate the possessive.

Prefer single quotes to double quotes, but single and double are used when there is nesting of quotes.

Units

Use SI units with standard abbreviations without points. The base units are: m, kg, s, A, K, cd, and mol. Note that, by international agreement, including France and the US, correct spellings are *gram* and *metre*.

Do not use a dash to denote an interval.

Write: between 20 and 25°C *not* 20–25°C.
Write: from 10 to 15 September *not* 10–15 September.

Capitals

Like *The Times*, resist a tendency to a Germanic capitalisation of nouns by avoiding capitals wherever possible. Too many of them break the flow of the eye across a sentence. They also make pompous what need not be. The general rule is that proper names, titles and institutions require capitals, but descriptive appellations do not. Thus government needs no capital letter, nor does committee nor department. The same goes for jobs that are obviously descriptive, such as prime minister, foreign secretary, or even president unless it is used as a personal title ('President Washington' but 'the president'). There are a few exceptions such as Black Rod, The Queen, God.

Laws are lower case (second law of thermodynamics) unless they are named after somebody (Murphy's law).

Integers: in text, write out 0 to 10 (as zero, one, … , ten) but for greater integers use figures (21). If an integer starts a sentence write it in words.

Plural: Data and media are still, in English usage at least, plural forms of nouns. The singular forms are datum and medium.

Things to avoid

- **Ornate words and phrases** such as convey (take), pay tribute to (praise), seating accommodation (seats), utilise (use), Fred underwent an operation (Fred had an operation), we carried out an experiment (we did an experiment).
- **Needless prepositions** tacked on verbs: check up, try out, face up to.
- **Vague words** like considerable, substantial, quite, very, somewhat, relatively, situation (crisis situation), condition (weather conditions), system.
- **Clichés** (last but not least, as a matter of fact).
- **Passive voice**
  — As is shown in Figure one (Figure one shows)
  — It was decided to … .
- **Obfuscation**
  I found the following on the World Wide Web. It is attributed to Mark P Friedlander.

  Learn to obfuscate
  Children, children, if you please
  Learn to write in legalese,
  Learn to write in muddled diction,
  Use choice words of contradiction.
  Sentences must breed confusion,

Redundancy and base obtusion,
With a special concentration
On those words of obfuscation.

When you write, as well you should,
You must not be understood.
Sentences concise and clear
Will destroy a law career.
And so, my children, if you please,
Learn to write in legalese.
So that, my dears, you each can be
A fine attorney, just like me.

Statistics

Several textbooks offer guidance about the presentation of data and of data analysis. In *A Primer in Data Reduction*, Andrew Ehrenberg (1982) wrote four chapters on communicating data: Rounding; Tables; Graphs; Words. These are worth reading before you write any papers or your report.

Rarely are measurements made to more than two or three digits of precision. Yet results of analysis are often shown to many more digits. Finney (1995) gives an example: 2.39758632 ± 0.03245019 'computed with great numerical accuracy from data at best correct to the nearest 0.1%'. Such numbers are crass and meaningless but computers automatically produce them. Would you then report them, pretending scientific precision, or would you round them to an understandable level that means something?

In his discussion of tables, Ehrenberg says that:

- rows and columns should be ordered by size
- numbers are easier to read downwards than across
- table layout should make it easier to compare relevant figures
- a brief verbal summary should be given for every table.

The briefest and best (in my view) of guides about the presentation of results is reprinted as an article from the *British Medical Journal*: statistical guidelines for contributors to medical journals (Altman *et al*, 1983). This has good advice for all research workers, not just those in the medical world, and I suggest that you obtain a copy. Here are a few of its points:

- Mean values should not be quoted without some measure of variability or precision. The standard deviation (SD) should be used to show the variability among individuals and the standard error of the mean (SE or SEM) to show the precision of the sample mean. You must make clear which is presented.
- The use of the symbol ± to attach the SE or SD to the mean (as in 14.2 ± 1.9) causes confusion and should be avoided. The presentation of means as, for example, 14.2 (SE 1.9) or 14.2 (SD 7.4) is preferable.
- Confidence intervals are a good way to present means together with reasonable limits of uncertainly and are more clearly presented when the limits are given: 95% confidence interval (10.4, 18.0) than with the ± symbol.
- Spurious precision adds no value to a paper and even detracts from its readability and credibility. It is sufficient to quote values of t, X^2, and r to two decimal places.
- A statistically significant association does not itself provide direct evidence of a causal relationship between the variables concerned.

Bibliography and references

Altman D G, Gore S M, Gardner M J and Pocock S J (1983) Statistical guidelines for contributors to medical journals. *British Medical Journal*, Vol. 286, pp. 1489–1493.
Barrass R (1978) *Scientists Must Write*. Chapman & Hall, London.
Crystal D, with cartoons by Edward McLachlan (1988) *Rediscover grammar*. Longman, Essex.
Cooper B M (1975) *Writing Technical Reports*. Penguin, London.
Ehrenberg A S C (1982) *A Primer in Data Reduction*. John Wiley, London.
Finney D J (1995) Statistical science and effective scientific communication. *Journal of Applied Statistics*, Vol. 22(2), pp. 293–308.
Kirkman J (1992) *Good Style: Writing for Science*. E & FN Spon, London.
O'Connor M and Woodford F P (1978) *Writing Scientific Papers in English*. Pitman Medical, London.
Partridge E (1962) *A Dictionary of Clichés*. Routledge & Kegan Paul, London.

29
Presenting your research: reports and talks

Paul Levy

Introduction

One of the more neglected areas of the research process is perhaps the most important: the report and the presentation. Clear and effective presentation of the findings of the research is vital if the researcher wishes the research to be properly understood. It is also important if the conclusions of the research are to have impact and be remembered by the readers.

What was the best piece of writing you have ever read and why? What was the worst piece of writing? What was the best piece of writing you have ever written yourself, and why do you regard it as the best? Whether it is a novel, a newspaper article, a research report, or a poem, good writing contains a number of distinguishing features.

Features of good writing

Good writing

- is well structured
- has a particular and appropriate style
- is satisfying for the readers
- contains an internal logic
- makes use of the richness of language.

Good writing is well structured

I like a film to have a beginning, a middle and an end, but not necessarily in that order.
Jean-Luc Godard

Some of the best exponents of structure in literature are crime novels, particularly classic whodunits! A structure is just as important in a research write-up as it is in an Agatha Christie! A well-structured report will have a clear structure which allows the readers to follow the entire research process from the early stages of formulating the research question, through the research design and choice of research methods, right through to the presentation of findings, analysis and conclusions. Here's a typical structure for a research report:

Title
Acknowledgements
Contents, list of figures, glossary of terms
Aims and objectives
Background to the research (possibly, including analysis of literature)

Research methodology
Presentation of research findings
Analysis of research findings
Conclusions (and possibly recommendations).

This structure is an amalgam of the most common to be found in research reports. There are many other structures that can be used (for a more detailed discussion on structure see the previous chapter).

Let us return to the crime novel for a few moments. What is good about a good crime novel? From the first few lines the reader is hooked in through a scene-setting which is both interesting and inviting. The plot begins to unfold, and more information is revealed about both events and characters. Through the use of language and suspense the reader is kept on the edge of his or her seat right to the end. At the end, the sign of a good novel is when the reader doesn't want the story to end, cannot put the book down and is anxious to read another book by the same author. In many cases the structure is described as neat or clever, there might be a twist at the end of the story, a sting in the tail. I am not suggesting that your research write-up should be structured like a crime novel! However, you can learn from the structures of good literature. A checklist for your writing might look something like this:

- Is the structure coherent?
- Does the structure support the readers in understanding the research process and logic?
- Does the structure make reading easier?

Good writing has a particular and appropriate style

People usually have a favourite author. You can pick up a book that a friend has recommended as 'absolutely brilliant' and not be able to get beyond the first page. You wonder what excuses you can give your friend who lent you the book with such enthusiasm. It's all a matter of style. And different styles appeal to different people.

The particular style of writing depends on the personality of both the writer and the reader. As a result, the more particular your style of writing, the more danger there is of enthusing some readers but switching off others (who might be assessing the work!). If there's one maxim worth remembering here it is this:

Remember who you are writing for.

How should you style your research write-up? It very much depends on who you are writing for. A write-up in a magazine may need a very different style to a formal research report or a chapter in an academic book. The same piece of research may need to be written up differently for different audiences. The magazine article may need a more informal, conversational style. A research report will need to be more formal. A book chapter might combine formal and informal.

In general research reports should tend towards the formal. Why is this? One of the key reasons relates to the nature of research itself. The research process carries with it the ethical responsibilities of scientific investigation. The researcher is concerned with discovery and the sharing of that discovery with a chosen audience. Clarity of thought on behalf of the researcher, and clarity of interpretation and understanding on behalf of the reader is critical. It is a serious and dangerous business trying to understand the world and its underlying processes! Because of this, the danger of misinter-pretation needs to be minimised. A clear, logical style is therefore appropriate. The formal style can aid understanding, not because it is necessarily the best style to use for that purpose, but simply because you are writing in a scientific tradition and the formal style has been in place in that tradition for over two centuries. As a result, it is embedded in the assumptions of the scientific community. It also has the advantage of ensuring:

- logic
- structure
- clarity
- precision.

In general then, a formal, impersonal style tends to be used in most research reports. For example:

the research involved the use of questionnaires and interviews ...

or

the researcher used questionnaires and interviews ...

is generally favoured over:

I used questionnaires and interviews in the research ... (see Chapter 28 for a contrary view).

The autobiographical can be appropriate in some research, particularly in inductive approaches where the researcher is very much involved with reflecting on the research as he or she goes along, and where this reflection is very much part of the research process. For example:

> At this point I began to discover a number of biases in my own interview style. This was a result of being in the research situation, not only as an observer, but also as a participant .... .

This autobiographical style can help to locate the reader right inside the research situation, and to relive the thought processes that helped the researcher to make key discoveries from the data. However, you should also be aware that this can be dangerous. With both the reader *and* the researcher *in* the research, the objectivity gained from taking a more distanced view can be lost.

In more traditional research approaches, such as in a typical research project based on a questionnaire survey, a more formal approach, using the third-person, is more appropriate.

Good writing is satisfying for the readers

Should reading a research report be a satisfying experience for the readers? Of course it should! There is no reason why research shouldn't be interesting and stimulating and equally no reason why readers of that research shouldn't be interested in, and stimulated by, what they read. A satisfying read will come from a number of sources:

- an interesting research topic
- a well-structured piece of writing
- relevant use of well-presented graphics
- a well-told research story
- a perceptive set of conclusions.

An interesting research topic isn't always as easy to find as we may like. Indeed you might be given a piece of research to do that doesn't seem interesting. However, it will be very hard to write up research in an interesting way if you aren't interested in the topic yourself. If this is the case, it is important to make a real effort to get interested. This may involve reading around the subject and perhaps finding a few related videos in the local library. It is possible to reach a threshold where working hard ceases to be hard work! You, the researcher, need to get involved if you want to involve the readers of the research later on.

A well-structured piece of writing is often conducive to a satisfying read.

Science is nothing but trained and organised common sense.

T H Huxley

In a long research report it is helpful to end chapters with a clear and simple summary of that chapter and to point the readers towards what is coming up in the next chapter. Similarly chapter beginnings may contain a paragraph which 'takes stock' of what has been read so far. Other structural techniques include:

- pose questions at the beginning or end of chapters which are picked up later
- pose big questions at the beginning of the report which are addressed one or several times throughout the report
- build a schematic diagram as the report progresses (such as a flow chart of the research process which is added to as the report proceeds)
- return in the last chapter to the aims set at the beginning of the report.

You may be able to think of others. A useful exercise here is to take a look at past research reports by different authors and look at them purely in terms of structure. Ask yourself:

- What was satisfying about the structure?
- What structural techniques were employed?

Relevant use of well-presented graphics can also support a satisfying read. This will be dealt with in the next chapter. Some readers respond better to the written word. Others are better engaged through the use of graphics and pictures. A further group prefer to hear arguments through the spoken word. It makes sense, therefore, to ensure that you vary your presentation approach to meet the varying needs of your audience!

A lot of people love to hear a good story. Storytelling is a traditional activity which can be found in most cultures in the world. For most people, listening to a good story can be a satisfying experience, particularly if the story has been well told! Storytelling can be a useful mechanism for presenting the chapter on research methodology. The story of the research can be an important support to understanding the research findings and core arguments. It is also a way of reminding the readers that the research process has been carried out by human beings with all the human problems that involves! The story of your research can be told informally in the traditional storytelling style of 'once upon a time ...' with autobiographical style, possibly conversational in tone. This might be appropriate for a magazine or for stand-up verbal presentation. On the other hand, a more formal style might be used for research reports where a story structure is used but presented using formal language. For example, the methodology chapter might be structured as follows:

1. The background and context to the research.
2. Why the topic was chosen.
3. What research methods were chosen and why.
4. The research process.
5. Problems encountered and overcome.
6. The outcome of the research.

For a bit of fun, here's that list again related to a typical story:

- The background and context to the research (*describing the lie of the land, the castle and the dark wood where the wicked witch lives*)
- Why the topic was chosen (*the quest for the magic key*)
- What research methods were chosen and why (*the choice of route and magic charms for the quest*)
- The research process (*the adventure*)
- Problems encountered and overcome (*battles on the way*)
- The outcome of the research (*and they all lived happily ever after!*).

Finally, a perceptive set of conclusions will round off the write-up. A sense of roundedness, of circularity, of coming back to the start, is a feature of a lot of good writing and can support the sense of a good read. Many research reports fail to draw sensible conclusions from the research data. For such research to be useful to the world, to avoid being placed in one of the many storage jars of the mind to be found on the dusty shelves of libraries, research should:

- draw practical conclusions from the research findings
- suggest areas for further research which lies beyond the scope of the current research project.

Good writing contains an internal logic

Poor research reports make it difficult for the readers to follow the logical thread of the research process. The write-up should enable the readers to follow the stages of the research from beginning to end. It is important to know from the start: are you

- testing a hypothesis (or more than one)?
- addressing one or more questions?
- exploring a broad or specific issue?
- measuring some variables?

or some combination of these.

The readers needs to be clear about:

- why you chose your research topic (the background)
- the environment in which the topic is important and/or interesting (the context)
- what type of research you are carrying out (the approach)
- what the key stages are (the methods and processes)
- how the results are presented and analysed) (findings and analysis)
- how your conclusions are drawn (conclusions drawing the threads together)
- how you have reached your recommendations (if you have).

The internal logic of the writing is concerned with creating a logical and coherent path for the readers through all of these stages.

It is always worth showing drafts of the write-up to a colleague or friend to get feedback on the internal logic of the report. What may seem obvious to you may not be to your readers! Here's a list for checking the internal logic:

- Do the conclusions at the end follow on from the aims set out at the beginning?
- Does the research process follow a logical path?
- Are any conclusions at the end not supported by the research data?
- Do the recommendations arise logically from the conclusions and, if not, what is the justification?
- Are there any contradictory arguments expressed anywhere in the writing?

Good writing makes use of the richness of language

A good novel, an inspiring poem, an exciting short story, all make use of the richness of language. A reader will not be impressed by:

- repetition of words ('it was a interesting party', 'the findings are interesting')
- use of clichés ('at the end of the day' ... , 'pros and cons ... ')
- over formal use of language ('the process was undertaken on a formalistic basis which nevertheless facilitated ... ').

Good writing in research makes use of plain English which makes full (and proper) use of language. The key is to maintain accuracy without losing the reader's interest. A number of resources are available to writers to support this process:

- A thesaurus and grammar checker (either on computer or manual).
- Wide reading: the best training for a writer is the writing of others; read other research papers, books, novels, poetry.
- A computer spell-checker to find repeated words. Combine this with a thesaurus to ensure variety.

Be aware of the two extremes of writing in research work – *the artist* and *the scientist*: (I am, of course being simplistic and unfair to the great variety of artists and scientists in the world!):

The artist is concerned with:
use of expression, colour, image, representation, interpretation, subjectivity, experience of natural processes.
The scientist is concerned with:
use of logic, objectivity, rationality, accuracy, measurement and control of natural processes.

Both extremes have something to offer the writer of research. The scientist dominates most research writing. However, you need only read the writings of Darwin, Einstein, Goethe, Gleick and Hawking to see the importance of the artist in such writing as well.

Present your results

One of the most crucial parts of the research writing process is the presentation of results and analysis.

You should present research results

- in clear language
- with helpful commentary
- with supporting use of graphics
- with proper references between the text and any graphics
- with appropriate reference to and use of appendices
- with guidelines to the readers on how the results were obtained and why they have been presented in the way they have
- with any relevant criticism of the results.

Present research results in clear language

It is all too often the case that the results section of a research report contains a long list of statistics hot off the computer! The key term is *reader friendly*. You should present results in as helpful a way as possible:

- pitch the presentation at the right level. It is always better to be on the safe side and assume the reader isn't as conversant with the subject as you are
- use simple, easy to understand terminology
- use helpful and simple graphics of summary data supported by more detailed data in appendices
- use formats such as 'bullet point' to list key findings, particularly in summary form at the end of a section or chapter.

Present research results with helpful commentary

Commentary sentences and paragraphs can guide the readers through the chapter, if the results are particularly complex. For example:

The chapter is divided into three sections. The first section will ...

or

Table 5 differs from Table 4 in that it includes the figure for 1995 ... .

A bullet-point summary of the main findings at the end of the chapter or section may help readers.

Present research results with supporting use of graphics

A picture really can paint a thousand words! In research writing the use of graphics will help the reader and will also keep the word-count down. Some ways to use fewer words are explained in the next chapter. Graphs, tables, diagrams or other types of graphics should support the argument which appears in the text. Remember:

- Never use a graphic which does not support something written in words.
- Never write long sentences when a short sentence supported by a simple graphic will do the trick.

Present research results with proper references between the text and any graphics

Always refer clearly to any graphics that have been employed. This need not be done with verbosity. It needs only a...

'(see Table 1) ...

or

'Diagram 1 shows an example of ... .

If you do not refer clearly to graphics (and it's very easy to forget), readers will easily become lost, may lose the thread of the argument, and, eventually, will become frustrated in their attempts to understand what the research is all about.

Present research results with appropriate reference to and use of appendices

Many research projects have appendices which are larger than the report itself. Appendices should be used sparingly and should not be a repository for the entire contents of the researcher's filing cabinet. Appendices are the support of a research report. They enable the readers to obtain further detail about:

- the research context (there may, for example, be a relevant extract from a report, some

laboratory or interview notes, a map of a factory floor, a company report, data about a specific product or process)
- the research methods (a full version of a questionnaire or interview schedule)
- the research findings (a full set of graphs showing research findings)
- other supporting information (such as a glossary of terms, list of references).

To be useful, such appendices need to be page numbered, included in the contents page, and properly referenced in the main text. For example:

Appendix 1 contains a full list of the questions contained in the research questionnaire

or

A fuller, more detailed account of the stages in the experiment can be found in Appendix 2.

Present research results with guidelines to readers on how the results were obtained and why they have been presented in the way they have

Even in a short research write-up, there ought to be at least a paragraph or two informing the readers of *how* the results were obtained. If the readers unerstands the research methods they may understand the results. If the research was interview based, for example, the use of the research methods helps to explain why the researcher is presenting the findings in the form of case examples. If the research method was questionnaire based, the presentation of findings in the form of graphs and tables is explained.

Present research results with any relevant criticism of the results

We should be suspicious of any research writing which appears to be presented as perfect. Research is a human process and is therefore open to error. Difficulties usually arise in one form or another and, even in the controlled conditions of a laboratory, the researchers have taken a philosophical approach to research which may be open to question.

All true research writing will therefore be self-critical, will present to the readers the problems as well as the successes. Good research writing will pre-empt the questions and criticisms which could be expected to arise in the mind of a typical reader. For example, in interview-based research, there is the danger of bias (in interviewer and interviewee); in questionnaire research, there is a danger that respondents misunderstand questions; in a controlled laboratory experiment there is the danger that environmental factors distort results.

In summary, always include a section on what went wrong, as well as what went right, what is problematic about the findings, as well as what is persuasive about them.

The process of writing

Different people write in different ways. Some people find writing easy and can throw a report together in an evening. For others it is a slow, painful process taking many days (and even nights!). Are there any general principles and tips to help a writer of research? Below is a set of guidelines and tips for writing research. At least some of the tips should be of use to a writer of any brand.

Tip 1: a process for writing

Here's a process for writing a piece of research. It can be adapted for different types of research:

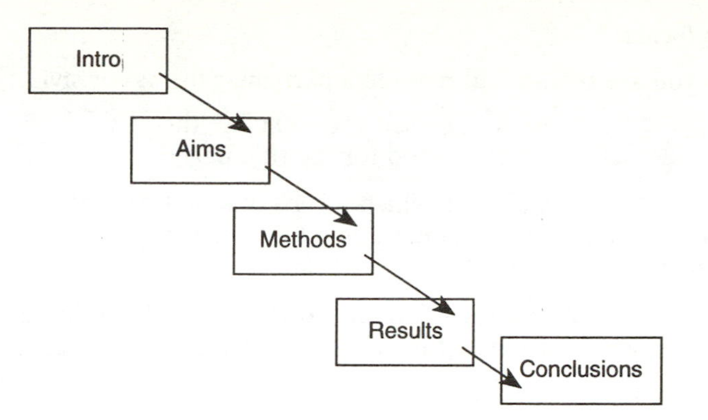

Figure 29.1 A flow chart.

Write the final paragraph first! This can be a strong motivator and can help a writer see where he or she is going in the writing. Get an overview of the whole piece of writing. One way is to draw a map of the report. This could take the form of a flow chart (see Fig. 29.1). Or, for a bit of fun, you might try something more imaginative. Picture the report as a road journey. This can help get an idea of the 'flavour' of the report (see Fig 29.2). Write the chapter or section headings first and then make notes under each heading. In that way you can build up a file which will eventually turn into the final draft. Keep a notebook with you at all times for writing down ideas and references. This could be invaluable to you later. Make several drafts. Always write the first draft by hand. When we type on a computer our fingers tend to run ahead of our thoughts. Three drafts should be enough: the first to get the ideas down on paper; the second to revise and rethink; and the final draft for a final check, to polish up, format and ensure good presentation. Find a supportive colleague or friend to check over your drafts.

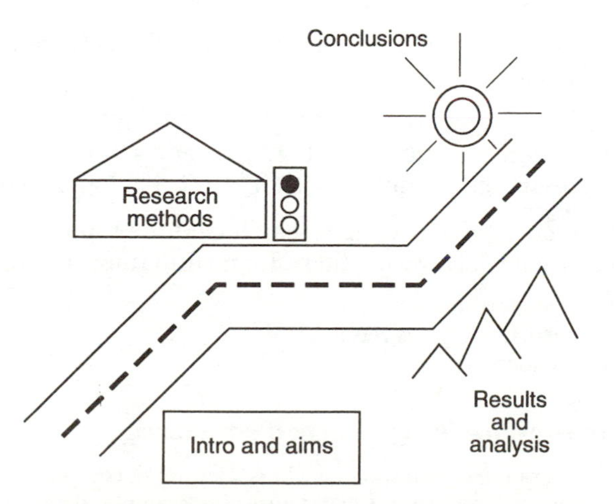

Figure 29.2 Thinking of your report as a journey.

Tip 2: ideas on format

Check to see if you are required to provide a particular format or style.

You can centre your text like this
which is good for the title pages

You can format, justifying to the left, which means that all text is aligned along the left-hand side of the page. This is standard in a lot of reports.
Or you can format,

justifying to the right which means that all text is aligned along the right-hand side of the page which might be of use on headed note paper, or for formatted questionnaires.

Or you can locate a block of text in a smaller space which can be useful for quotations or for placing text next to a diagram or a table. Desktop publishing software is very good for doing this!

Finally, you may wish to format the text so that the text is double justified or blocked equally across the page. This makes the report look neater than left justification.

Tip 3: ideas on the report style

Few people agree on what is most appropriate style for a research report. It really depends on the audience. Or it may be that the style is specified for you by someone else.

Book style takes the form of text divided into chapters, with chapters divided into paragraphs.

Report style also makes use of chapters which tend to be quite short, often called 'Sections' and which make extensive use of bullets points or numbering:

1. like this
2. and this
3. and this

* or like this
* and this

- or even this
- and this.

A further style of business report which is also used in technical reports involves the use of numbering. For example:

<u>1.0 Each major section is numbered 1.0, 2.0, 3.0 and so on</u>
1.1 while each sub-section is numbered 1.1, 1.2, 1.3 and so on
1.1.1 and each sub-sub-section is numbered 1.1.1, 1.1.2, 1.1.3 and so on.

The book style is used more for academic research, and the report style is used more in business and technical research. However, there are no firm rules. The main guidelines are:

- be sensitive to the readership
- check to see if a particular style is required
- ensure maximum readability.

Tip 4: use the resources available to you to best advantage

- Check spelling and grammar with spell-checkers. Beware of computer spell-checkers which can encourage laziness. They cannot distinguish, for example, between 'an' and 'and'.

- Make full use of graphics and, if using computers, use a clear font:

  This is Times, and this is Univers both clear and acceptable. *This is Script* a lot of fun, but unreadable and unacceptable!

  A typeface with serifs, like Times, should be used for long texts. Sans serif types, like Univers and Helvetica, tire the eyes and should be used sparingly, perhaps for emphasis or for headings.
- If you use a computer printer, ensure that the printout is of reasonable quality, and beware of bad photocopiers.
- Use **bold** to emphasise and *italics for quotations* and for paraphrasing.
- Use colleagues for proofreading and criticism.

Tip 5: plan your writing in detail

Set **aims**. Three should be enough. An aim is a general statement of direction of the research. For example: 'the aim of this research is to explore the link between ...'.

Set **objectives**. These should arise logically from the aims. They are more specific than aims in that they are explicit about what actions are required to meet the aims. For example: 'the objective of this report is to measure the differences and similarities between ...'.

Present your research standing up

For some, even to stand up in front of a small audience is a terrifying experience. Others seem to love it. Love it or hate it, researchers are often asked not only to give a written version of their research work, but also to stand up and present it verbally. PhD students defend their theses at a *viva voce*. Professional researchers may be asked to present their findings to a client. Students are often asked to give individual or group presentations to their peers and to the lecturers.

Whether a verbal form of presentation is requested or not, it may be of benefit to the research process to consider it. A verbal presentation can complement the written word through:

- direct, live contact with the researcher
- the researcher's demonstrated enthusiasm for the research
- the opportunity for questions and discussion.

In general the verbal presentation of research isn't too different from other kinds of verbal presentation such as business presentations. As in these other types of presentation, the presenter needs:

- to plan properly and rehearse
- confidence and enthusiasm in voice and posture
- a structured approach to the material
- appropriate and supporting use of visual aids.

The presenter needs to plan properly and rehearse.

Few people can stand up and present effectively without any planning or rehearsal. An audience will usually be able to identify lack of preparation through:

- an unconfident speaker
- visual aids which look rushed or simply lifted from the research report
- contradictions in the main argument
- a lack of cohesion between the written and verbal material
- poor time management during the presentation
- lack of preparation for questions and discussion at the end.

One of the tips for writing mentioned earlier was the importance of drafting. A similar principle applies to stand-up presentations.

When you prepare the material

- mark the research report with a highlighter pen to identify key areas for inclusion in the talk
- make notes on cards which can be sequenced to provide a guide for the final version of the talk
- use report contents as a guide to structure.

When you rehearse the talk

- practice in front a mirror, and again in front of a friend or colleagues
- audio- or videotape the talk to get an idea of what you look and sound like
- time the presentation and practice until you are in control of the time available to you
- sit in on a colleague's presentation and try to identify what was good about it, and what was not so good, so as to draw lessons which you can apply to your own talk.

When you prepare the visual aids

- handwrite the materials, such as slides, before committing them to print or foil
- use large pieces of flip-chart paper with pens or crayons to design your slides
- ensure you know how to use the technology: do you always put slides the right way up on an overhead projector? Do you know which way up to put slides in the carousel of a 35 mm projector?

The presenter needs confidence and enthusiasm in voice and posture

He who whispers down a well
About the wares he has to sell
Will never make as many dollars
As he who climbs a tree and hollers

(Anon)

No matter how impressive the visual aids are, no matter how important the research findings are, lack of enthusiasm on the part of a speaker will send an audience to sleep. Enthusiasm doesn't simply mean jumping around a lot, being loud, or arriving inside a huge wedding cake. Particularly in research presentations, the voice and eye contact can be used to:

- create enthusiasm at the beginning of the presentation to communicate the importance and interesting nature of the research
- emphasise key points throughout the presentation
- highlight conclusions and recommendations.

Here are some tried and tested keys to maintain confidence:

- 'Laughter is inner jogging.' Enjoy it!
- Confidence comes with practice.

- It is never as bad as you think it will be.
- The audience is not the enemy.
- Always prepare properly.
- Relax, and don't forget to breathe!
- Proceed at a pace you are comfortable with.
- Prepare for questions, don't be afraid to respond 'I don't know'.

Last of all, take a look at yourself in the mirror. Practice your talk. What can you do to improve the basic skills of verbal presentation?

- voice projection
- voice clarity
- voice tonal range
- body posture
- time management
- listening and observation
- control of body language
- enthusiasm.

Pick one area for improvement. Use the presentation as an opportunity to develop yourself.

The presenter needs a structured approach to the material

Unstructured presentations are anathema to research presentations. Research is a structured process and a good presentation should mirror this. One of the simplest ways of structuring a research presentation is to follow (at least in part) the written report structure. Here's a suggested simple structure for a slide presentation:

Slide 1 – research project title, your name, department, university, sponsor
Slide 2 – aims and objectives of the presentation
Slide 3 – context/background to the research
Slide 4 – research methodology
Slide 5 – findings (graphical or text)
Slide 6 – analysis
Slide 7 – conclusions
Slide 8 – recommendations.

You might also include slides containing:

- summary of findings
- an abstract or executive summary at the beginning
- a pertinent and/or amusing quotation
- a cartoon which captures the theme of the research
- photographs
- a set of *questions for further research* at the end.

The presenter needs appropriate and supporting use of visual aids

There is nothing worse than a one-hour talk where the only available source of information is the speaker's monotone voic! There is also nothing worse than a speaker who says very little but puts slide after slide on the overhead projector, crammed with too much information, too small to read anyway.

Visual aids should act as a support to a presentation not a substitute for it! There are a range of visual aids available which can be used to present the research effectively, which have different advantages and disadvantages.

Overhead projector

Advantages: simple to use; effective; easy to prepare; slides are relatively cheap, can be used in conjunction with a computer, good for presenting research data graphically.
Disadvantages: projectors can overheat; can be noisy; light can get in the speaker's eyes.
Tips: check that the slides can be seen from all parts of the room; and have a spare bulb ready.

Slide projector

Advantages: can show colour photographs; allows remote control; if slides are professionally made can look very impressive; good for most kinds of data if prepared on computer.
Disadvantages: the world is littered with stories of slide projectors breaking down; slides must be preloaded; light can get in speaker's eyes.
Tips: check the equipment beforehand; have a spare to hand; and check that your slides are all the right way up and in the right order.

White/blackboard

Advantages: allows speaker to write things up; build up research ideas; diagrams.
Disadvantages: messy; requires legible handwriting; unprofessional; in a fixed position unless portable one is available.
Tips: use a clean board; erase any previous words written up by someone else; practise writing beforehand; be clear about what you are writing up; write legibly.

Flip chart

Advantages: dynamic; movable; good for small group presentation; very little can go wrong with it.
Disadvantages: too small; relies on speaker's writing; requires pages to be torn off; good for building-up diagrams and recording questions and ideas.
Tips: use in conjunction with an overhead projector for writing-up questions; and before the presentation, draw any items you want to include in light pencil to act as a guide for neat writing and drawing during the presentation!

Video

Advantages: allows pictures to be shown of the live research situation, can create interest for audience.
Disadvantages: danger of technical difficulties, some people don't enjoy television.
Tips: check the equipment beforehand, try to get a player with remote control, and check the room to see that the screen is big enough.

Other visual aids

Other visual aids for research presentations include:

- live demonstrations
- use of supporting notes handed out to the audience
- use of audio-cassette-based material
- posters and exhibition boards
- drama-based presentations (use actors if you can).

The important point for all of these aids is to ensure:

- they are appropriate to the research
- they are suited to the audience type, size, and expectations
- you can confidently use them.

Present the results of different research methods

Different research methods indicate different types of presentation. This will vary with the type of research project and audience. Here are some guidelines for the main research methods:

Presentation of questionnaires and statistics

Tips:

- use graphics wherever possible
- remove jargon and technical words that will alienate audience members (perhaps support with a glossary of terms handout)
- do not overload the audience with too many figures
- use supporting handouts
- select only the major findings for presentation
- allow time for questions of clarification, either as you go along, or at the end
- use only one graph or table per slide, and make it big enough to read.

Presentation of interviews

Tips:

- do not put too much information on a slide
- use italic style for direct quotations
- create summary tables to help audience gain an overview of the research findings
- supply a supporting handout with a list of interview questions, a summary of interviewee characteristics, or a research process summary showing how the interviews were done.

Presentation of case studies

Tips:

- use storytelling where possible
- use pictures to give a visual flavour
- use summary tables to compare and contrast different cases.

Presentation of experiments (such as laboratory work)

Tips:

- use storytelling where possible
- use graphics and photographs for visual flavour
- allow time for questions of clarification
- use practical demonstrations if possible and appropriate.

A last word ...

The presentation of research, like all presentation, is a skill that can be developed. The presentation or research must be designed and delivered to ensure:

- understanding
- clarity and accuracy
- interest and enthusiasm for both speaker and listener.

Much more could be written about the skills, techniques, equipment, materials and support that can be used to achieve a high standard of presentation. The reading list provided at the end of this chapter is for the speaker who would like to develop these skills a little further. Good luck!

Bibliography

Reading on presentation skills

Bell G (1990) *The Secrets of Successful Speaking and Business Presentation*. Heineman.
Leigh A and Maynard M (1994) *Perfect Presentation*. Arrow Business Books.
Peel M (1995) *Successful Presentation in a Week*. Headway: Hodder & Stoughton.

Reading on writing skills

Bartram P (1994) *Perfect Business Writing*. Arrow Business Books.
Bell J (1994) *Doing Your Research Project* (2nd edn). Open University Press, Chapters 11 and 12.
Bowden J (1994) *How to Write a Report* (2nd edn). How To Books Ltd.
Cooper B M (1990) *Writing Technical Reports*. Penguin Books.
Daniels D and Daniels B (1993) *Persuasive Writing*. HarperCollins.
Goodworth C (1990) *The Secrets of Successful Business Report Writing*. Heineman.

30 Graphical presentation

Paul Levy

Introduction

In this chapter we look at the graphical presentation of information. This is a huge topic and there is not enough space to explore it in depth. Here are some basic guidelines for those who are new to research and wish to put some graphical life into their research reports.

Information expressed visually can be a guide for the writer as well as for the reader. Graphics act as points of light in the writing, beacons, signalling the development of the research from aims, through methods, to findings, analysis and conclusions. Even if you omit them from the final draft they can help you while you are writing.

Some of the graphics used in research reports are:

- graphs
- pie charts
- tables
- figures
- photographs
- cartoons.

Graphs

If you have collected a lot of numbers, and if these numbers are, at least in part, necessary for a reader to understand the research, then a graph of one kind or another will be appropriate. Any book on graph making will talk about the different kinds of graph: bar charts, histograms, line graphs. What is common to them all is that the data, through graphic presentation, is easier to read and understand than in text form.

As Fig. 30.1 illustrates (from the first question of a questionnaire), the clear and simple presentation is the key.

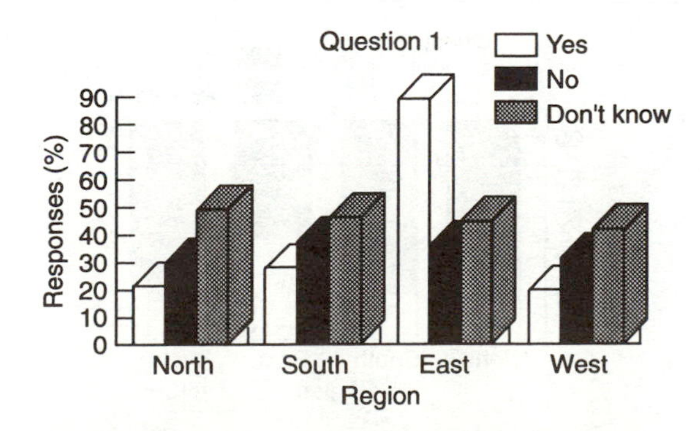

Figure 30.1 A graph with three results side by side.

In Fig. 30.1 the responses to a question which requires either a 'yes', 'no' or 'don't know' response are shown for four respondent groups from different parts of a country.

The reader is able to get a rough idea of the responses by looking at the graph, without having to read the numbers in detail. These numbers can be contained in an appendix, or in the report itself. But the visual impact is the important thing, particularly when the graph is being presented in a stand-up presentation. The researcher may have a particular point to make about the responses to the questionnaire, and the point can be made on the back of the visual impression the graph creates. In the case of Fig. 30.1 there may be a point to be made about the 'yes' response from the group of respondents from the 'east'.

In Fig. 30.2 there is a second version of the same graph. In this case, by putting the three types of response into one bar for each group, we can get a visual sense of the proportion of responses from each group.

In Fig. 30.3, we have an example of the wonders of technology: a simple graphic version of Fig. 30.1 generated three-dimensionally. Impressive though it may look, the ability to compare the responses is lessened through the distortion provided by the poor perspective of the three-dimension.

Finally, in Fig. 30.4, the *proportion* graph shown in Fig. 30.2 is presented in a different way, allowing the reader a little more accuracy in assessing the proportions. The reader can pick out the different responses and relate them more quickly to the percentages on the scale provided.

The choice of the type of graph depends on what you believe should have visual impact. What is the aim of the graph? What response should it elicit in the reader?

The following guidelines may help:

- check the purpose of the graph from the readers' point of view
- do not use too many graphs
- label the axes properly
- include exact figures on the graph if this is required
- keep the amount of data on one graph manageable.

Figure 30.5 shows a graph presenting two kinds of data on one chart. This enables the reader to see a relationship between several variables. Once again, it is important to be clear about the purpose of such a graph.

Graphs have a range of uses for different kinds of research. They can:

- show changes in a research variable over time
- show relative proportions
- establish relationships between research variables.

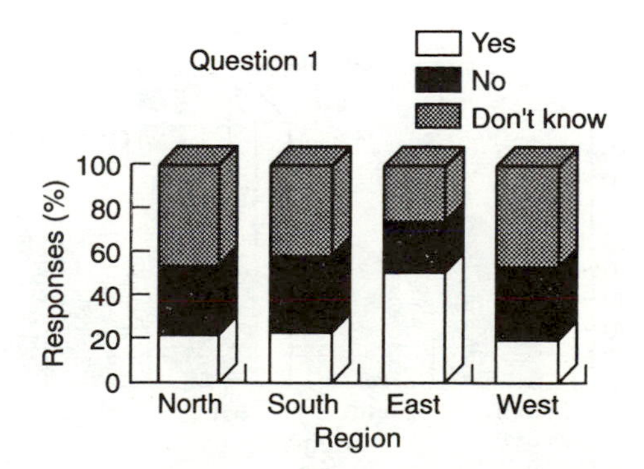

Figure 30.2 The same graph as in Fig. 30.1 showing relative proportions.

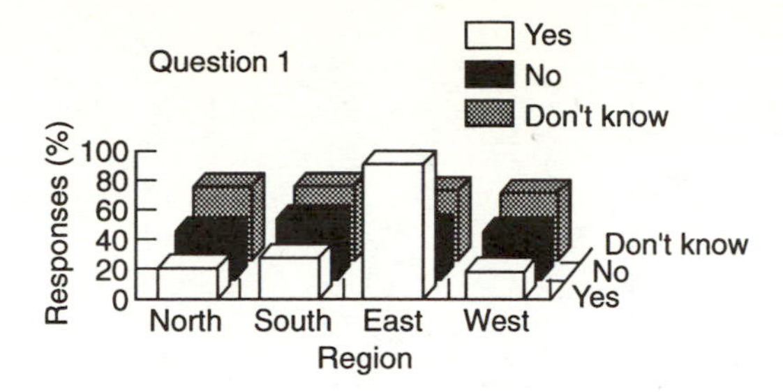

Figure 30.3 A three-dimensional version of Fig. 30.1, lacking clarity for the reader.

Pie charts

Pie charts can have impact in a stand-up presentation but should not be used in written reports since the few numbers they represent are more easily read in a simple table.

Pie charts can be used:

- to show the relative 'share' of different elements
- as an alternative to the graphs shown in Fig. 30.1.

As with graphs, they can be generated easily by computer programs. They are harder to construct by hand. Figure 30.6 shows the simple visual power of a pie chart in showing the results of a question from an attitude survey.

Figure 30.7 is a poor example of a pie chart. It has the same data as were used in Fig. 30.1. In this case, the percentages are missing and the pie pieces have been separated. Clarity is lost and it is very hard to see what the chart is trying to say. However, if in a stand-up presentation, we only wish to make the simple point that salad is the preferred lunchtime food, then this chart does the trick.

Finally, in Fig. 30.8, we have a three-dimensional pie chart. Once again salad stands out clearly as the preferred food, and the graphic overall looks impressive. There is, however, a small danger that the distortion of perspective distorts the reader's view of the data. Compare each of the pie slices in Figs 30.7 and 30.8. What do you think?

In general, pie charts can make for good, impressive visuals. As with graphs some guidelines should be noted:

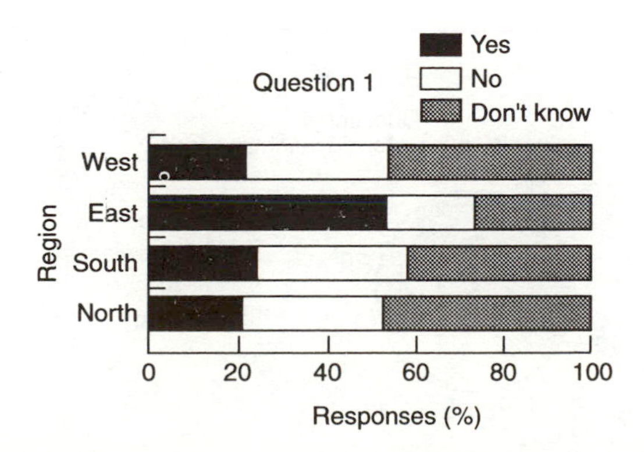

Figure 30.4 Another version of Fig. 30.1 presented differently.

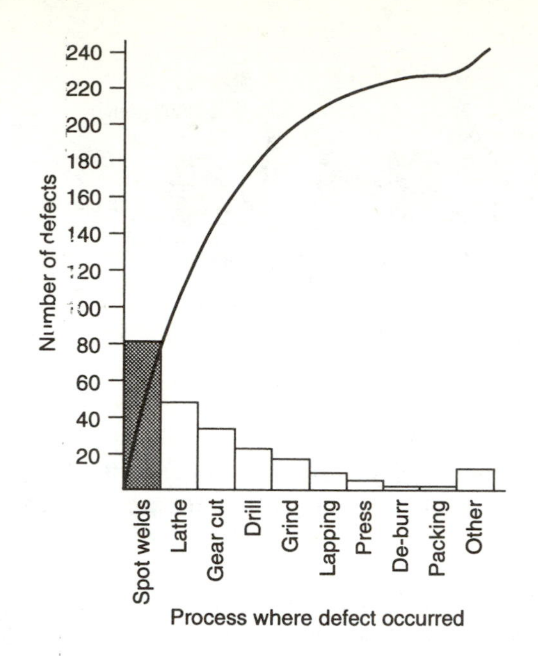

Figure 30.5 An example of a graph showing two kinds of data.

- label the pie chart properly
- be clear about the purpose of using the chart
- too many pie pieces can make the chart look overloaded with information and hard to read
- be careful of the impact of using three-dimensional effects
- pie charts can be most effective in stand-up presentation
- pie charts should not be used in written reports.

Tables

Tables can be used in conjunction with other, more visual, charts or they can be used alone. With other charts, the table can be used to hold the detailed data while the chart gives a general visual impression. On its won, a table holds research data which can be read easily and understood by the reader.

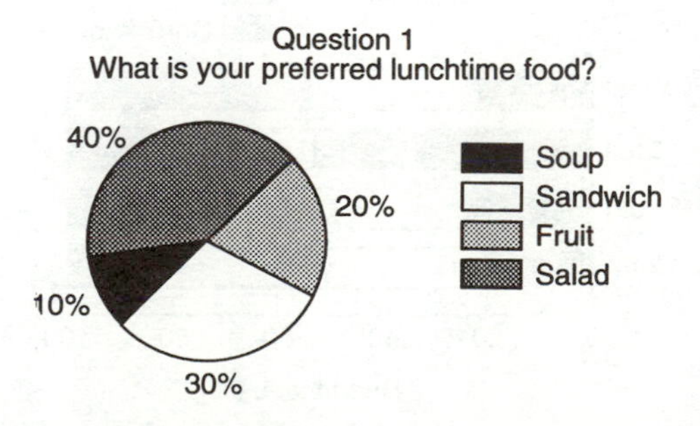

Figure 30.6 A pie chart showing the responses to a question from a questionnaire.

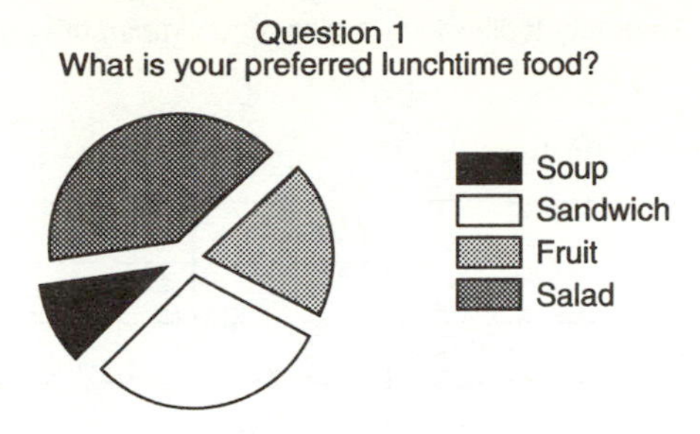

Figure 30.7 A pie chart without numbers attached, giving a rough idea of proportion but lacking precision for the reader.

Tables are *successful* in research writing when they are:

- simple
- clear
- not too overloaded with data
- clearly linked to the text and other graphics.

Tables are *unsuccessful* in research writing when they:

- are used a 'dumping grounds' for piles of numbers
- try to say too much at once
- 'stand out' alone and are not integrated with the text.

Table 30.1 is a simple table without any data. It has four columns for data and allows the reader a quick view of the results of a piece of survey work. It is poorly labelled and would rely on proper referencing within the text.

In summary, tables depicting research data need:

- supporting commentary
- proper labelling
- simple and clear presentation
- links to other types of graphic.

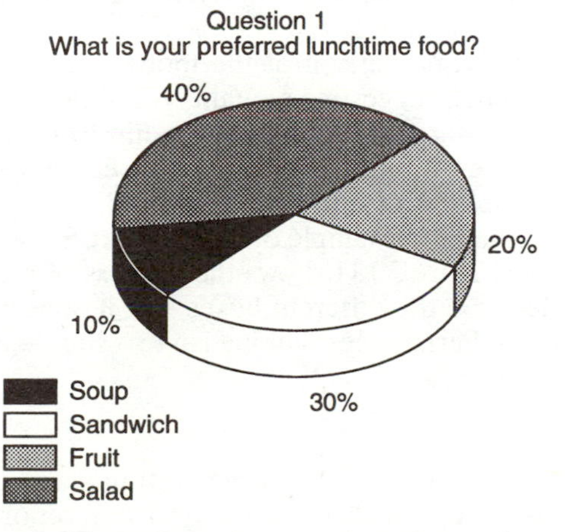

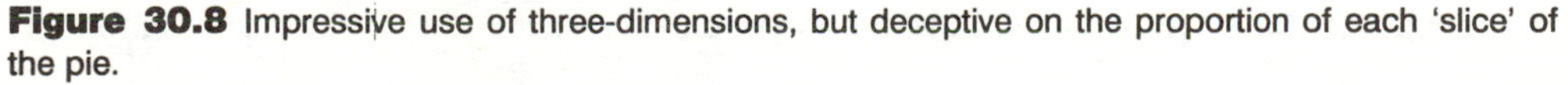

Figure 30.8 Impressive use of three-dimensions, but deceptive on the proportion of each 'slice' of the pie.

Table 30.1 An empty table which is clear and simple, but poorly labelled

C I Travel Ltd
Survey of favourite holiday destinations

Destination	Tally	Total	Cumulative total	%
Costa del Sol				
Tenerife				
Florida				
Paris				
Kos				
Blackpool				
Bognor Regis				

Figures

Figures can take many forms. Some of their uses in research are:

- to show an experiment
- depict a process
- map an organisation
- link a set of ideas.

A figure will support written research if it:

- helps to clarify an argument or a context
- demystifies a process
- helps to link critical ideas together
- improves the visual appeal of a piece of writing.

Figures 30.9 and 30.10 show one researcher's attempts to describe an organisation's philosophy known as 'continuous improvement'. In Fig. 30.9 we have a sense of a number of ingredients (shown on the left) all being important in the creation of the *Continuous Improvement* (CI, on the right). This is clear. What is not so clear is the relationship between each of the elements on the left.

In Fig. 30.10 we have a second version of the model. In this case the elements are depicted as 'cogs in a machine', suggesting a quality of motion in the diagram. They are now viewed by the reader in terms of their relationship to each other: they are interdependent – one cannot move without another. If the researcher's intention was to make this point, then Fig. 30.9 is less successful than Fig. 30.10!

Finally, in Fig. 30.11, we have an example of a flow chart. Flow charts are useful if you wish to describe a process. Figure 30.11 shows the process of making a cup of tea. The diagram lacks a key to describe the different boxes, but it does give a visual impression of all of the stages required. Perhaps tea making isn't so simple after all!

Photographs

Photographs are powerful visual aids. They work well in stand-up presentations (used mainly with slide projectors) but can also be used in written reports. Once again we have to deal with the basic question: what is the purpose of the visual aid? If the research report

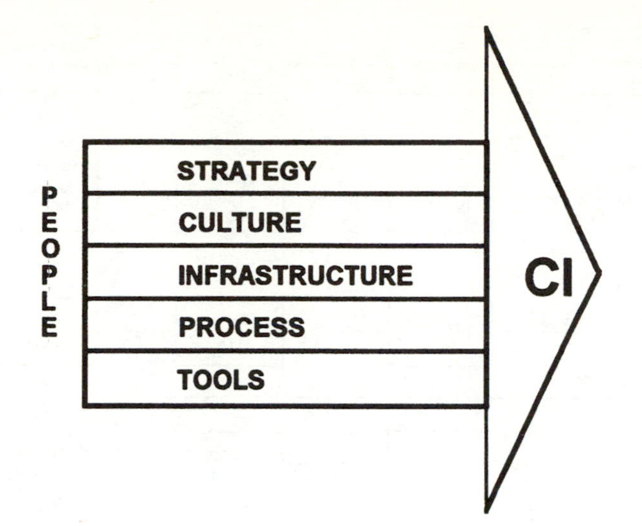

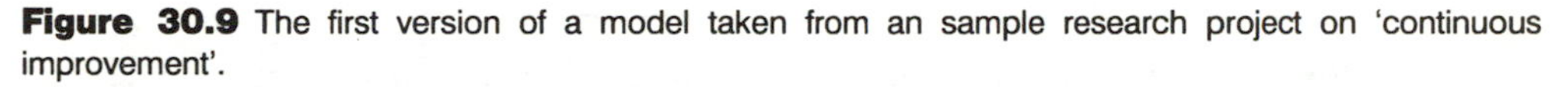

Figure 30.9 The first version of a model taken from an sample research project on 'continuous improvement'.

is to be academically credible then the use of photographs should not be used to create the impression that the write-up is like a piece of journalism, no matter how appealing the pictures are!

Photographs can and should be used where a real picture is critical to the reader's understanding of the research. For example, in a biological research write-up, photographs taken from microscope slides may be appropriate. Or in a piece of engineering research, a picture of a laser machine being studied may be appropriate. What are less appropriate, for example, are snapshots of the researcher in action, or a picture of the interviewer and interviewee shaking hands.

Guidance for using photographs

- Be clear about who owns the copyright on photographs.
- Do not use poor photocopies of photographs.
- Be clear on why a photograph is being used.

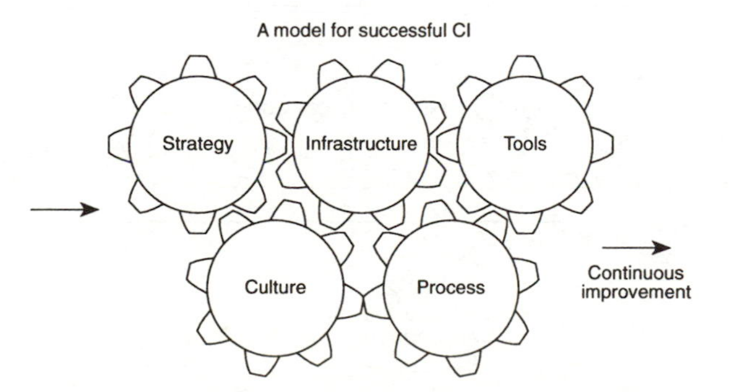

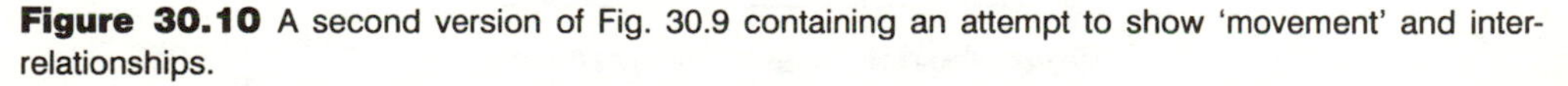

Figure 30.10 A second version of Fig. 30.9 containing an attempt to show 'movement' and inter-relationships.

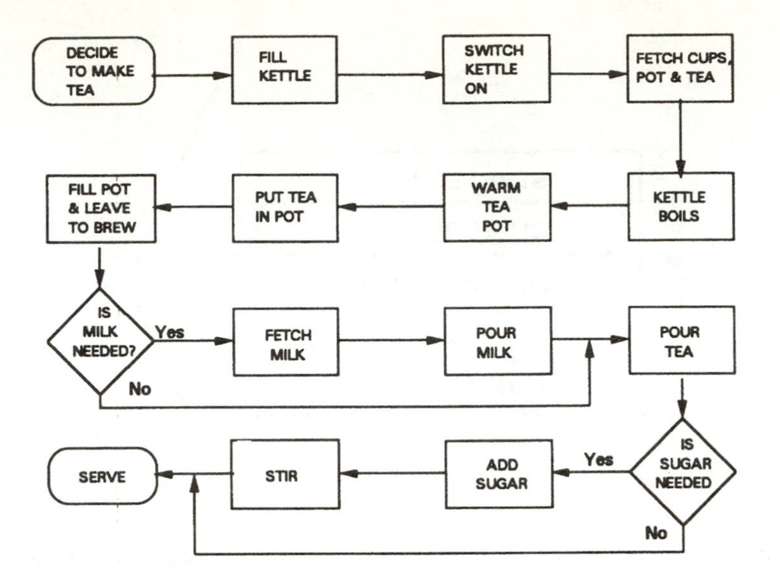

Figure 30.11 A flow chart making a complex process visual for the reader.

Cartoons

Cartoons can be used cautiously in stand-up presentations. But be wary about using them in a written research report, unless they illustrate a point effectively and are not likely to annoy the reader. The two rules shown in Fig. 30.12 may help.

Computer versus hand drawing

It is now largely standard for research reports and graphics to be word processed or computer generated. However, drawing programs on many computers are user-unfriendly and researchers don't always have the time to spend on learning such programs. Some strategies are available:

- get support from a computer literate colleague
- use neat handwriting (stencils can be used for slides)
- write the graphic by hand and use a computer scanner.

Figure 30.12 Rules of using cartoons.

Remember: The quality of the graphic may be seen by audiences and readers to represent the quality of the research.

Colour

If colour pens or printers are available, use them! At the simplest level, colour is appealing to look at. At a deeper level colour is a powerful tool for presentation. Colour can be used to:

- create identity for different sections of the research, such as blue for methods, green for findings
- depict emotional data, such as red for problems, green for actions, orange for warnings.

When you draft your report, experiment with different colours. Too much black and white can create the impression of greyness to the reader.

And finally ...

And finally, there are the logos; icons; sections inside boxes; proper page numbers; a clear contents page; a list of figures and tables; an impressive title page; reader-friendly report covers; the final spell-check, and then ... presenting it to the reader! Phew!

Reference

Tufte E (1983) *The Visual Display of Quantitative Information*. Graphics Press, Cheshire, CT.

Part Eight

THE FUTURE

31 Protecting and exploiting technology

Edward Nodder and Fiona Dickson

Technology transfer has become a common expression in academic circles but what does it mean? It is easy to envisage a tangible piece of technology, such as a transgenic animal or a new EXAFS detector, being physically handed over to a third party. The law, however, has also created property rights in intangible matter arising out of the creative intellect. This too can be transferred to a third party in a variety of ways, often in return for monetary consideration.

The intangible matters alluded to are generically known as **intellectual property rights** (or IPRs) and form the key to any technology transfer programme.

In this chapter we discuss the following issues:

- Why is technology transfer of interest to academics?
- What intellectual property rights exist and in what do they subsist?
- How can intellectual property rights be protected?
- Who owns the intellectual property rights in any particular technology?
- How can intellectual property rights be transferred?

The discussion which follows is intended to provide a generous overview of considerations relevant to technology transfer programmes. It is based on the law at December 1994. The law, however, is constantly developing and complex. Accordingly, before applying any information in this chapter to particular circumstances, professional advice must be taken from a lawyer or patent agent as appropriate. Additionally, the comments relate to English law. For other jurisdictions, advice should be sought from lawyers or patent agents qualified to practice there.

Sources of funding

One of the reasons why an academic institution may become involved in a technology transfer programme is to generate funding for future research without having to rely on external sources. By way of background, it is useful to look at some of the sources currently available for funding academic research.

Whilst some academic institutions may be fortunate enough to receive donations from benefactors, alumni or endowments, many have to look for research funding from other sources. The most obvious are the Research Councils, of which there are now six (see Chapter 4): biotechnology and biological sciences; economic and social sciences; engineering and physical sciences; medical; natural environment; and particle physics and astronomy.

The publication, however, of the government's White Paper *Realising our Potential* marked a change in direction for centrally funded research in this country. The White Paper notes that:

> *While the Research Councils should focus on the value of proposed research in terms of scientific excellence and timeliness, they should take more fully into account the extent to which outcomes could be taken up by potential users.*

The aim is therefore to foster stronger links between science and engineering communities, industry and the research charities as a way of ensuring the efficiency and effectiveness of centrally funded research.

In line with the policy objectives of the White Paper, higher educational institutions have engaged in collaborative research projects with commercial partners. Sponsorship by industry of an academic partner to carry out research on its behalf is also becoming increasingly common. Alternatively, some institutions provide consultancy services to industry.

The LINK scheme, a cross-government initiative, aims to bridge the gap between the country's science and engineering base and industry. There are currently over 30 LINK programmes. Each supports a series of projects. Every project requires at least one science and engineering base partner (which can be an academic institution) and one industrial partner. The government provides a grant of half of the total project costs, while industry provides the other half.

The European Commission also has a number of schemes for promoting collaborative research across the EU provided the research has industrial application. Examples include projects such as ESPRIT (the European Strategic Programme of Research and Information Technology) and EUREKA (the European high technology programme). Such funding programmes, however, are subject to a standard-form contract issued by the Commission. As such, the contract terms may not be suited to the particular requirements of an academic institution. This is in contrast with privately arranged projects with industry where there is usually the opportunity to negotiate terms which take account of any special circumstances associated with academic institutions.

Beyond the research phase lies the development of the technology with a view to realising its commercial potential. At that stage, other funding opportunities may arise. A more formal joint venture could be considered with an industrial partner, for example, through a company jointly owned by the institution and the industrial partner. Alternatively, private venture capitalists could be approached who will assess whether or not the project is likely to prove a worthwhile investment.

Compared with other methods, the advantage of a technology transfer programme as a way of funding research is that it can, at least in theory, be self-perpetuating. Exploitation of the IPRs arising from one research project could be used to fund future ones. However, there must be a demand for the technology concerned.

What are the IPRs that can be exploited through technology transfer and in what do they subsist?

Intellectual property rights

Under English law, the most important forms of intellectual property that are relevant to academic research institutions are:

- patents
- copyright
- industrial designs (registered designs and unregistered design rights).

Know-how (unpatented confidential technical information) is commonly included in this general category, although strictly speaking it is not a property right.

Special rights have also been created in line with technological developments, such as the advent of computerisation. These include semiconductor topography rights and (in the US) mask-works. Rights can also be obtained in some countries, such as Germany, through petty patents or utility models, available where the inventive step requirements of patentability are not satisfied.

Patents

A patent is a monopoly right granted by the state in return for adequate disclosure to it, and in turn the public, through a written specification of a new invention which the inventor could otherwise have kept secret. The invention may, for example, be a product such as a novel designer drug or a process such as the method for manufacturing that drug.

As noted, patents are only available for novel inventions. To be new, the invention must not have been used in or in any way disclosed to the public (in the UK or elsewhere) by the inventor or anyone else before its priority date, (normally the date the patent application is filed). There must also be an inventive step. In other words, when looked at through the eyes of a person skilled in the particular area of technology concerned, the invention must not be obvious when compared with everything that is known at the priority date claimed. Patentable inventions must also be capable of industrial application.

However, patents are not available for all inventions. For example, scientific theories, mathematical methods and mere discoveries are not patentable *per se*, although their application to a product or process may be if there is a technical effect. The patentability or otherwise of computer programs is the subject of topical debate. In the absence of a new technical effect, patent protection for computer software has not been granted in Europe, copyright being the chosen medium (see below). The agreement reached on intellectual property rights (TRIPS) in the last round of talks on the GATT did not, however, expressly exclude computer software from patentability. That is akin to the position under US law where it has been possible to patent software provided it is applied in some manner to physical elements or process steps. It remains to be seen whether European patent laws will have to be revised to accord with TRIPS.

A patent grants to its registered proprietor the exclusive right to exploit the product or process claimed for up to 20 years from the filing date and the right to prevent others from doing so unless authorised by the registered proprietor. As such, they can be extremely valuable rights, especially when entering a new market. Time is often needed to establish a new product and recover research and development (R&D) costs before competitors arrive on the market. A possible disadvantage, though, of patenting an invention rather than merely keeping it secret is the cost involved in obtaining and, once granted, maintaining the patent. In addition to paying renewal fees, this could include defending a claim that the patent is invalid. As such, they are not always the most appropriate form of protection for an invention.

If it is possible that a patentable invention may have been made it is crucial not to use it in public or disclose it to any other person before a patent agent has drafted and filed the patent application. Otherwise, the validity of the patent may otherwise later be brought into issue. Care should therefore be exercised when publishing or presenting papers at a conference or to colleagues at a seminar to ensure that the invention is not accidentally disclosed, or before demonstrating the invention on departmental open days. Should a potential industrial partner wish to evaluate technology belonging to an institution that is potentially patentable, it is important to ensure that before representatives of that prospective industrial partner see or have disclosed to them any such technical information, they are told and have acknowledged such demonstrations and/or disclosures are being made to them in confidence. They should be required to sign a confidentiality agreement confirming this. Professional advice should be taken on the form of such an agreement; many academic institutions, however, already have their own standard form agreement.

Know-how

Know-how is a general term used to describe technical information, often unpatentable, which is confidential to its originator. Formulae, recipes and process descriptions are common examples. Its commercial value resides in its confidential nature. Clearly, this gives rise to a conflict between, on the one hand, academic freedoms and the requirement to publish the results of research work, and on the other, the necessity to keep know-how confidential to preserve its value.

Keeping technical information confidential has some advantages over patent protection. No formalities are required and hence there are no application and renewal fees to pay. Furthermore, unlike patents which are territorial in nature, no such restrictions apply to confidential information. The commercial value of such information, though, is destroyed if it loses its confidentiality either because it leaks out or because someone else independently develops the same know-how and publishes it. Once a product is on the market it may also be possible to discover its formula or its manufacturing method by reverse engineering. Accordingly, whilst in theory the life of know-how is perpetual, in practice it is indefinite. This is a major disadvantage compared with patent protection.

The following ground rules can help to prolong the life of know-how. Marking documents containing such information '*confidential*' and permitting access to the information only on a need-to-know basis are obvious steps. Storing confidential documents in a secure place, minimising the number of copies made, and protecting access to information stored on a computer with a security code are further practical steps that should be taken.

Disclosure of know-how often accompanies the grant of a licence to exploit a patent on a commercial scale. The recipient of that information should be required to keep it confidential and not to disclose it to anyone else and not to use it for any purpose other than that specified. In certain circumstances, such duties can be implied by law. The prudent course of action, however, is to ensure such obligations are recorded in legally binding documents, signed before the information is imparted, and in which the relevant information is clearly identified. There remains the risk of those obligations being ignored and the information leaking into the public domain. It is therefore important that practical steps such as those outlined above are also imposed upon the recipient.

Confidentiality is frequently an issue when an academic research body collaborates with an industrial partner. For the former it is often important to publish results wherever possible, but for the latter this may prejudice the patentability or confidentiality, and ultimately the commercial value, of the technology whose development they have financed. Usually a compromise will be reached under which the research body is permitted to publish important scientific discoveries, perhaps after scrutiny by and the approval of the industrial partner. The publication of theses by research students must also be taken into account, bearing in mind any relevant university regulations.

Copyright

Copyright protects the particular expression of an idea rather than the idea underlying the expression. It subsists, for instance, in original written materials (laboratory notebooks, instruction manuals, graphs and preparatory design materials for a computer program are common examples) provided certain qualification requirements are met (such as by the author being a UK resident). It is also the legal medium for protecting databases and, as noted above, computer software.

At present, copyright in a work subsists in general for the life of its author plus 50 years. However, by mid-1995, the UK and other European Member States are required to have harmonised the term of copyright protection to the life of the author plus 70 years.

One advantage of copyright in the UK is that there are no registration or other formalities for it to subsist. It arises automatically on creation of the work. It is therefore an ideal medium through which to maintain an advantage over competitors. By, for example, making relatively minor adjustments to a computer program a new copyright work could be created.

The owner of the copyright in a work will be the person who created it, unless it was made by an employee during the course of his employment, in which case it will usually be the employer. Establishing ownership of copyright, however, can sometimes be difficult. This may be facilitated by the addition of a copyright notice as follows:

© [insert name of copyright owner] [insert year of creation of work]

It is also important to keep all original works and records of the dates on which all copyright works were produced.

Copyright in the UK gives its owner the exclusive right to do certain acts, such as copy the work or issue copies of the work to the public. It is an infringement of copyright to do any of the acts the copyright owner has the exclusive right to do, unless that owner gives permission. Special provisions apply for making back up copies of and decompiling computer programs.

Like a patent, copyright is also a national right but, by virtue of various international conventions, protection may extend beyond the country in which copyright protection originally arises.

Designs

Design right in the UK protects an original design which relates to any aspect of the shape or configuration of a three-dimensional article, such as a design for a part of a machine. There are exceptions, however, for those features of shape or configuration which 'must fit' or 'must match' with another article. It is an unregistered right which arises provided the design qualifies for protection by reference, for example, to the citizenship of the designer and it has been recorded in a design document or an article has been made to the design. It may give rise to protection for up to 15 years.

If the design has 'eye appeal' (such as the design of a new watch face) then the registered design protection may be available. Registration gives the proprietor of the design certain exclusive rights for up to 25 years. To be registrable, the design must be novel and must be applied to an article by an industrial process. Protection is not available though, for example, if the design depends on the appearance of another article of which it forms an integral part (such as a car body panel).

There are currently proposals for harmonising design laws across Europe which may modify the protection available in the UK.

Handling IPRs in practice

All research staff should have a basic knowledge of the intellectual property rights discussed and the technologies in which they are likely to arise. Maintaining detailed and accurate laboratory notebooks and records (including dates) of any inventions made is very important.

Procedures should be implemented to prevent any accidental disclosure of intellectual property rights, in particular, technical information which needs to be kept confidential. All publications, theses, posters, abstracts, and seminar material should ideally be screened by the academic institution concerned before sending for publication or being presented and amended accordingly.

Identifying what intellectual property rights are available for exploitation is a key step in any technology transfer programme. This is usually done by means of a technology audit. In the same way as accountants audit books, there are specialist technical auditors who can identify what intellectual property rights may subsist in the technology concerned, evaluate them and advise on their protection and exploitation.

To maximise the effectiveness of a technology transfer programme, technology audits should be carried out regularly (annually is sensible). Otherwise, the exploitable value of that technology may be lost. Delay carries the risk that a competitor will reach the market or apply for patent protection first, or that the technology may be superseded by something much more valuable.

Preliminary points before exploiting technology

Having identified some IPRs which may be suitable for exploitation, the following preliminary matters should be considered.

Who owns the intellectual property rights?

It is important to establish the ownership of the IPRs concerned as, generally speaking, it is only the owner of those rights who has the right to exploit them or grant or transfer that right to others.

In the case of an academic institution this is often a complex issue. There are several possible candidates ranging from the academic institution itself to the research supervisor, the graduate student, any collaborative partner or any other funding body. Legal rules set out for each type of IPR who owns that particular right in law. In certain cases, these rules may be set aside by agreement. In addition, many academic institutions also have policies on the ownership and exploitation of IPRs by their academic staff and research students, but employment contracts, terms upon which research grants are awarded, or collaborative agreements would need to be checked in conjunction with these.

How are the intellectual property rights owned?

They may be owned by the academic institution alone, or perhaps jointly with a collaborative partner or even an academic. Patents, for example, may be exploited by each joint owner without recourse to the other, but a licence under such a patent can only be granted by all joint owners, unless it is otherwise agreed.

By whom are the intellectual property rights to be exploited?

Often an individual or a group within an academic institution is given responsibility for technology transfer activities. Sometimes, a limited liability company will be formed, the share capital of which is owned by the institution. Commercial contracts are then entered into by that company (to whom the intellectual property rights will have been transferred) rather than the institution. This is an important way of reducing the risk of prejudicing the institution's tax and charitable status.

What is the nature of the technology to be exploited?

Whether the technology represents a major breakthrough or merely an improvement in its particular field will determine its commercial value and often the appropriate basis for its exploitation.

Common ways of exploiting technology

Two common routes used by academic institutions are assignment and licensing.

Assignment

An assignment is the legal term for an outright transfer of the property right concerned (or the interest held in that right), including the right to sue infringers. Parts of that property right may even be carved out for assignment in certain cases. To be valid, assignments of patents, protected designs or copyright must all be in writing, but there are different rules on execution. It is advisable to seek professional help on the form and scope of the assignment, especially where the technology concerned may be required as a basis for further research. Otherwise, once assigned, the institution will not be able to exploit those particular rights itself without permission from the assignee. Assignment may be appropriate if the institution has no way to exploit the rights itself, the purchaser is prepared to pay full market value for the technology and needs the title to those rights for its purposes. This is often the case for a pharmaceutical product where considerable expenditure must be committed to clinical trials and evaluation of the drug before it can be marketed.

Licensing

The alternative route to an assignment is usually to grant a licence. A licence, however, does not transfer any property right as such to the licensee. Rather, it allows the licensee to use the IPRs concerned in accordance with the terms agreed, the use of which would otherwise be an infringement.

Licences can be granted expressly or by implication and on an exclusive, non-exclusive or sole basis. A written agreement is, of course, preferable to define the scope of the licence and to ensure adequate control over the licensing arrangement. Biotechnology and software contracts, in particular, require special provisions to take account of the unique nature of these technologies.

An exclusive licence is the closest to an assignment as it permits only the licensee to use the IPRs licensed. The licensor is also precluded. The licence may be restricted to certain fields of use. For example, an exclusive licence under a chemical patent could be limited to the production of the substance claimed for use in only one of several possible applications. This could appropriate only where an institution may wish to use the IPRs being licensed for further research, exclusive manufacturing and sale rights, being granted to the licensee. To enable the institution to protect its income stream, the licence agreement could contain provisions enabling the institution to terminate the licence if the licensee fails to exploit the technology to its full potential and look for another licensee, or perhaps to alter the licence to a non-exclusive basis.

If the licensing institution also wishes to be able to exploit the technology then a sole or non-exclusive licence should be considered. A non-exclusive licence is usually more appropriate for a technology that has many applications. The institution will be free to use the technology itself as well as to grant non-exclusive licences to others. Under a sole licence only the licensor and licensee are entitled to use the relevant technology.

More elaborate ways of dealing with intellectual property rights are possible. For example, intellectual property rights can be used as consideration for the issue of shares in an exploitation company or mortgaged in the same way as a house as security for borrowing.

The consideration for the transfer of the IPRs concerned is usually a matter for negotiation. It could, for example, be a lump sum payable on the grant of rights, or more commonly a percentage royalty on sales of the product resulting from their exploitation.

Combinations of these routes and more sophisticated arrangements are always possible.

Finally, national and European competition rules and taxation consequences should always be considered when deciding how to exploit IPRs.

Conclusion

Provided appropriate steps are taken, academic institutions can endeavour to exploit the technology arising from their research efforts to produce income for further research, thus reducing the need to rely on external sources of funding. However, technology transfer programmes require specialist assistance at all stages, whether it is in identifying and protecting intellectual property rights or in their subsequent exploitation.

Bibliography

Jacob R and Alexander D (1993) *Guidebook to Intellectual Property: Patents, Trademarks, Copyright and Designs* (4th edn). Sweet & Maxwell.

Bainbridge D I (1992) *Intellectual Property*. Pitman.

CVCP (1992) *CVCP Guidance: Sponsored University Research: Recommendations and Guidance on Contract Issues*. Published by the Committee of Vice Chancellors and Principals (CVCP) of the Universities of the UK.

UK Patent Office, various publications including:

- *Basic Facts: Patents*
- *Basic Facts: Copyright*
- *Basic Facts: Designs*
- *Patent Protection*
- *Design Registration*
- *How to prepare a UK Patent Application.*

Useful Contacts

The UK Patent Office
Concept House
Cardiff Road
Gwent MP9 1RH
Tel: 01633 814000

Lists of firms of solicitors specialising in intellectual property law and practice can be found in the following publications:

(i) *The Legal 500* (1995 edn)
Published by Legalease
28–33 Cato Street
London W1H 5HS

(ii) *Chambers & Partners Directory of the Legal Profession* (1994–1995)
Published by Chambers & Partners Publishing
74 Long Lane
London EC1A 9ET

Chartered Institute of Patent Agents
Staple Inn Buildings
High Holburn
London WC1V 7PZ.
Tel: 0171-405 9450

LINK Secretariat
Third Floor
151 Buckingham Palace Road
London SW1Y 9SS
Tel: 0171-215 6671

TII (European Association for the Transfer of Technology, Innovation and Industrial Information)
3 Rue des Capuchins
L-1313 Luxembourg
Tel: 00-352 463 035

CVCP
29 Tavistock Square
London WC1H 9BZ
Tel: 0171-387 9231

32
Career opportunities

Ralph Coates

Introduction

In this chapter I present some views and information about sectors of employment together with some useful tips to help you locate and apply for appropriate posts.

A postgraduate degree (at doctorate or master's level) does not in itself guarantee you employment. However, your skills, abilities and some of the specialised knowledge derived from your postgraduate study will most certainly assist when matching yourself to both the requirements of an employer and the demands of a job. Recognition of what you have to offer is crucially important so that you can effectively explore the appropriate job market and project yourself competitively within it.

Sectors of employment

Teaching and education

In the area of higher education a postgraduate degree coupled with evidence of a good record of research interests and experience is normally a prerequisite for lecturing posts in most institutions of higher education.

Postdoctoral fellowships of between three and five years duration are available in universities throughout the world. Many are funded by an appropriate research council and will provide opportunities for researchers to continue working in their specialist areas.

There are occasionally some short-term appointments of one to three years as demonstrators or research assistants.

If however you are contemplating a teaching career in schools then despite your higher degree you are strongly recommended to take a *Postgraduate Certificate in Education* (PGCE) course. All state schools and increasingly public schools and sixth-form colleges require teachers to have this teaching qualification. Without it your future promotion opportunities could be adversely affected.

International agencies

The **Commonwealth Development Corporation** (CDC) recruits a small number of staff each year on agriculture-, engineering- and science-based projects related to the developing world. Postgraduate qualifications together with experience are required.

The **Consultative Group on International Agricultural Research** (CGIAR) is an association of countries, international and regional organisations, and private foundations dedicated to supporting a network of 13 agricultural research centres and programmes around the world. These centres offer the opportunity to carry out research in developing countries with excellent facilities provided by the World Bank and other agencies.

The **European Commission**, the largest of the institutions of the EU, occasionally has vacancies. In particular the **Joint Research Centre** employs postgraduate and postdoctoral researchers in physical and computer sciences, engineering and related disciplines to work on joint projects.

Overseas Development Administration (ODA) recruit experienced staff for tours of duty in many countries in Africa, Latin America, the Middle East, and South-East Asia. The disciplines required include agriculture, engineering and the sciences. The ODA also handles recruitment for several other international organisations.

Civil service (including government laboratories) and industrial research centres

Each year the civil service (and its many research establishments) and major industrial companies (with established and active research centres) will recruit people with a sound background of higher education in research techniques coupled with some proven experience.

A report in 1993 by the Institute of Manpower Studies suggests that 'there is no evidence of any significant broadening of the PhD science labour market beyond its traditional scientific research employment areas, except possibly in mathematics. However the market is made up of a number of different types of employers, only a small minority of whom are those which can demonstrate a discrete demand for newly qualified PhD science graduates'.

I suspect that the same is true for newly qualified PhD engineering graduates.

Business and industrial companies

In the 1993 report by the Institute of Manpower Studies the view was expressed that 'based on present economic trends, overall demand by industrial, commercial and public sector employers for PhD graduates is likely to remain fairly steady to the end of the decade with possibly a slight increase overall'.

Many employers in this sector do not make any significant distinction between those with a master's degree and an applicant with a bachelor's level of qualification. Although the particular study topic undertaken at the postgraduate level may, in some cases, be regarded as useful to have as an area of knowledge. The doctorate level of degree may appeal to some employers because they will recognise the nature of the individual study as evidence of many of the skills required for the jobs they have available.

The European and overseas connection

The notion of working outside of the UK has considerable appeal to many postgraduates and with the widening of the EU together with a policy of internationalism on the part of many companies it is likely that more opportunities will arise. However such posts will require proven and relevant experience, perhaps with some previous overseas contact, as well as a postgraduate qualification. A facility with languages will undoubtedly be advantageous if you intend to live and work abroad.

Skills employers seek

Employers expect postgraduates to have trained minds, with the abilities to apply their intellects to business or technical situations and to continue learning throughout their careers. Some vacancies require postgraduates with specific expertise. Whatever qualifications sought for the vacancy to be filled, employers also expect the postgraduate applicant to have acquired a wide range of personal skills.

The following list of core skills and definitions have been distilled from:

- work by members of the Careers Advisory Board and Careers Service of the University of Newcastle
- a survey of employers' 1994 graduate recruitment brochures and application forms by M W Rae, formerly graduate recruitment manager for a major employer
- a Graduate Skills Survey 1995, by Fiona Hewitt, Careers Service, University of Newcastle
- an Enterprise in Higher Education (EHE) 1994 graduate survey conducted by the University of Nottingham EHE Unit
- the Association of Graduate Recruiters (AGR) graduate salaries and vacancies 1994 summer update survey
- the skills module of PROSPECT (HE), the computer-assisted guidance system for students in higher education.

Communication

The ability to operate three distinct modes of communication:

- written communication
  – write well-expressed, concise and grammatical documents (especially reports and letters), aprropriate to the needs of the readers
- interpersonal communication
  – listen attentively and seek to understand what other people say
  – ask probing questions, consider differing views
  – negotiate with people to reach agreement on a point of view or course of action
  – speak in clear and succinct language
- oral presentation
  – make lucid and confident oral presentations, appropriate to the audience.

Teamwork

The ability to:

- recognise and respect the attitudes, actions and beliefs of the other members of a group
- establish a good rapport with others and work effectively to meet an objective or complete a task
- contribute to the planning and coordination of a group's work
- assist the working process of a group by helping to resolve conflicts, recognising the strengths of others and encouraging them to contribute
- when appropriate, take a leadership role, setting direction and winning the commitment of others.

Planning and organising

The ability to:

- take a long-term view and set challenging but achievable objectives
- decide priorities for attaining targets
- make a plan and arrange resources to carry it out
- draw up a work schedule and meet deadlines
- effectively manage personal time and handle a range of activities simultaneously.

Problem solving

The ability to:

- assimilate, analyse and evaluate complex information, identify key issues and principles, and draw well-reasoned conclusions
- think critically, learn from mistakes, challenge established assumptions and make well supported judgements
- take a broad view, seeing less obvious connections and interdependencies
- think conceptually and creatively. Generate ideas that pay-off in practice
- implement action based on the assessment of all available data.

Initiative

The ability to:

- be a *self starter*, take appropriate action unprompted
- set demanding personal goals and overcome difficulties to achieve them
- pursue an activity to a high standard and rise to challenges
- take well researched decisions quickly.

Organising

The ability to:

- take a long-term view and set objectives
- decide priorities for achieving targets
- make a plan and organise resources to follow it through
- draw up a work schedule and meet deadlines
- effectively manage personal time and handle a range of activities simultaneously.

Adaptability

The ability to:

- respond readily to changing situations and priorities
- recognise potential for improvement
- re-apply known solutions to new situations
- think creatively, to generate ideas that pay off in practice
- initiate change and make it happen: *pro*act not *re*act
- manage stress and to remain effective under pressure.

Employers may also look for other skills, most commonly:

- **Numeracy** – The ability to understand and use numbers accurately.
- **Computer literacy** – the ability to use word processing and spreadsheet software, and databases, competently.
- **Languages** – the ability to speak, write or learn foreign languages.

Finally, each person has a *unique profile of skills and knowledge*. Employers may well expect postgraduates to have reached informed career choices by assessing how well their own knowledge and skills match those required by the job or career sought. Awareness of personal strengths and weaknesses will be looked for in the candidates. You will be asked to explain your career choices and to give reasons for the specific job application.

Where to find those advertised posts

You will most certainly find science-, engineering- and technology-related vacancies regularly advertised in the major national newspapers and the scientific and technical journals some of which are:

- *The Independent* (Monday)
- *The Daily Telegraph* (Tuesday and Thursday)
- *The Evening Standard* (Wednesday)
- *The Guardian* (Thursday)
- *The Times* (Friday)
- *The Observer* (Sunday)
- *The Sunday Times*
- *The Sunday Telegraph*
- *New Scientist*
- *Nature*
- *Biologist*
- *Chemistry in Britain*
- *Chemistry and Industry*
- *The Chemical Engineer*
- *New Civil Engineer*
- *Professional Engineering*
- *IEE Recruitment*
- *Materials World.*

For teaching and lecturer posts you will need to refer to *The Times Educational Supplement* (TES) and *The Times Higher Educational Supplement* (THES).

The appointments department of the Association of Commonwealth Universities also circulates vacancy details to the universities careers services.

Prospects Today (available from most university careers services) is a fortnightly publication advertising vacancies for graduates and some of the posts will specify opportunities for those with postgraduate qualifications.

If you are particularly seeking to work abroad then *Overseas Jobs Express* is a fortnightly publication carrying advertisements for jobs based overseas some of which will require postgraduate qualifications.

Recruitment agencies, many of which advertise in the national newspapers and technical journals, are worth contacting particularly if they work for employers in a specialised work sector.

Taking action

Advertised posts

Response to an advertised post is fairly straightforward. You get some indication of what the job involves and what the employer is expecting of the applicant. Moreover, you get a clear instruction of how they want you to make application: by letter and CV, or with an application form.

Whatever the style of approach, always remember it is your first, and maybe your only opportunity to impress the selector enough to be invited to an interview. So, write about yourself in a positive way. Your application must have impact, be of quality and match the job specification. Remember, you are a quality product in a highly competitive market. Your potential will be judged largely on your track record of performance and achievement in the major periods of your life so far (such as university, school, working experiences and even some of your extracurricular activities). So, this is not the time to be over modest!

Once you are over this first hurdle and have been invited to an interview it is important that you are able to keep your storyline going and live up to the impression you have so far created.

Speculative applications

In some cases you may consider it appropriate to make speculative applications to employers, organisations or institutions to tell them of your existence and hope to influence them into considering you as a possible candidate either immediately or at some future date. To make the necessary impact you must have a good perception of the employer's business and work activities and recognise where you believe you can make an appropriate and effective contribution by applying your skills, abilities and expertise.

If you are contemplating this more open job search you may find it helpful to talk to members of academic staff in your university department about your plan of action. They may have information about the type of work you are seeking and may also have some useful contacts to enable you to make that first approach.

Your application, usually by letter and CV, must have impact and illustrate your self-confidence. Try to find the name and designation of the person to whom it would be most appropriate to send your letter and CV.

Speculative applications are certainly worth trying in a carefully controlled way.

Conclusion

Finally, for further information about any of the employment sectors or help in dealing with applications get in touch with your university careers service.

Index

abbreviations 249–50
abstracts 49
accessibility sampling 131
Ackroyd, S 172
adaptability 292
adaptive sampling 134
advertised posts 292–3
advice 36–7, 67–8
aeroplane fuselage problems 215
agricultural research
  experimentation principles 114
  field experiments 107–8
  transformations 113–14
  treatment structure 108–12
algebra 55
algorithms 55, 195, 229
Allan, G 77
Altman, D G 87, 91, 93, 251
amplifiers 162
analogue input–output 160, 163–4, 165
analogue-to-digital converter 162
Anderson, D 141
Annual Abstract of Statistics 158
appendices 247, 259–60
Apple Corporation 53, 64
Appleby, J C 216, 217
applications, speculative 293–4
Applied Statistics 202
ARIMA models 235
ASCII 56
Association of Commonwealth Universities 293
Attenborough, M 207
augmented designs 104–5
autoregressive models 234–5
averages 183

Babbage, C 34
Bàlàs 215
Barnett, S 214
batch mode analysis 197
Bell, J 39, 40, 172, 174
Bellman, R E 225
Bernoulli distribution 184
Beveridge, W I B 6
Beynon, R 77
Bhattacharya, R N 209, 228
bias
  in measurement 144–5, 147–8
  safeguarding against 88
  sampling frame 116, 129–31, 139
BIDS-II 50
BIDS-ISI service 49–50
binomial distribution 185
Biotechnology and Biological Sciences Research Council 22, 23
Birch, J 77
bit-map image 58
blackboard 266
block structures 112–13
blocking 88, 107–8
BMDP 197–8, 203
BMDP2V 197–8, 203
body language 173
Bond, D 138
Bondi, C 207
Bonferroni correction 92
book searches 48–9
Borrie, J A 214
box and whisker plot 149, 151
Bras, R L 228
British Academy 25
British Council 25
British Medical Journal 251
British Standard 5497 96
Bronowksi, J 5
Buchanan, D 172, 174
Burdess, J S 211, 214
Burghes, D N 207, 223
Burrows, C R 211

Buzan, T 42
Byte 202

CAMAC 167
Cameron, R G 214
capitals, style 250
capture–recapture sampling 134
card index 41
career opportunities 289–94
careers services 69
Carroll, R J 146
cartoons 276–7
case studies, presentation 267
Cassel, C M 128
catastrophe theory 208, 223, 224, 225
catheter, fixed to valve body 105
CD-ROM 49, 50, 61
CD-ROM drive 63
census 142, 152–5
Central Statistical Office 68
Chalmers, I 93
Chamber of Commerce 68
Chambers & Partners Directory of the Legal Profession 288
chaos theory 224
chaotic dynamics 208–9
charities, research funding 26–7
Chartered Institute of Patent Agents 288
Chatfield, C 234–5
Checkland, P 43
chi-squared test 93, 149, 187
Church, A 147
civil service 290
class-intervals, frequency tables 149
Classical Mechanics and Control (Burghes and Down) 207
classificatory scale 146–7, 150
Clements, A M 207
Cleveland, W S 150
clinical trials
  analysis 91–2, 93
  design 87–91
  methodology 87
  protocols 90–1, 92
  quality 93
  random allocation 87–93, 186
clustering 138
Cochran, W G 113, 132, 145
collaborative research 20, 25–6, 27, 282
Committee of Vice Chancellors and Principals 19, 288
Commonwealth Development Corporation 289
communication skills 291
company information, financial pages 69
composite designs 104–5
computer, in statistics 195–203
computer editing 124
computer literacy 292
computer magazines 62
computer programs, patents 283
computer-aided design 61
computers 52–4, 61, 63–4
  algebra 55–6
  crashes 63
  data collection 54–6
  graphics 58–9
  purchase 60
  in research 52–9
  response times 60
  in statistics 181–2, 195–203
  survey analysis 123–4
  text as data 57–8
  text processing 56–7
  see also databases
conclusions 246, 257
concomitant variables 100
conduct, codes 30–2
conferences 68
confidence interval 186, 251
confidentiality 118–19, 284
consultancy 21
Consultative Group of International Agricultural Research 289
contingency table 187
control group 87, 144
control variables 100–4
controller, rotor
Cooke 229
cooking results 35–6
copyright 62, 284–5
Cormack, R M 132, 134
correspondence 44
council tax lists 140
covariates 100
Cox, D R 184, 233
Cox, G M 113
Cramer, H 149
creativity 3, 5
cross-classification 187
Curnow, R N 114
curve, as measurement 146–7, 196

dams 208, 238–9
data
  continuous/categorical 182
  from interviews 174–5
  graphics 58–9
  incomplete 91–2
  instrumentation 160
  multimodal 183
  numerical/non-numerical 182
  plug-in acquisition 164–5, 167
  text 57–8
  transformations 113–14
data analysis 4, 122–4, 175
  batch mode 197
  discriminatory 192–3
  intention-to-treat 92
  non-linear multivariate 196
  numerical 55
  regression 189–91
  soft system 43
  spectral/wavelet 209, 239
  variance 108–10, 113, 129–31, 184, 193
data collection 9
  computers 54–5
  forms 33–4
  induction 9
  instruments/procedures 119–21
  pilot study 122
  software 65
  surveys 118–19
data entry, professional 124
data tables 246
databases 41, 47–8, 50, 65
Davis, I L 221
Davis, T P 146
Decennial Supplement on Occupation Morality 157
deduction 9
Deming, W E 132
demo disks, computer magazines 62
Demographic Statistics – Population and Social Conditions 158
demonstrators 289
Den Hartog, J P 211
Department of Trade and Industry 68
dependent variable 189
Derive 56
design right 285
desktop publishing 57, 64
deterministic models 208, 209
  finite difference method 219–22
  finite element method 216–18
  new developments 222–6
  vibration control model 211–16
diagrams 41–2
diary 39, 41, 248
dictaphone 41
differential equations 208, 216
digital input–output 160
Directory of Grant Making Trusts 26
discriminant analysis 192–3
disk maintenance program 61
dispersion 183–4
dissertation 243
distribution, statistics 129, 184–7, 196, 233, 235–6
Dobson, A J 147
document, computer 53
documentation, research 38–45
Doole 215
Doré, C J 91, 93
DOS 63
Down, R R 207
drawing 43
Drazin 215
Dyke, G V 107
dynamic programming 225–6
dynamics, chaotic 208–9

Easterby-Smith, M 40, 42
Economic and Social Research Council 22, 23, 71
editing, surveys 123, 124
Edwards, D 207, 228
Ehrenberg, A S C 57, 251
eigenvalues 213
elasticity 216
electoral register 140–1
electrical characteristics 160
electronic journals 51
Ellenberg, J H 93
Ellios, P W 211
Elliott, D 135, 142
Elton, L 78
employment, sectors 289–90
engineering, computer simulation study 214
Engineering Mathematics Exposed (Attenborough) 207
Engineering and Physical Sciences Research Council 22, 23

errors
  edit 123
  in measurement 147–9
  random 228
  root mean square 130
  in sampling 116
  standard 130
  statistical accuracy 129–31
  type one 98
  type two 99
ESPRIT 282
estimators 128, 130
ethics 30
  and commercial pressure 32–4
  research 29–37
  software copying 62
  tape recording 40–1
  trial design 89, 90, 91
EUREKA 282
Europe: Funding from the fourth Framework Programme for Research and Technological Development (Office of Science and Technology) 25–6
European Commission 26, 68, 290
European Union, research funding 25–6
EuroStat 158
Everitt, B 200
exciters 162
experimental design 95–105
  comparative 95–6, 97–8
  descriptive 95, 97
  factorial 96, 101–4, 111–12
  presentation 267
  principles 114
  response 96, 100–5
  statistical analysis 96–100
explanatory variables 100
eye contact, presentation 264–5

factorial experiments 96, 101–4, 111–12
Falconer, K 224
Faust, C R 209, 221, 222
fax 62
feedback control 215–16
fellowships 21, 289
field experiments 107–8, 137–8
field models 209
figures 274–5
filters 162
FINEL 216
finite difference method 219–22
finite element model 208, 215, 216–19
Finney, D J 251
Firth, J M 239
flip chart 266
flooding, risk estimation 234–8
floppy disks 53, 63
flow chart 261, 276
fluid flow, multiphase 221–2
font styles 262
forecasting 194
forging 35
format 261
Forward Look of Government-funded Science, Engineering and Technology (HMSO) 24
Foster, K 140
Fourier, J-B 220
Fourier series 208
fractal geometry 209, 224
Framework IV Programme 25–6
fraud 29, 34–6, 90
free language searching 48
frequency table 149–50
Frets, G P 182
Friedlander, M P 250–l
Frost, P 77
ftp, anonymous 63
funding 15–17, 20
  charities 26–7
  European Union 25–6
  forms of 21
  government 22–5
  industry 27–8
  technology transfer 281–2

Gantt chart 18
Gardner, M J 93
gas platform 218
gatekeepers 118
GATT, TRIPS 283
GAUSS 200, 203
Gaussian distribution 186, 233
Geiger, H 149
General Register Office (Northern Ireland) 156, 158
General Register Office (Scotland) 155, 156, 158
general-purpose interface bus 165–6
Genstat 54, 108, 114
Gifi school, non-linear multivariate analysis 196

Gilks, W R 239
Glim 54, 114
Godard, J-L 253
Goltardi, G 219
Goodman 142
goodness-of-fit 190
Goupilland, P 239
government
  departments/ministries 68
  publications 68
  research funding 22–5
  and universities 20
  White Papers ix, 3, 20, 27, 281–2
Government Actuary's Department 156
graduates, social life 79–80
grammar checker 258
graph theory 196
graphical presentation 259, 269–77
graphics 54, 58–9, 65, 259
graphs 269–71
Grebenik, E 170
groundwater pollution 221–2
Grove, D M 146
Guide to Mathematical Modelling (Edwards and Hamson) 207

Haberman, R 207
Hall, C 77
Hamilton, L 200
Hamson, M 207, 228
hard disk 61, 63
hardware 52, 53, 60–2
head sizes 182–3
Hearn, G E 239
heat conduction, one-dimensional 219–20
heat-flow equation 220
Hedges, B 141
Helsinki Declaration 30–1
Hickman, M 140
histogram 149, 184, 185
HMSO Publications Centre 24, 157
hoaxes 35
Hobbs, P V 234
Hogan 215
Hohn, M E 132
Holmes, R 213
Holt, D 135
hospitals, information 69
Howard, K 173, 174
Huba, G J 193
Huberman, A M 174, 175
Hughes, J A 172
Huxley, T H 256
hypotheses, testing 186–7

IBM 53
image processing 58, 209
Imam, I 218
independence, statistical 185, 187
independent variables 189
indexes, journal 49
induction 8–9
industrial liaison officer 28
industry, research funding 27–8
inference 9, 128
information retrieval 47–51, 52
information services 47
informed consent 90
Institute of Manpower Studies 290
instrumentation
  basics 161–4
  data 160
  research 164–7
  safety 161
  signal types 160–1
integration 10
intellectual discovery, methods 5
intellectual property rights 281, 282–6, 287
intention-to-treat analysis 92
interlibrary loans 50, 159
international agencies 289–90
Internet 51, 62, 78, 200
interviews
  analysing data 175
  presentation 267
  process 172–4
  questions 172
  rationale 169–70
  recording data 174
  response 119
  survey 137
  types 170–1
initiative 292
IQ testing 147
Isaaks, E H 239
ISDN-standard card 62
Isham, V 233
isolators 162
Ito calculus 209
Jacobs, O L R 208, 225
JANET 49

Jankowitz, A D 10
Jeffrey, A 207, 208, 212
Joglekar, G 189
John, J A 114
Joint Research Centre 290
Jones, R T 211
journal articles 49–50, 51
Journal of the European Commerce 26

Kalton, G 129, 130, 133, 137, 141
Kane, E 173
Kant, I 30
Keith, L 132
Kelvin, W T 37
keywords 48, 49, 244–5
Kiaer 129
Kingsley, C 30
Kirkwood, M 218
Kish, L 130, 132, 133, 140
know-how 284
Koestler, A 6
Kreyszig, E 207
Kriging 209, 239

laboratory notes 40
laboratory staff 69
labour turnover 7 (note)
languages 290, 292
LaTeX system 57
Latin square 112–13
least squares regression 104
lecturing 289
Levy, P 243
Lewis, P A W 184
librarians 47, 50
libraries 47, 49, 68
licensing 284, 287
Lievesley, D 138, 139, 140, 142
Likert scales 54
linear discriminant analysis 192–3
linear models 114
linearisers 163
LINK scheme 24, 27–8, 282, 288
local services 68–9
Locatelli, J D 234
logistic regression 191–2
Lunn, A D 200
Lynn, P 138, 139, 140, 142
Lyons, W 78
McCullagh, P 114, 147
Macintosh computers 53, 64
Mackay, R 209
McNeil, D R 200
Manchester University, Microdata Unit 155
Mandel, J 146
Mandlebrot, B 224
Mantoglou, A 233
mapping 41–2
Markov chain model 208, 209, 235, 238
Marshall, C 169, 171
materials properties 100
materials/equipment supply companies 69
Mathematics 56
mathematical modelling 207–9
  deterministic 208, 209, 211–16
  matrices 212
  stochastic 209, 228–39
  vibration control 211–16
Mathematics for Engineers and Scientists (Jeffrey) 207
Matheron, G 233
MATLAB 214, 239
matrices, for modelling 212
matrix, transition 235
Maunch, J 77
Maxwell, A E 147
May, G S 91
Mead, R 107, 114
mean 183, 251
means table 111
measurement
  bias 144–5, 147–8
  classificatory scale 146–7
  errors 147–9
  expert advice 146
  instrumentation 161–3
  investigations 145–6
  precision 148–9, 251
  repeats 114, 147–8
  survey 116–17
Medawar, P 36
median 149, 183
medical research, ethics 29, 31–2, 33
medical research charities 26
Medical Research Council 22, 23
Meirovitch, L 211
Mellor, D 232, 233–4
memoranda 44
mentoring 77–8
Metcalfe, A V 105, 132, 149, 211, 214, 239
Michell, J 147
Microdata Unit, University of Manchester 155

Miles, M B 174, 175
mind-mapping 41–2
Minitab 54, 235
mode 183
modem 61–2
Mohr, L B 129, 130
monitors 154, 157
morality, and culture 30
Moran, P A P 238
Moser, C A 129, 130, 133, 137, 170
Moses, I 73
Mote, C D Jr 211
motion equations 212–13
MRC 87
MTB model, rainfall 230, 232, 233–4
Mullings, L 171
multimedia computers 61
multiplexing 163
multistage sampling 133–4, 138, 140
Murrell, G 77
Murthy, D N P 207

National Postgraduate Committee 75, 79
Natural Environment Research Council 22, 23
Nelder, J A 114, 147
networks 51, 53, 61–2, 65
Neumann, John von 5
neural networks 196
Newland, D E 239
Newton, R 73, 77
Nietzsche, F W 30
non-linear models 195, 196
non-parametric tests 187
non-probability sampling 131
non-response 123, 135, 142
normal distribution 186
note-taking 39, 174
null hypothesis 186–7
numeracy 292
numeric data 54–5
Numerical Recipes (Press) 229

obfuscation 36, 250
observations, recording 39–40
Office of Population Censuses and Surveys
    141, 155, 156, 157
Office of Science and Technology 22
optical character readers 58
oral presentation 263–6, 291
organising skills 292
OS2 64
outcome measures 90, 92
outliner 56–7
overhead projector 266
Overseas Development Administration 290
Overseas Jobs Express 293

PAFEC 216
pain resistance 187
Pappus method, discovery 5
parametric tests 186, 187
Particle Physics and Astronomy Research
    Council 22, 23
Partington, J 72, 82
Patent Office 288
patents 283
Patton, M Q 169, 170, 172, 173, 174, 176
Payne, R W 108, 113
PC computer 53
PCI bus 63
Peaceman, D W 220, 221
Pease, A 173
Personal Computing 202
PERT diagram 18
Peters, J 173
Peto, R 87
pharmaceuticals, ethics 32–4
PhD
  books on 77
  completion 80–2
  examination 82–4
  organisation 75–6
  planning 73–5
  requirements 72–3
  self-management 79–80
  stylistic conventions 81–2, 248–9, 250, 253–8
  uncompleted 71
  working conditions 75–6
  writing up 74–5
  see also research proposal; thesis
Phillips, E 71, 75, 77, 80
photographs 44, 274–5
pie charts 271–4
pilot study 117, 118, 122
placebos 87, 89
plane stresses 216–18
Plato 30
plug-in data acquisition 164–5, 167
Plutchik, R 40
Pocock, S 87, 90
Poincaré, H 223
Poisson distribution 185

politics, and statistics 32
pollution modelling 221–2
Polya, G 6
population
  census 142, 152–5
  estimates and projections 157–8
  intercensal estimates 155–7
  international statistics 158
  reference volumes 157–8
  sampling 128, 140–2
  screening 141
post-modernism 30
post-stratification 135
postal surveys 118, 120
Postcode Address File 140–1
postcode sectors 138
postdoctoral fellowships 289
Postgraduate Certificate in Education 289
postgraduate charter 79
Poston, T 208, 223
PostScript 58
power curve 99
precision 148–9, 251
prediction 189–91
Preece, R 77
presentation
  case studies 267
  experiments 267
  graphical 259, 269–77
  interviews 267
  oral 263–6, 291
  questionnaires 121, 267
  references 259
  research 253, 263–6
  results 258–60
  statistics 251, 267
  writing style 253–8
Press, W H 229
principal components analysis 193
printer 61
probability sampling 131–4, 138, 196, 228
problem solving 291
prognosis 192–3
programme funding 21
programs 52, 53–4
project funding 21
project management 18
pronouns, first/third person 248–9, 250, 255
proofreading 56
Prospects Today 293
protocols 90–1, 92
pseudo-random number generator 224
psychoactive drug study 193
public libraries, reference sections 159
Pugh, D 71, 75, 77, 80
purposive sampling 131

quartiles 183
Quenouille, M H 114
questionnaires 119, 120–1, 267
questions 121, 172
quota sampling 131

rainfall model 230–4
Ralph, C J 132
RAM memory 63
random errors 228
random sampling 131–4, 190–1
randomisation 87–8, 128, 186
randomised complete block design 108, 112, 113
range
  analogue input 164
  data 183–4
  measurement 148
Realising Our Potential Awards scheme 22
Realising Our Potential (White Paper) ix, 3, 27, 281–2
records
  anonymised 154–5
  graphic 41–3
  observations 39–40
  organisation 76
  photographs 44
  tape/video 40–1
recruitment agencies 293
recursive partitioning algorithms 195
reference management 50–l
reflection 10
regression
  analysis 189–91
  least squares 104
  logistic 191–2
  simple/multiple 191
regression coefficient 190
repertory grid 42
reports 253–4, 261–3
resampling methods 195
research ix, 3–4
  advice 36–7, 67–8
  applied 222
  commercial pressure 32–4

documentation 38–45
ethics 29–37
instrument systems 164–7
planning 38–9
presentation 253, 263–6
research assistants 289
Research Councils 22, 281–2
research and development, competitive tendering 23
research proposal
contract 19
costs 17–18
description 16–17
funding 15–17, 19
infrastructure 21
objectives 16
process 7–8, 11
writing 15–16
see also PhD; thesis
research studentships 21, 27
researcher network 8, 78
response experiments 96, 100–5
response variables 100, 189
results, presentation 258–60
'rich picture' 43
Ripley, B 200
risk estimation, flooding 234–8
Roberts, F S 147
Rodriguez-Iturbe, I 228
Rogers, C 9
rolling project support 21, 25
root mean square error 130
Rossman, G B 169, 171
rotor, vibration control 211–16, 218
Royal Society 24–5
Royal Statistical Society 31
Rudestram, K 73, 77
Russell, B 30
Rutherford, E 149

safety 161
Sahinkaya, M N 211
Salman, P 71
sample size 33, 89, 99–100, 127, 134
sampling 117–18, 127–35, 183–5
accessibility 131
adaptive 134
bias 116, 129–31, 139
capture–recapture 134
data collection 119–21, 141
error 116
inference 128
multistage 133–4, 138, 140
non-probability 131
populations 140–2
probability 131–4
purposive 131
quota 131
random 131–4, 190–1
randomisation 87–93, 128, 186
stratification 132
systematic 133
unequal fractions 133
weighted 133, 134–5, 142
sampling frame 128, 139–40, 142
sampling rate 145, 163–4
Särndal, C E 128
SAS Software Ltd 54
SAS Institute 203
SC (Statistic Calculator) 200–2, 203
scales, statistics 182
Scanlan 215
scanning 58
scatter plot 145
Scott, J N 132
screening, population sampling 141
searching techniques 47–8
seat-belt studies 146
Seber, G A F 132
seminars 67, 68
Senn, S 89
SERC 78
SGML 58
Sharp, J 174
Siegel, S 147
signal conditioners 162
Silverman, D 175
simulation 214–16
skewness 184
skills in demand 290–2
Skinner, C 77
slide projector 266
Smith, D K 225
Smith, G D 221
Smith, S 140
Smith, T M F 135
smoothing, non-parametric 196
snowballing techniques 142
Social Trends 68
soft system analysis 43

software 52, 53–4, 62–3, 203
graphics 58
instrumentation 167
numeric data 54–5
numerical work, non-data 55–6
plug 'n' play 62
search facilities 39
for statistics 196–202
system maintenance 65
text 57–8
types 64–5
SORT Group 93
sound card 63
space rocket velocity 207–8
space/time models 208, 209
Spector, P 200
spectral and wavelet analysis 209, 239
spell-checker 56, 249, 258, 262
SPIDA 200, 203
split-plot design 113
SPlus 203
Splus 199–200, 203
sponsor 15–19
Sponsored University Research: Recommendations and Guidance on Contract Issues (CVCP) 19
spreadsheets 53, 64
SPSS 54, 198, 203
Srivastava, R M 239
Stablein, R 77
staffing, for research proposal 17
standard deviation 148, 184, 251
standard error 130
Stata 200, 203
statistical languages 196, 198–202
statistical packages 196, 197–8
statistics 36–7, 181–2
accuracy 129–31
agricultural research 108
basic measures 182–4
computer use 195–203
ethics 31
independence 185, 187
inferential 9
official 32
political factors 32
presentation 251, 267
scales 182
software 65, 196, 197–203
Statistics and Computing 202
Statistics in Engineering (Metcalfe) 207
Stewart, H B 209, 215, 224
Stewart, I 208, 209, 220, 223
stochastic models 195
dams 238–9
flood risk 234–8
rainfall model 231–4
waiting times 228–30
storage systems, water 238–9
stratification 132–3, 139
studentship schemes 21, 27
style
first/third personal pronouns 248–9, 250, 255
good writing 253–8
reports 262
style-checkers 56
subject librarian 47
supervisor 66–7, 74, 78–9
surveys
computer analysis 123–4
data collection 118–19
edit errors 123
interview 137
item non-response 123
measurement 116–17
planning and conducting 117
postal 118, 120
precoded answers 122–3
quantitative 115
self-completion 120–1
time chart 117
system maintenance, software 65
systematic sampling 132

T-shirt selling model 223–5
t-test 89, 93, 187
tables 272–4
tape recording 39, 40–1, 174
target population 137
teaching 289
Teaching Company Scheme 28, 68
teamwork 291
technical support 69
technology, exploiting 286–8
Technology Foresight (White Paper) 20
technology transfer 281
tensile strength graph 100
Tertullus method, discovery 5
text, as data 57–8
text processing 56–7
theory building 8–9

thermocouple 160
thesaurus 258
thesis
  acknowledgements 248
  appendices 247
  bibliography 247
  conclusions/recommendations 246
  contents list 245
  on database 50
  diary 248
  discussion 246
  drafting 80–1
  glossary 247
  introduction 245
  keywords 244–5
  management 76–8
  method, results and analysis 246
  notation 248
  publishing 243
  references 247
  structure 243–8
  style 248–51
  subject choice 245
  summary 244
  title 244
  see also PhD; research proposal
Thom, R 223
Thompson, J M T 209, 215, 224
Thompson, S K 128, 131–2, 133, 134
Thomson, W T 212, 214
TII (European Association for the Transfer of Technology, Innovation and Industrial Information) 288
time chart, surveys 117
time series models 194, 209
Times Educational Supplement 293
Times Higher Educational Supplement 293
timing input–output 161
Tong, H 209
transducers 161–2
transformation
  data 113–14
  logistic 192
  wavelet 239
transition matrix 235
trial design: see clinical trials
triangulation 9
trimming results 35
TRIPS 283
Truesdell, C 220
trusts, research funding 26–7
Tsang, W W 234, 236
Turabian, K 81
Turner, M J 216
Twain, Mark 36
two-sided test 98

UFC 21
UK Patent Office 288
UK Research and Higher Education European Office 26
uncertainty principle 90
United Nations, Population Division 158
United Nations Demographic Yearbook 158
units 250
Use Your Head (Buzan) 42
user groups 53
utilitarianism 30

variables
  concomitant 100
  confounded 144
  control 100–4
  dependent 189
  in experiments 96
  explanatory 100
  independent 189
  in prediction 189
  response 100, 189
variance, analysis 108–10, 113, 129–31, 184, 193
VDU 61
velocity, space rocket 207–8
Venables, W 200
Venutelli, M 219
VESA standard 63
vibration control 211–16
video card 61
video recording 39, 266
virtual reality 58–9
visual aids 264, 265–6
viva voce 82, 83
VME extensions for instrumentation 166
VXI 166–7

waiting times, stochastic model 228–30
wavelet transformation 239
Waymire, E C 209, 228, 233
Weaver, W Jr 212
Weibull distribution 235–6
weighing designs 147
weighting 133, 134–5, 142

Wellcome Trust 26, 27
Welsh, J 75
WGR model, rainfall 233
whiteboard 266
Whyte, W F 173
Wilson, J L 233
wind-powered generators 189–91
Windows 63
Winfield, G 71
Wood, A D 207, 223
Woodford, C F 216
word-processing, companies and products 53–4, 56–7, 64–5
Works 64
World Wide Web 51
Wragg, E C 172
writing
  active/passive voice 248–9, 250, 255
  planning 263
  process 260–3
  reports 262
  style 253–8
  see also presentation
Yates, F 131
Youden, W 147
Young, K 71, 78

Zienkiewicz, O C 216